Digital Signal Processing

Zahir M. Hussain · Amin Z. Sadik ·
Peter O'Shea

# Digital Signal Processing

## An Introduction with MATLAB and Applications

Springer

Prof. Zahir M. Hussain
School of Electrical and Computer Engineering
RMIT University
Latrobe Street 124, Melbourne
VIC 3000
Australia
e-mail: zmhussain@ieee.org

Peter O'Shea
School of Engineering Systems
QUT
Gardens Point Campus
Brisbane 4001
Australia
e-mail: pj.oshea@qut.edu.au

Amin Z. Sadik
RMIT University
Monash Street 19 Lalor, Melbourne
VIC 3075
Australia
e-mail: azsadik@ieee.org

ISBN 978-3-642-15590-1 e-ISBN 978-3-642-15591-8

DOI 10.1007/978-3-642-15591-8

Springer Heidelberg Dordrecht London New York



*Cover design:* eStudio Calamar S.L.

Printed on acid-free paper

Springer is part of Springer Science+Business Media (www.springer.com)

*This book is dedicated to our loving families.*

# Preface

Signal Processing (SP) is a subject of central importance in engineering and the applied sciences. Signals are information-bearing functions, and SP deals with the analysis and processing of signals (by dedicated systems) to extract or modify information. Signal processing is necessary because signals normally contain information that is not readily usable or understandable, or which might be disturbed by unwanted sources such as noise. Although many signals are non-electrical, it is common to convert them into electrical signals for processing. Most natural signals (such as acoustic and biomedical signals) are continuous functions of time, with these signals being referred to as analog signals. Prior to the onset of digital computers, Analog Signal Processing (ASP) and analog systems were the only tools to deal with analog signals. Although ASP and analog systems are still widely used, Digital Signal Processing (DSP) and digital systems are attracting more attention, due in large part to the significant advantages of digital systems over their analog counterparts. These advantages include superiority in performance, speed, reliability, efficiency of storage, size and cost. In addition, DSP can solve problems that cannot be solved using ASP, like the spectral analysis of multicomponent signals, adaptive filtering, and operations at very low frequencies.

Following the recent developments in engineering which occurred in the 1980s and 1990s, DSP became one of the world's fastest growing industries. Since that time DSP has not only impacted on traditional areas of electrical engineering, but has had far reaching effects on other domains that deal with information such as economics, meteorology, seismology, bioengineering, oceanology, communications, astronomy, radar engineering, control engineering and various other applications.

This book is based on the Lecture Notes of Associate Professor Zahir M. Hussain at RMIT University (Melbourne, 2001–2009), the research of Dr. Amin Z. Sadik (at QUT & RMIT, 2005–2008), and the Notes of Professor Peter O'Shea at Queensland University of Technology.

Part I of the book addresses the representation of analog and digital signals and systems in the time domain and in the frequency domain. The core topics covered are convolution, transforms (Fourier, Laplace, Z, Discrete-time Fourier, and

Discrete Fourier), filters, and random signal analysis. There is also a treatment of some important applications of DSP, including signal detection in noise, radar range estimation for airborne targets, binary communication systems, channel estimation, banking and financial applications, and audio effects production. Design and implementation of digital systems (such as integrators, differentiators, resonators and oscillators are also considered, along with the design of conventional digital filters. Part I is suitable for an elementary course in DSP.

Part II (which is suitable for an advanced signal processing course), considers selected signal processing systems and techniques. Core topics covered are the Hilbert transformer, binary signal transmission, phase-locked loops, sigma–delta modulation, noise shaping, quantization, adaptive filters, and non-stationary signal analysis.

Part III presents some selected advanced DSP topics.

We hope that this book will contribute to the advancement of engineering education and that it will serve as a general reference book on digital signal processing.

May 2009

Prof. Zahir M. Hussain
Amin Z. Sadik
Peter J. O'Shea

**Prerequisites**:

Basic knowledge in calculus, programming, and circuit theory is recommended.

**Objectives**:

The book aims to facilitate the development of expertise in analyzing and synthesizing signals, both natural and synthetic. It provides various tools which can reveal the critical information contained in the time and frequency structure of signals of interest. The book also provides advanced applications and topics in signal processing, with MATLAB experiments to give practical experience in implementing analog and digital signal processing systems.

**References**:


1. Carlson, G.: Signal and Linear System Analysis. Wiley, New york (1998)
2. Williams, A.B., Taylor, F.J.: Electronic Filter Design Handbook: LC, Active, and Digital Filters, 2nd edn. McGraw-Hill, New York (1988)
3. Ambardar, A.: Analog and Digital Signal Processing, Brooks/Cole, Monterey (1999)
4. Oppenheim, A.V., Schafer, R.: Discrete-Time Signal Processing, Prentice-Hall, Englewood Cliffs (1989)
5. Haykin, S., Veen, B.V.: Signals and Systems. Wiley, New York (1999)
6. Proakis, J.G. Manolakis, D.G.: Digital Signal Processing: Principles, Algorithms and Applications, Macmillan, New York (1996)
7. Proakis, J.G. Salehi, M.: Contemporary Communication Systems, Brooks/Cole, Monterey (2000)
8. Hsu, H.: Signals and Systems, Schaum's Outline Series, McGraw-Hill, New York (1995)

# Contents

## Part I Theory and Selected Applications

# Acronyms, Symbols and Abbreviations

| | |
|---|---|
| $\forall$n | For all $n$ |
| $C_n$ | $n$th-Order Chebychev polynomial |
| $\delta(t)$ | Dirac delta function |
| $\mathcal{E}$ | Expectation functional |
| $\mathcal{F}$ | Fourier transform (FT) |
| $G_m$ | Maximum filter gain |
| $G_c$ | Filter gain at cutoff |
| $G_o$ | Filter gain at DC |
| $\mathcal{H}$ | Hilbert transform |
| $\mathcal{L}$ | Laplace transform (LT) |
| $\mathcal{L}_d$ | Double Laplace Transform (DLT) |
| $\rho$ | Time–frequency distribution (TFD) |
| $\mathbf{h}(n)$ | Vector of an FIR impulse response coefficients at sample time $n$ |
| $\mathbf{w}(n)$ | Vector of an adaptive FIR filter coefficients at sample time $n$ |
| $\omega_c$ | Filter cutoff (radian) frequency |
| $z^*$ | Complex conjugate of $z$ |
| $Z$ | $z$-Transform |
| AC | Alternating current |
| ADC (or A/D) | Analog-to-digital converter |
| a.k.a. | Also known as |
| AM | Amplitude modulation |
| ASP | Analog signal processing |
| AWGN | Additive White Gaussian noise |
| BIBO | Bounded-input bounded-output |
| B-BPF | Butterworth band-pass filter |
| bps Bits | Per second |
| B-HPF | Butterworth high-pass filter |

| | |
|---|---|
| B-LPF | Butterworth low-pass filter |
| BPF | Band-pass filter |
| BP | Band-pass |
| BS | Band-stop |
| BSF | Band-stop filter |
| BW | Bandwidth |
| CFS | Complex Fourier series |
| C-BPF | Chebychev band-pass filter |
| C-HPF | Chebychev high-pass filter |
| C-LPF | Chebychev low-pass filter |
| D/A (or DAC) | Digital to analog converter |
| DC (or DC) | Direct current |
| DCO | Digital controlled oscillator |
| DCT | Discrete cosine transform |
| DFS | Discrete Fourier series |
| DFT | Discrete FT (finite length N) |
| DLT | Double-sided Laplace transform |
| DM | Delta modulator |
| DSB/ DSBTC | Double side-band transmitted carrier |
| DSP | Digital signal processing |
| DTFT | Discrete-time FT |
| ECG | Electrocardiogram |
| EEG | Electroencephalogram |
| EM | Electromagnetic |
| EOG | Electrooculogram |
| ESD | Energy spectral density |
| FDM | Frequency division multiplexing |
| FFT | Fast FT (algorithm to compute DFT) |
| FIR | Finite impulse response |
| FM | Frequency modulation |
| FT | Fourier transform |
| FS | Fourier series |
| HP | High-pass |
| HPF | High-pass filter |
| HT | Hilbert transform |
| Hz | Hertz |
| IDFT | Inverse discrete FT |
| IFFT | Inverse fast FT (algorithm to compute IDFT) |
| IIR | Infinite impulse response |
| I/O | Input/output |
| ILT | Inverse Laplace transform |
| ISI | Inter-symbol interference |
| IZT | Inverse $z$-transform |
| LHS | Left-hand side |
| LMS | Least mean-square |

| | |
|---|---|
| LP | Low-pass |
| LPF | Low-pass filter |
| LSB | Lower sideband |
| LT | Laplace transform |
| LTI | Linear time-invariant |
| MF | Matched filter |
| mse | Mean-square error |
| Mux | Multiplexer |
| NBC | Natural binary code |
| PAM | Pulse–amplitude modulation |
| PCM | Pulse code modulation |
| PDF | Probability density function |
| PLL, DPLL | Phase-locked loop, digital PLL |
| PM | Phase modulation |
| P/S | Parallel-to-serial converter |
| PSD | Power spectral density |
| RF | Radio frequency |
| RHS | Right-hand side |
| ROC | Region of convergence |
| Rx | Receiver |
| SDM | Sigma–delta modulator |
| SH (or S/H) | Sample-and-hold |
| SLT | Single-sided Laplace transform |
| SNR | Signal-to-noise ratio |
| snr | Signal to noise ratio $= E_b = N_o$ (digital comms) |
| SP | Signal processing |
| S/P | Serial-to-parallel converter |
| sps | Symbol per second |
| SSBSC | Single side-band suppressed carrier |
| TDM | Time division multiplexing |
| TFD | Time–frequency distribution |
| TFS | Trigonometric Fourier series |
| Tx | Transmitter |
| USB | Upper sideband |
| Var | Variance |
| VCO | Voltage-controlled oscillator |
| WKT | Wiener–Kinchin theorem |
| w.r.t | With respect to |
| ZT | $z$-Transform |

# Part I
# Theory and Selected Applications

# Chapter 1
# Analog Signals and Systems

## 1.1 Definitions, Classifications, and Overview

### *1.1.1 Definitions*

A signal is a parcel of information (natural or synthetic) expressed as a function of time and perhaps other variables such as the dimensions $x$, $y$, and $z$. This information can be conveyed in various ways, e.g. by a sequence of amplitude values corresponding to regularly spaced intervals of time.

A system is a physical or mathematical entity, typically realized in hardware or software, which performs operations on signals to extract or modify information. A low-pass filter is a system that removes high frequency content from a signal.

### *1.1.2 Representation of Signals and Systems*

Signals and systems can be ***represented*** or characterized in various ways. A common way is to express or approximate them with mathematical models—such models facilitate relatively simple implementation of the relevant signals and systems in hardware and/or software.

### *1.1.3 Examples of Signals*

Some sample signals are provided below.

1. The angular displacement $\theta(t)$ of an oscillating pendulum (see Fig. 1.1). This signal corresponds to simple harmonic motion provided that the pendulum does not deviate too far from the equilibrium or rest position [1].

Z. M. Hussain et al., *Digital Signal Processing*,
DOI: 10.1007/978-3-642-15591-8_1, © Springer-Verlag Berlin Heidelberg 2011

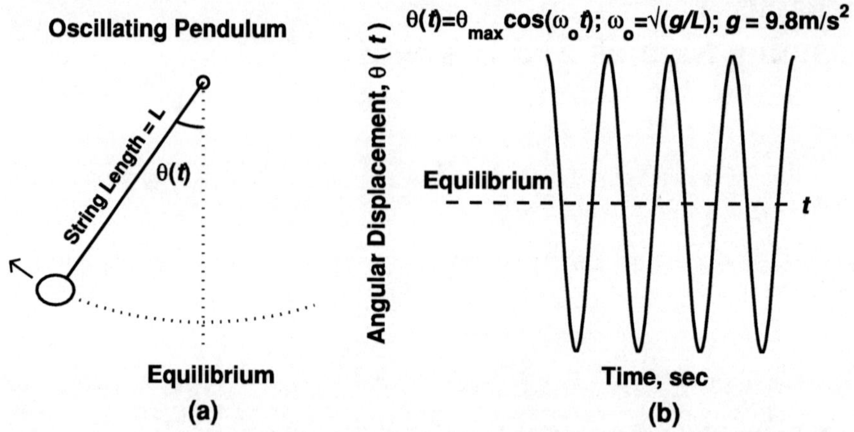


**Fig. 1.1** **a** Oscillating pendulum, **b** the displacement signal for the oscillating pendulum

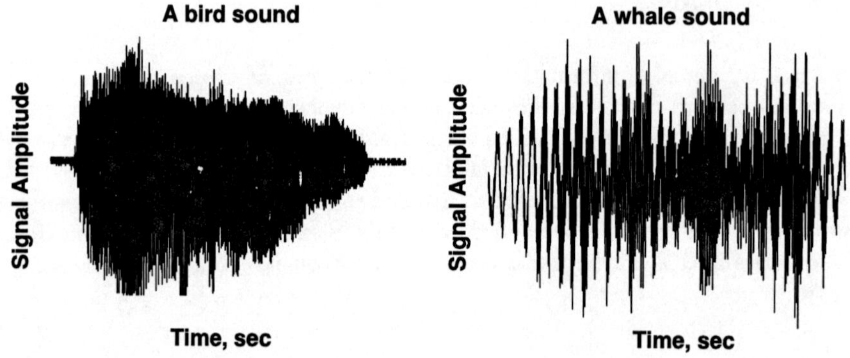


**Fig. 1.2** Sounds of animals

2. Profit as a function of time and market variables.
3. Trend of a stock price as a function of time.
4. Brain (EEG), heart (ECG), or eye (EOG) signals.
5. Speech, music, or other naturally occurring sounds (see Fig. 1.2).
6. Video signals.
7. Atmospheric pressure as a function of time.
8. Unemployment ratio as a function of social, political, and economic variables.
9. Electronic image signals sent from a satellite or space station.

## *1.1.4 Classification of Signals*

A signal can belong to one or more of the following categories.

1. **Analog, discrete, and digital signals**:

   *An analog signal* is represented as a continuous function of time, and at any point in time can have a continuum of values.
   *A discrete-time signal* is represented only at discrete time instants, but at those time instances can have a continuum of values.
   *A digital signal* is a discrete-time signal which is quantized to specific (discrete) levels (see Fig. 1.3).
   Analog signals are typically processed only by physical systems (hardware), while digital signals can be processed by hardware or software systems.

2. **Real and complex signals**: complex representation of signals and systems is a useful mathematical abstraction which can help to simplify the analysis of signals and systems. This kind of representation is particularly useful for finding or analyzing information pertaining to delays (timing) and phase relationships. For example, suppose that there is an AC circuit with a voltage source, $v(t) = \sin(\omega t)$ Volts, connected to a series combination of a resistor and a capacitance. Suppose also that the resistor has a resistance of $R$ Ohms and that the capacitor has a capacitance, of $C$ Farads). One can define a complex impedance (i.e., a generalization or abstraction of conventional resistance) for the resistor capacitor combination. This impedance is given by $Z = R - j/(\omega C)$. The advantage of using this kind of complex representation is that the phase relationships between the current and the voltage in the circuit can be easily obtained using Ohm's law: $V = IZ$. It is easy to show that the current leads the voltage by an angle of $\tan^{-1}|\text{Imag}(Z))/\text{Real}(Z)|) = \tan^{-1}[1/(\omega\, RC)]$.
3. **Periodic and non-periodic signals:** a periodic time signal repeats its value at time instants which are multiples of the "period" of the signal. For example, the signal $x(t) = \cos(2\pi f_o t) + \sin(2\pi f_o t)$ is periodic with a period $T_o = 1/f_o$ (corresponding to the inverse of the lowest frequency present in the signal). The signal, $x(t) = e^{-t}$, on the other hand, is non-periodic.
4. **Deterministic and random signals:** a deterministic signal is a time function with a known and perfectly predictable structure. For example, $x(t) = \sin(\omega_o t)$ is deterministic—its exact value is known at any point in time. A random signal cannot be perfectly predicted. One can only estimate (or "guess") its value based on its statistical properties. For example, the thermal noise, $n(t)$, that occurs naturally in many electrical devices is a random signal, as it cannot be predicted exactly. This kind of unpredictability often presents a practical design and analysis challenge for engineers.
5. **Single-channel and multi-channel signals:** single channel signals are associated with a single source, whereas multi-channel signals are associated with multiple sources. For example, a black- and-white TV picture is a single-channel signal, whereas a color TV picture is a three-channel signal.
6. **Power and energy signals**: *Power* is defined as the time average of energy. A **power signal** is a signal with finite power over the time interval $(-\infty, \infty)$, i.e.,

$$P = \lim_{T\to\infty} \frac{1}{T} \int_{-T}^{T} |x(t)^2| dt < \infty. \tag{1.1}$$

Hence, the energy $E = \lim_{T\to\infty} \int_{-T}^{T} |x(t)|^2 dt$ should be *infinite for power signals*. An **energy signal** is a signal with finite energy over the time interval $(-\infty, \infty)$, i.e.,

$$E = \lim_{T\to\infty} \int_{-T}^{T} |x(t)^2| dt < \infty. \tag{1.2}$$

It necessarily follows that $P = 0$ for energy signals. An example of a power signal is $x(t) = \sin(\omega_o t)$, and the power of this signal is $P = 0.5$ W. For this same signal $E = \infty$. An example of an energy signal is $x(t) = e^{-|t|}$. For this signal $E = 1$ J and $P = 0$.

7. **Mono-component and multi-component signals:** if only one (possibly time-varying) frequency component is present in a signal then that signal is said to be mono-component. If more than one component is present then the signal is referred to as multi-component. For example, the signal $x(t) = \sin(\omega_o t) + \cos(5\omega_o t)$ is a 2-component signal; its spectrum (Fourier transform magnitude) is shown in Fig. 1.4 for $f_o = 1$ Hz. The two components are represented by spikes at the relevant frequencies.

### *1.1.5 Analog and Digital Signal Processing*

Most signals in nature are analog. To process those signals using digital systems, analog-to-digital (A/D) conversion is first required. Once the analog signal has been digitized it is typically processed by a digital signal processing (DSP) unit. Subsequent to that, digital-to-analog (D/A) conversion is applied to recover the

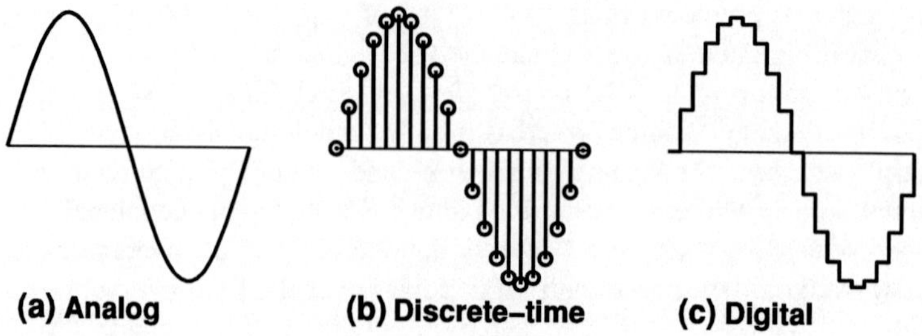


**Fig. 1.3** A sinusoid in different versions: **a** analog, **b** discrete-time, and **c** digital version

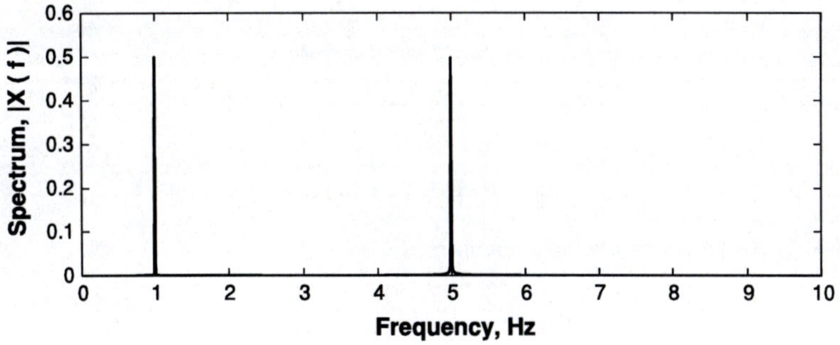


**Fig. 1.4** Spectrum of the two-component signal, $x(t) = \sin(\omega_o t) + \cos(5\ \omega_o t)$

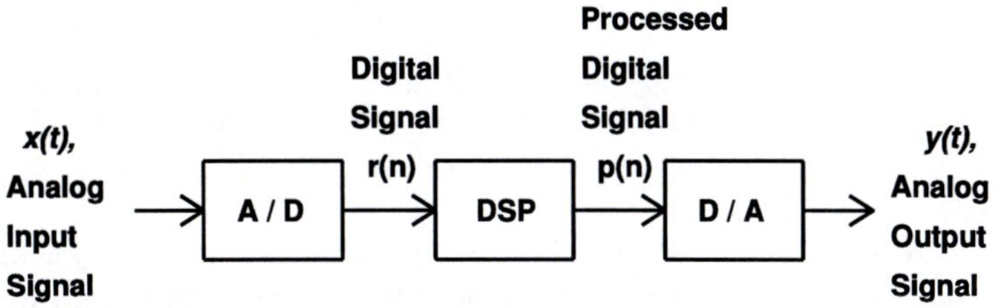


**Fig. 1.5** Block diagram of a generic signal processing system

modified analog signal. A diagrammatic view of a general digital signal processing system is shown in Fig. 1.5.

### *1.1.6 Digital Signal Processing Versus Analog Signal Processing*

Digital signal processing has a number of advantages over analog signal processing (ASP), namely:

1. DSP is less susceptible to noise and power supply disturbances than ASP.
2. DSP can more accurately and reliably represent signals and systems.
3. Storage of signals is easier in DSP.
4. DSP is more flexible and versatile, especially in changing the system parameters to accommodate changing environments (e.g., in adaptive filtering).

### *1.1.7 System Modeling*

A single input–single output system can be represented mathematically in the time domain as an operator or transformation $\mathcal{T}$ on the input signal, as shown in Fig. 1.6.

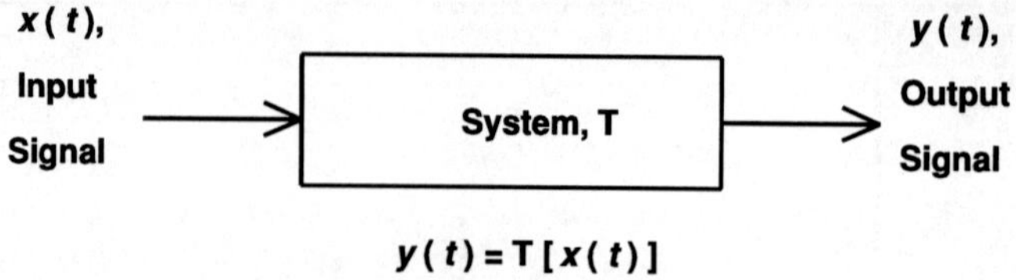


**Fig. 1.6** Signal processing system as an operator

### *1.1.8 Classification of Systems*

A system can belong to one or more of the following categories.

1. **Analog, discrete-time,** and **digital** (discrete-time quantized) systems. Analog systems operate on continuous-time signals, discrete-time systems operate on discrete-time signals and digital signals operate on discrete-time quantized signals.
2. **Time-varying (non-stationary) and time-invariant (stationary) systems**: in a time-invariant system, shifts in the input produce corresponding shifts in the output. That is, if a system input $x(t)$, gives an output $y(t)$, then an input of $x(t - t_o)$ will give an output of $y(t - t_o)$. This can be expressed more formally as:
   If
   $$y(t) = \mathcal{T}[x(t)]$$
   then
   $$y(t - t_o) = \mathcal{T}[x(t - t_o)]$$
   where $t_o$ is a constant positive time-shift.
3. **Causal and non-causal systems**: the output of a causal system at time, $t$, is only dependent on values of the input up to and including time $t$. The output is *not* dependent on input values beyond $t$. Practically realizable systems must be causal— otherwise they would need to be able to predict the future to generate outputs.
4. **Static (memoryless) and dynamic (with memory) systems**: a system whose output does not depend on either a previous or future value of the input signal $x(t)$ is called **memoryless**, i.e., $y(t)$ is a function only of $x(t)$. In a dynamic system the output depends on inputs at either past or future values of time. An **inductor** which has voltage as input and current as output is an example of a system which is dynamic. The voltage across the inductor is $v(t) = L{\cdot}di/dt$ and the current is $i_L = (1/L)\int_{-\infty}^{t} v_L(t)dt$. Hence the inductor has a memory. The same argument is applicable to a **capacitor**, which has current as input and

voltage as output, since $v_C = (1/C)\int_{-\infty}^{t} i_C(t)dt$. On the other hand, a **squarer** is a memoryless system, since $y(t) = [x(t)]^2$.

5. **Stable and unstable systems**: if a system takes a bounded input signal (i.e., $|x(t)| < \infty$) and always produces a bounded output signal $y(t)$, the system is said to be **bounded-input, bounded-output (BIBO)** stable.
6. **Linear and non-linear systems**: a system which produces an operation $\mathcal{T}$ on a signal is called **homogeneous** if it satisfies the scaling property:

$$\mathcal{T}[c \cdot x(t)] = c\mathcal{T}[x(t)],$$

where $c$ is an arbitrary constant. A system is referred to as *additive* if it satisfies the additivity condition:

$$\mathcal{T}[x_1(t) + x_2(t)] = \mathcal{T}[x_1(t)] + \mathcal{T}[x_2(t)].$$

A **linear** system satisfies the **superposition property**, which is the combination of scaling (homogeneity) and additivity:

$$\mathcal{T}[a \cdot x_1(t) + b \cdot x_2(t)] = a\mathcal{T}[x_1(t)] + b\mathcal{T}[x_2(t)],$$

where $a$ and $b$ are arbitrary constants.

**Example 1** The system represented mathematically by $y(T) = x(T) + 2$ is not linear. To see this more clearly, assume that $x(t) = a \cdot x_1(t) + b \cdot x_2(t)$, where $a$ and $b$ are constants. Then it follows that:

$$\mathcal{T}[a \cdot x_1(t) + b \cdot x_2(t)] = \mathcal{T}[x(t)] = x(t) + 2 = a \cdot x_1(t) + b \cdot x_2(t) + 2,$$

whereas

$$\begin{aligned} a \cdot \mathcal{T}[x_1(t)] + b \cdot \mathcal{T}[x_2(t)] &= a[x_1(t) + 2] + b[x_2(t) + 2] \\ &= a \cdot x_1(t) + b \cdot x_2(t) + 2a + 2b. \end{aligned}$$

**Example 2** The system represented mathematically by $y(t) = \ln[x(t)]$ is non-linear since $\ln[c \cdot x(t)] \neq c \cdot \ln[x(t)]$.

**Example 3** The system $y(t) = dx(t)/dt$ is linear since it can be shown to satisfy both homogeneity and additivity.

### *1.1.9 Linear Time-Invariant Systems*

Linear time-invariant (LTI) systems are of fundamental importance in practical analysis firstly because they are relatively simple to analyze and secondly because they provide reasonable approximations to many real-world systems. LTI systems exhibit both the linearity and time-invariance properties described above.

## 1.2 Time-Domain / Frequency-Domain Representations

There are two approaches to analyzing signals and systems: the time-domain approach and the frequency-domain approach. The two approaches are equivalent, with both domains being connected by the well known Fourier transform [2].

This chapter focuses on *analog* signals and LTI systems. There are similar representations and relationships for *discrete-time and digital* signals, and such systems will be covered later in Chap. 2.

Within the time domain, LTI systems are typically characterized and analyzed with the following alternative approaches:

1. representation with constant-coefficient differential equations,
2. formulation with constant-coefficient state-space matrix equations,
3. characterization with impulse responses.

This book will focus on the impulse response approach for the time-domain representations of LTI systems.

### *1.2.1 Basic Functions and Relations*

#### 1.2.1.1 The Convolution Integral

The convolution of two functions $h(t)$ and $x(t)$, denoted by $h(t)*x(t)$, is defined by:

$$y(t) = h(t) * x(t) = \int_{-\infty}^{\infty} h(\lambda) \cdot x(t-\lambda) d\lambda \tag{1.3}$$

As seen in the above equation, convolution performs integration on the product of the first function and a *shifted and reflected* version of the second function.

It will be seen later that the output of any LTI system can be obtained by computing the convolution of the input signal and the impulse response of the system.

Properties of the Convolution Integral

Convolution is:

1. Commutative: $h(t) *x(t) = x(t) *h(t)$
2. Associative: $h(t) *[x(t) *v(t)] = [h(t) *x(t)] * v(t)$
3. Distributive: $h(t) *[x(t) + v(t)] = h(t) *x(t) + h(t) * v(t)$.

The above properties are important in predicting the behavior of various combinations of LTI systems.

#### 1.2.1.2 The Dirac Delta Function

The Dirac delta function, denoted in the time-domain by $\delta(t)$, is invoked frequently in signal analysis. It is a *generalized function*, not an ordinary mathematical function, and rigorous study of this function is complicated. It can be defined in conjunction with an arbitrary continuous function $x(t)$ by the integral:

$$\int_{-\infty}^{\infty} x(t)\delta(t - t_o)dt = x(t_o),$$

where $t_o$ is a constant.

In plain engineering language, $\delta(t)$ is an even, tall, narrow spike of infinite height with zero width concentrated at $t = 0$. Hence, $\delta(t - t_o)$ is concentrated at $t = t_o$. A loose engineering definition can be given by:

$$\delta(t - t_o) = \begin{cases} 0, & t \neq t_o \\ \infty, & t = t_o. \end{cases}$$

In practical diagrams, one normally represents $\delta(t)$ by an arrow of unit length, while a scaled delta function $a\ \delta(t)$ is represented by an arrow of height $= a$ (see Fig. 1.7).

Properties of the Delta Function

The delta function has the following properties:

1. **P1**: $\int_{-\infty}^{\infty} \delta(t)dt = 1$ (unit area), or more generally,

$$\int_{a}^{b} \delta(t - t_o)dt = \begin{cases} 1, & a < t_o < b \\ 0, & \text{otherwise.} \end{cases}$$

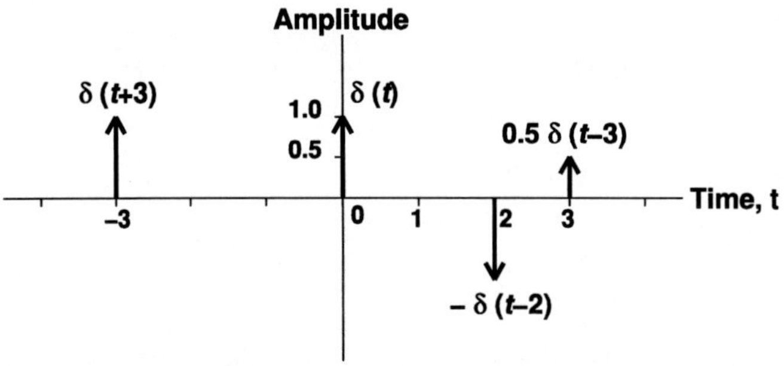


**Fig. 1.7** Representation of the delta function and shifted versions thereof

2. P2: $\delta(t) = \delta(-t)$ (even).
3. P3: $x(t) * \delta(t) = x(t)$, or, more generally, $x(t) * \delta(t - t_o) = x(t - t_o)$, where $t_o$ is a constant. That is, the convolution of a function $x(t)$ with $\delta(t)$ yields $x(t)$.

Alternative Representations of the Delta Function

The Dirac delta function can also be defined as the limit of several even functions that satisfy the above properties in the limit. These definitions include:

1. The limit of the weighted rectangular pulse (box), $\Pi_{2a}(t)$ (see Fig. 1.8, left):

$$\delta(t) = \lim_{a \to 0} \frac{1}{2a} \Pi_{2a}(t) = \lim_{a \to 0} \frac{1}{2a} \begin{cases} 1, & |t| \leq a \\ 0, & |t| > a. \end{cases}$$

2. The limit of the weighted absolutely-decaying exponential:

$$\delta(x) = \lim_{a \to 0} \frac{1}{2a} e^{-\frac{|x|}{a}}.$$

3. The limit of the weighted triangular pulse (see Fig. 1.8, right):

$$\delta(t) = \lim_{a \to 0} \frac{1}{2a} \Lambda_{2a}(t) = \lim_{a \to 0} \frac{1}{a} \begin{cases} 1 - \frac{|t|}{a}, & |t| \leq a \\ 0, & |t| > a. \end{cases}$$

#### 1.2.1.3 The Unit Step Function

The unit step function is defined as:

$$u(t) = \begin{cases} 1, & t > 0 \\ 0, & t < 0. \end{cases}$$

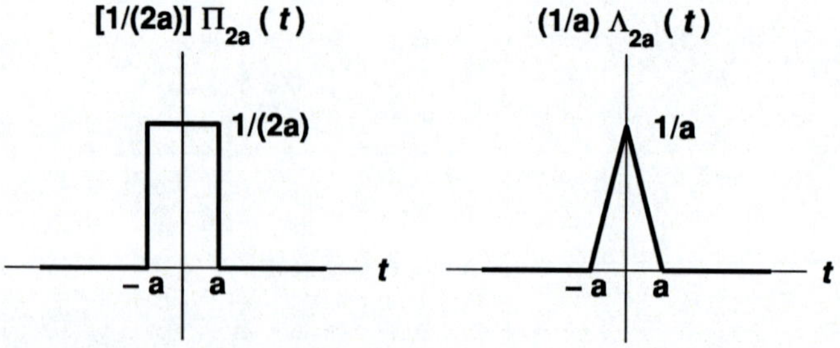


**Fig. 1.8** Weighted rectangular and triangular pulses

This function has a discontinuity at $t = 0$, where its value is *undefined*. If $u(0)$ is chosen as 0.5, $u(t)$ is called the **Heaviside** unit step. The above definition is equivalent to the following integral relation:

$$u(t) = \int_{-\infty}^{t} \delta(t)dt. \tag{1.4}$$

Hence, it follows that $\delta(t) = du(t)/dt$ (see Tables, Formula 7).

## 1.2.2 Time-Domain Representation

This section discusses representation of analog signals and systems in the time domain.

### 1.2.2.1 Mathematical Time-Domain Representation

An analog signal is represented in the time domain By simply defining its values for all time. An LTI analog system is usually characterized in the time domain by defining its impulse response for all values of time. The impulse response is the output of the system when the input is the Dirac delta function, $\delta(t)$. Typically the impulse response is denoted by $h(t)$, as shown in Fig. 1.9.

It can be shown that the input/output (I/O) relationship for the system is described by the convolution of the impulse response $h(t)$ and the input signal $x(t)$ [3]. This is expressed mathematically below:

$$y(t) = h(t) * x(t) = \int_{-\infty}^{\infty} h(\lambda)x(t - \lambda)d\lambda. \tag{1.5}$$

*Note*: for causal systems $h(t) = 0$ for $t < 0$, otherwise instantaneous anticipating quantities like $h(-\tau)x(t + \tau)$, $\tau > 0$, can appear in the above integral when $\lambda = -\tau < 0$, this in turn implies that the system needs to predict future values of the input in order to form the output.

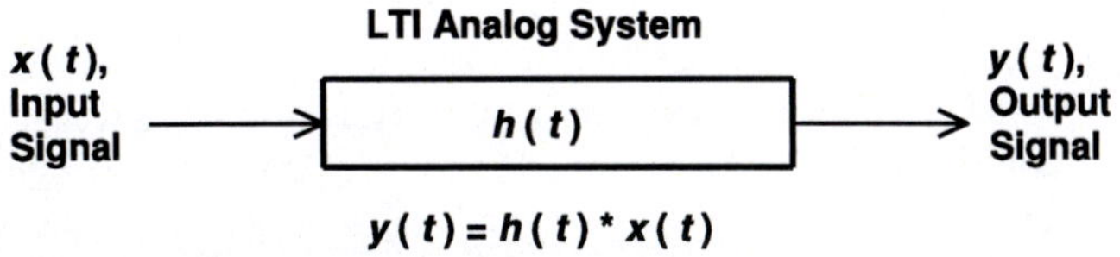


**Fig. 1.9** Time-domain representation of an LTI system

Practical Measurement of the Impulse Response

It can often be difficult to generate an analog impulse function accurately in hardware. It is typically easier to generate the unit-step function. One can show that the impulse response of the system $h(t)$ is related to the unit-step response according to (see Tutorial 8):

$$h(t) = d\rho(t)/dt. \tag{1.6}$$

#### 1.2.2.2 Stability of Analog LTI Systems in the Time Domain

An analog system is BIBO stable if its impulse response is absolutely *summable*, i.e., if

$$\int_{-\infty}^{\infty} |h(t)| < \infty. \tag{1.7}$$

**Example 1** Consider the system described by the impulse response

$$h(t) = \begin{cases} e^{-at}, & t \geq 0 \\ 0, & t < 0, \end{cases}$$

where $a$ is a positive constant. This system is causal (since $h(t) = 0$ for $t < 0$) and stable since $\int_{-\infty}^{\infty} |h(t)| = 1/a < \infty$. This system can be representative of a series capacitor-resistor circuit.

*Exercise*. Find the output of the above system when the input is $x(t) = \cos(t)$.

**Example 2** The system $e^{-|t|}$ is non-causal since $h(t) \neq 0$ for $t < 0$, but it is stable since $\int_{-\infty}^{\infty} |h(t)| = 2 < \infty$.

Although a time domain approach can be used for predicting the stability of systems, it tends to be difficult to do so for complicated systems. The frequency domain approach using the Laplace transform turns out to be more practical. Frequency domain analysis is therefore considered next.

### *1.2.3 Frequency-Domain Representation*

This section considers the representation and analysis of analog signals and systems in the frequency domain. Analysis in this domain is achieved with the help of suitable transformations, which yield equivalent results to those that would be obtained if time domain methods were used. The frequency domain, however, can reveal further characteristics of signals and systems which are useful in dealing with practical systems. As suggested already, for example, stability analysis is easier to perform in the frequency domain.

Ideally, transformations which are used in frequency domain analysis should be **unitary**, i.e., they should be invertible, they should be energy preserving and they should preserve the four key arithmetic operations (in the general sense).

The most important transformations in frequency domain analysis are:

1. **The Fourier Transform:** which is a transformation from the time domain to the frequency domain.
2. **The Laplace Transform (LT):** which is a transformation from the time domain to the generalized (complex) frequency domain.

Both the Fourier and Laplace transforms are **unitary**. That is, they are invertible and energy is preserved in transiting from one domain to the other.

The Fourier transform of a signal is normally referred to as the **spectrum** of the signal.

*The Fourier series (FS)* is a specific transform that is used *only* for periodic signals.

The transforms above can be used to analyze all signals and systems under some basic conditions that are met in almost all practical applications.

#### 1.2.3.1 Fourier Series Representation of Periodic Signals

A **periodic** time signal $x(t)$ with period $T_o$ repeats its values every $T_o$ seconds (see Fig. 1.10). The Fourier series (FS) is a **decomposition** of a periodic time signal into a linear combination of sine and cosine functions. The frequencies of these functions are multiples of the fundamental frequency of the signal, $f_o = 1/T_o$. Such a representation of sines and cosines is called a **trigonometric Fourier series.**

Using Euler's formula $e^{\pm j\theta} = \cos(\theta) \pm j\sin(\theta)$, one can obtain the "exponential" or "complex" Fourier series from the trigonometric series. (As a point of interest, Euler's formula provides a connection between algebra and geometry, and can be proved using a Taylor series expansion of $e^{j\theta}$, $\cos(\theta)$, and $\sin(\theta)$ around $\theta = 0$).

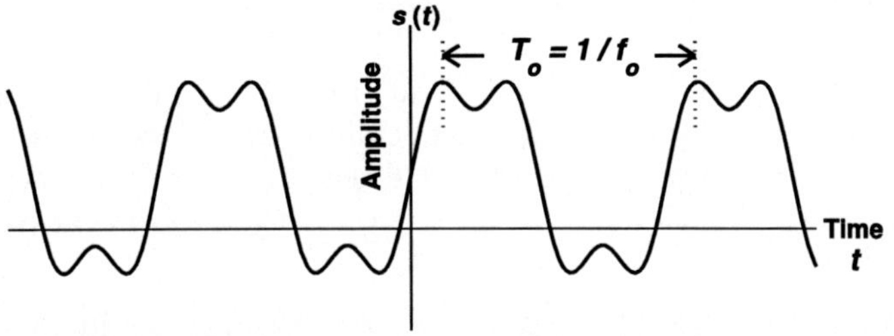


**Fig. 1.10** A periodic signal

Note that although the Fourier series reveals the frequency content of the signal, it is **not strictly speaking a frequency transform** as the representation is still in the time domain.

Trigonometric Fourier Series

If $x(t)$ is a periodic signal with fundamental period $T_o$, then it can be expanded as follows:

$$x(t) = a_o + a_1\cos(\omega_o t) + a_2\cos(2\omega_o t) + \cdots + b_1\sin(\omega_o t) + b_2\sin(2\omega_o t) + \cdots$$
$$= a_o + \sum_{n=1}^{\infty}[a_n\cos(\omega_n t) + b_n\sin(\omega_n t)], \tag{1.8}$$

where $\omega_o = 2\pi f_o = \frac{2\pi}{T_o}$, and:

$$a_o = \frac{1}{T_o}\int_0^{T_o} x(t)dt, \text{ (the constant term)} \tag{1.9}$$

$$a_n = \frac{2}{T_o}\int_0^{T_o} x(t)\cos(n\omega_o t)dt, \tag{1.10}$$

$$b_n = \frac{2}{T_o}\int_0^{T_o} x(t)\sin(n\omega_o t)dt. \tag{1.11}$$

Special Cases

1. If $x(t)$ is odd, then $x(t)\cos(n\omega_o t)$ is odd, hence $a_o = a_n = 0$, and the Fourier series is a series of sines without a constant (zero frequency) term.
2. If $x(t)$ is even, then $x(t)\sin(n\omega_o t)$ is odd, hence, $b_n = 0$ and the Fourier series is a series of cosines.

**Example** Consider the signals $x(t)$ and $s(t)$ depicted in Fig. 1.11. The signal $x(t) - 1/2$ is odd, so one can use results related to odd functions in deducing the Fourier series. The signal $s(t)$ is even. The fundamental period of both signals is $T_o = 2$. Using the formulae in (1.9)–(1.11), the Fourier series of these two signals are found to be:

$$x(t) = \frac{1}{2} + \frac{2}{\pi}\left[\sin(\omega_o t) + \frac{1}{3}\sin(3\omega_o t) + \frac{1}{5}\sin(5\omega_o t) + \cdots\right], \tag{1.12}$$

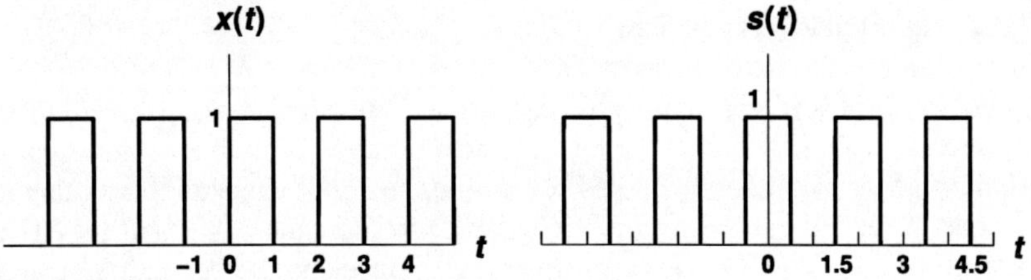


**Fig. 1.11** Two square waves

and

$$s(t) = \frac{1}{2} + \frac{2}{\pi}\left[\cos(\omega_o t) - \frac{1}{3}\cos(3\omega_o t) + \frac{1}{5}\cos(5\omega_o t) + \cdots\right], \tag{1.13}$$

where $\omega_o = \frac{2\pi}{T_o} = \pi$.

Complex Fourier Series

Using Euler's formula, one can write the trigonometric Fourier series in complex exponential form. This form is:

$$x(t) = \sum_{n=-\infty}^{\infty} X_n e^{+jn\omega_o t} = \sum_{n=-\infty}^{\infty} X_n e^{+j2\pi n f_o t} \tag{1.14}$$

where,

$$X_n = \frac{1}{T_o}\int_0^{T_o} e^{-jn\omega_o t}dt = \frac{1}{T_o}\int_0^{T_o} e^{-j2\pi n f_o t}dt, \quad n = 0, 1, 2, \ldots \tag{1.15}$$

Relationship Between CFS and TFS Coefficients

The coefficients of the CFS and the TFS are related as follows:

$$X_o = a_o, \tag{1.16}$$

$$X_n = \frac{1}{2}(a_n - jb_n), \tag{1.17}$$

and,

$$X_{-n} = \frac{1}{2}(a_n + jb_n). \tag{1.18}$$

*Exercise:* Verify the above relations using **Euler's formula**.

#### 1.2.3.2 The Fourier Transform

The Fourier Series (FS) is applicable only to periodic signals, with period $T_o$ and fundamental frequency $f_o = 1/T_o$. One can adapt the FS to be able to analyze *non-periodic* signals by setting $T_o$ in the FS definition to $T_o \to \infty$. With this setting the FS tends to the *Fourier transform* (FT). The time domain signal and its FT are often referred to as a *Fourier transform pair*, since the two quantities can be obtained from one another by transformation/inverse transformation:

$$X(f) = \mathcal{F}\{x(t)\} = \int_{-\infty}^{\infty} x(t)e^{-2\pi ft}dt \tag{1.19}$$

$$x(t) = \mathcal{F}^{-1}\{X(f)\} = \int_{-\infty}^{\infty} X(f)e^{+2\pi ft}dt \tag{1.20}$$

The Fourier transform (FT) reveals the ***frequency content*** of an arbitrary a-periodic (or non-periodic) signal, and is often referred to as the ***spectrum***. If the Fourier transform is applied to the system impulse response, one obtains the ***frequency response*** or ***transfer function*** of the system. This function describes the ratio of the system output to the system input as a function of frequency.

The Fourier transform $X(f)$ of a real signal $x(t)$ is generally complex. To plot the Fourier transform spectrum, one typically plots a magnitude spectrum and a phase spectrum. The magnitude spectrum is a plot of $|X(f)|$ versus frequency $f$, while the phase spectrum is a plot of $\angle X(f)$ versus frequency.If the signal which has been Fourier transformed happens to be the impulse response of the system, the magnitude and phase spectra are referred to as the system magnitude response and the system phase response respectively.

Although the Fourier transform was initially introduced to analyze ***non-periodic*** signals, it can be used to analyze ***periodic signals*** as well. For such signals one obtains Fourier transform expressions containing Dirac delta functions.

*Note*: The CFS pair (1.14)–(1.15) is similar in structure to the FT pair (1.19)–(1.20). For periodic signals with a period of $T_o$, it is easy to see that $X_n = \frac{1}{T_o} X_{1p}(f)|_{f=nf_o}$, where $X_{1p}(f)$ is the FT of one period of $x(t)$, hence, $X_{1p}(f)/T_o$ is the *envelope* of the CFS coefficients.

**Example 1** Consider the signal $x(t) = e^{-2t}u(t)$ (a decaying exponential). Its FT is:

$$X(f) = \int_0^{\infty} e^{-2t}e^{-j2\pi ft}dt = \int_0^{\infty} e^{-t(2+j2\pi f)}dt = \frac{1}{2+j2\pi f}.$$

$$\text{Amplitude spectrum} = |X(f)| = \frac{1}{\sqrt{4+4\pi^2 f^2}}.$$

$$\text{Phase spectrum} = \angle X(f) = \tan^{-1}\left(\frac{2\pi f}{2}\right).$$

Implementation in MATLAB:

In MATLAB, Fourier transforms can be implemented with the help of the *Fast Fourier Transform (FFT) algorithm.* While the FFT is, strictly speaking, applicable only to digital signals, it can be used to give good approximations to FTs for analog signals as well. An example of a signal and its spectra are shown in Fig. 1.12.

```
%Implementation of Fourier Transform

clear,clc,clf % clear all variables, command line, and figure
%----------------------

L=5; % Total signal time (in seconds)

Ts=0.001; % Time step-size (in seconds)

% The less Ts the better is the implementation accuracy.

t=0:Ts:L-Ts; N=length(t); % Time vector

fs=1/Ts; % Sampling Frequency

F=fs/N; % Frequency step-size (in Hertz)

f=-fs/2:F:fs/2-F; % Implementation frequency range

%----------------------

% The time-domain signal

a=2; x=exp(-a*t);

%----------------------

% Spectra

X=fftshift(fft(x))/fs; % FT centered at f=0 Hz.

AX=abs(X); % FT Magnitude
```

```
PX=angle(X);  % Phase, restricted to (-pi,pi).

% You can also use PX=atan2(imag(X),real(X));

%----------------------

% Checking with theoretical formulas

Xth=1./(a+j*2*pi*f); AXth=abs(Xth);  % Theoretical FT (see Tables)

PXth=angle(Xth); % Theoretical phase. Should be identical with PX.

%----------------------

figure(1),clf

set(gcf,'units','centimeters'), set(gcf,'position',[1 1 22 5])

subplot(1,3,1)

plot(t,x,'k','linewidth',2), hold on,

plot([-L L],[0 0],'k',[0 0],[0 1.2],'k','linewidth',1), hold off

axis([-5 5 0 1.2])

xlabel('\bf Time, sec','fontsize',12)

ylabel('\bf Signal Amplitude','fontsize',12)

text(0,1.0,'\bf {\itx}({\itt})','fontsize',14)

subplot(1,3,2)

plot(f,AX,'k','linewidth',2), hold on

plot([-12 12],[0 0],'k',[0 0],[-4 4],'k')

axis([-5 5 0 .6])

xlabel('\bf Frequency, Hz','fontsize',12)

ylabel('\bf FT Magnitude','fontsize',12)

text(0,.5,'\bf |{\itX}({\itf})|','fontsize',14)

subplot(1,3,3) plot(f,PX,'k','linewidth',2),hold on

plot([-12 12],[0 0],'k',[0 0],[-4 4],'k'),hold on
```

```
axis([-10 10 -pi pi])

xlabel('\bf Frequency, Hz','fontsize',12)

ylabel('\bf FT  Phase (rad)','fontsize',12)

text(0,2.5,'\bf \angle{\itX}({\itf})','fontsize',14)
```

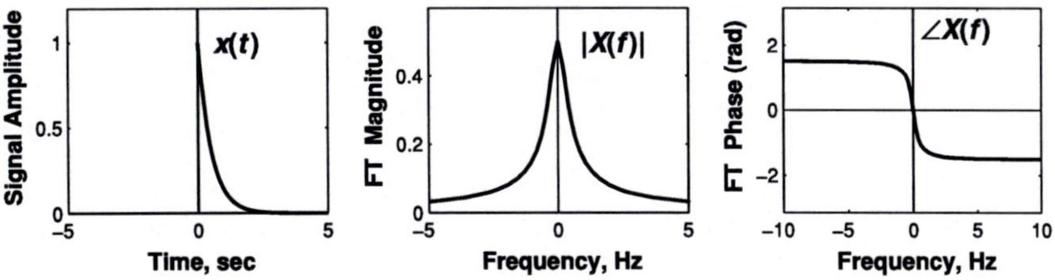


**Fig. 1.12** A decaying exponential signal with its magnitude and phase spectra

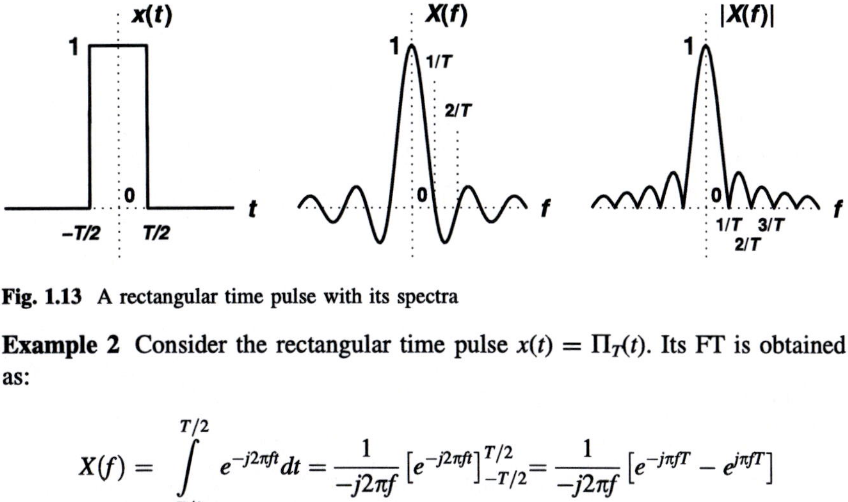


**Fig. 1.13** A rectangular time pulse with its spectra

**Example 2** Consider the rectangular time pulse $x(t) = \Pi_T(t)$. Its FT is obtained as:

$$X(f) = \int_{-T/2}^{T/2} e^{-j2\pi ft} dt = \frac{1}{-j2\pi f}\left[e^{-j2\pi ft}\right]_{-T/2}^{T/2} = \frac{1}{-j2\pi f}\left[e^{-j\pi fT} - e^{j\pi fT}\right]$$

$$= \frac{1}{\pi f} \cdot \frac{e^{j\pi fT} - e^{-j\pi fT}}{2j} = \frac{\sin(\pi f\mathrm{T})}{\pi f} \ \textbf{(Using Euler's Formula)}$$

$$= T\frac{\sin(\pi f\mathrm{T})}{\pi fT} = T\mathrm{sinc}(\pi fT).$$

The signal, along with its FT and magnitude spectrum, are shown in Fig. 1.13.

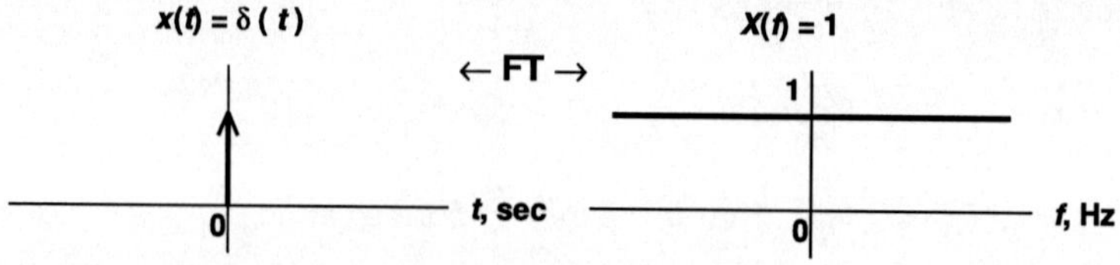


**Fig. 1.14** The time-domain delta function and its spectrum

**Example 3** If $x(t) = \delta(t)$, then its spectrum is given by:

$$X(f) = \int_{-\infty}^{\infty} \delta(t)\underbrace{e^{-j2\pi ft}}_{g(t)}dt = g(0) = 1.$$

(using the definition of the delta function from Tables, Formula 16) (see Fig. 1.14). It is seen that for the infinitely narrow time domain delta function the corresponding spectrum is infinitely wide. In general, if the time duration of a signal is narrow, its frequency spread tends to be wide, and vise versa.

## Properties of the FT

The properties of the FT are detailed in the **Tables** at the end of this book. The reader should prove properties himself/herself as a means of deepening their understanding of the Fourier transform (These proofs can be found in standard references such as [3]). Some of these properties are discussed below.

1. *Duality of the FT:*

$$\text{If } x(t) \stackrel{\mathcal{F}}{\longleftrightarrow} X(f) \quad \text{is a FT pair, then:}$$

$$X(t) \stackrel{\mathcal{F}}{\longleftrightarrow} x(-f) \quad \text{is a FT pair}$$

Note that the time and frequency variables are *exchanged* in the above relations. If $x(t)$ is *even*, then the duality relations become even simpler:

$$X(t) \stackrel{\mathcal{F}}{\longleftrightarrow} x(f)$$

**Example 1** Previously it has been shown that

$$\Pi_T(t) \stackrel{\mathcal{F}}{\longleftrightarrow} T\,\text{sinc}(fT).$$

Using the duality property it follows that:

$$B\,\text{sinc}(Bt) \stackrel{\mathcal{F}}{\longleftrightarrow} \Pi_B(f)$$

Hence, a time-domain sinc function will be transformed into a frequency rectangular box function in the frequency domain.

**Example 2** It has previously been seen that:

$$\delta(t) \stackrel{\mathcal{F}}{\longleftrightarrow} 1$$

By duality it follows that:

$$1 \stackrel{\mathcal{F}}{\longleftrightarrow} \delta(f)$$

2. *Time shift:*

$$x(t - t_o) \stackrel{\mathcal{F}}{\longleftrightarrow} X(f)\, e^{-j2\pi f t_o}$$

3. *Frequency shift:*

$$\text{If } x(t) \stackrel{\mathcal{F}}{\longleftrightarrow} X(f), \text{ then } \quad x(t)e^{j2\pi f t_o} \stackrel{\mathcal{F}}{\longleftrightarrow} X(f - f_o).$$

Example 3 Since

$$1 \stackrel{\mathcal{F}}{\longleftrightarrow} \delta(f) \quad \text{(as shown in Example 2 above)}$$

then

$$e^{j2\pi f_o t} \stackrel{\mathcal{F}}{\longleftrightarrow} \delta(f - f_o),$$

Hence, the Fourier transform of a complex exponential is a frequency-shifted delta function.

Fourier Transform of Sinusoids

From Euler's formula one can write:

$$e^{j\theta} = \cos(\theta) + j\sin(\theta) \tag{1.21}$$

$$e^{-j\theta} = \cos(\theta) - j\sin(\theta) \tag{1.22}$$

$$\therefore \cos(\theta) = \frac{1}{2}\left(e^{j\theta} + e^{-j\theta}\right); \ \sin(\theta) = \frac{1}{2j}\left(e^{j\theta} - e^{-j\theta}\right),$$

$$\therefore \cos(\omega_o t) = \cos(2\pi f_o t) = \frac{1}{2}\left(e^{j2\pi f_o t} + e^{-j2\pi f_o t}\right),$$

$$\therefore \mathcal{F}\{\cos(\omega_o t)\} = \frac{1}{2}\delta(f - f_o) + \frac{1}{2}\delta(f + f_o),$$

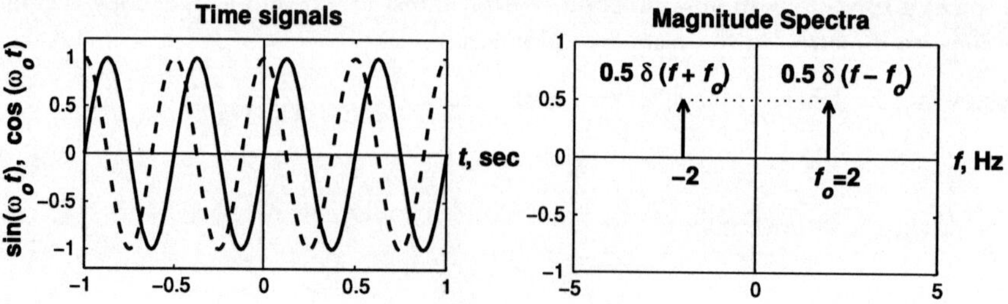


**Fig. 1.15** Sine and cosine (with the same $f_o$) have identical magnitude spectra

using the result from the previous example. Similarly,

$$\mathcal{F}\{\sin(\omega_o t)\} = \frac{1}{2j}\left[\delta(f - f_o) - \frac{1}{2}\delta(f + f_o)\right].$$

Hence, the magnitude spectra of $\sin(\omega_o t)$ and $\cos(\omega_o t)$ are identical, as shown in Fig. 1.15. The phase spectra, however, would be different.

### Fourier Transform of Periodic Signals

A periodic signal $x(t)$ can be represented by a Fourier Series Expansion according to:

$$x(t) = \sum_{k=-\infty}^{\infty} X_k e^{+j2n\pi f_o t},$$

where $\{X_k\}$ are the FS coefficients. Taking the Fourier transform of both sides yields:

$$\begin{aligned}\mathcal{F}\{x(t)\} &= \mathcal{F}\left\{\sum_{k=-\infty}^{\infty} X_k e^{+j2n\pi f_o t}\right\}\\ &= \sum_{k=-\infty}^{\infty} X_k\, \mathcal{F}\{e^{+j2n\pi f_o t}\} = \sum_{k=-\infty}^{\infty} X_k\, \delta(f - kf_o).\end{aligned}$$

Hence, the FT of a periodic signal $x(t)$ with period $T_o$ is a sum of frequency impulses at integer multiples of the fundamental frequency $f_o$ (i.e., at $f = kf_o$), weighted by the FS coefficients.

**Example** The complex Fourier series expansion and Fourier transform of the square wave shown in Fig. 1.16 are given respectively by:

$$x(t) = \sum_{k=-\infty}^{\infty} \underbrace{\left[\frac{1}{2}\operatorname{sinc}\left(\frac{k}{2}\right)\right]}_{X_k} e^{j\frac{1}{2}\pi k t} \quad \text{and} \quad X(f) = \sum X_k\, \delta\left(f - \frac{k}{4}\right)$$

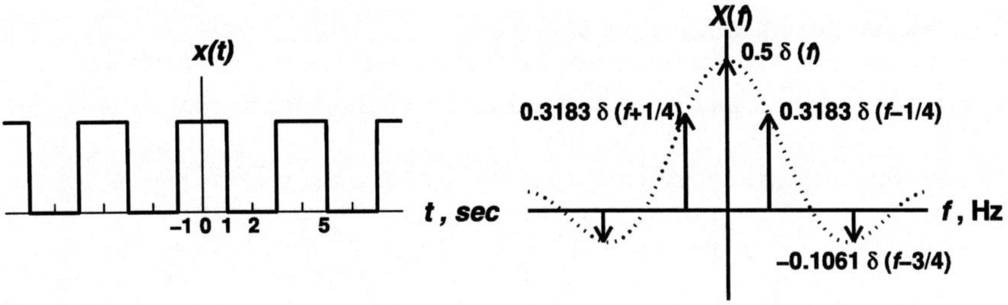


**Fig. 1.16** A square wave and its Fourier transform (*full line*) and the envelope of the Fourier transform (*dotted line*)

Here $T_o = 4$ sec, $f_o = 1/4$ Hz. The envelope of the $X_k$ coefficients in the frequency domain is obtained by substituting $f$ for $kf_o = k/4$, hence $(1/2)\mathrm{sinc}(k/2)$ becomes $(1/2)\mathrm{sinc}(2f)$.

The above rectangular pulse is useful in many applications. Its general form is given by: $\sum_{k=-\infty}^{\infty} A\,\Pi_T(t - nT_o)$, and its Fourier coefficients are $X_k = (AT/T_o)\{\mathrm{sinc}(kf_oT)$, where $T$ is the duration of the "ON" state. The envelope of these coefficients is given by $E(f) = (AT/T_o)\mathrm{sinc}\ (fT) = X_{1p}(f)/T_o$, where $X_{1p} = \mathcal{F}\{\Pi_T(t)\} = FT$ of one period.

**MATLAB:** The rectangular pulse train can be simulated in MATLAB as follows:

```
r=T/To*100;

fo=1/To;

wo=2*pi*fo;

x=A*0.5*(1+square(wo*(t+T/2),r));
```

### 1.2.3.3 The Laplace Transform

The Laplace transform (LT) is a generalization of the Fourier transform in which one decomposes a signal into the sum of *decaying* complex exponentials, rather than simply complex exponentials. The incorporation of amplitude as well as frequency information into the basis functions of the LT introduces a number of advantages. Firstly, it enables the LT to deal with a wider class of signals—including many signals which have no FT (such as the unit ramp $t^n u(t)$). Secondly, the LT formulation is more naturally suited to analyzing the stability of a system. Because of these advantages, the LT is the main tool for representing and analyzing analog (continuous-time) feedback systems, where stability is of extreme importance. There are two definitions of the LT as detailed below.

## The Double-Sided Laplace Transform

The double-sided Laplace transform (DLT) definition incorporates an integral which is evaluated across all possible values of time (both negative and positive):

$$X_d(s) = \mathcal{L}_d\{x(t)\} = \int_{-\infty}^{\infty} x(t)e^{-st}dt,$$

where $s = \sigma + j\omega$ is the **complex frequency** variable. Note that:

$$\mathcal{L}_d\{x(t)\} = \mathcal{F}\{x(t)e^{-st}\}.$$

The inverse DLT, $\mathcal{L}_d^{-1}\{X_d(s)\}$, can be used to recover $x(t)$ as follows:

$$x(t) = \mathcal{L}_d^{-1}\{X_d(s)\} = \frac{1}{2\pi j}\int_{\sigma_1-\infty}^{\sigma_1+\infty} X_d(s)e^{st}ds,$$

where $\sigma_1$ is any arbitrary value of $\sigma$. Note that $\mathcal{L}_d^{-1}$ requires integration in the complex plane.

## The Single-Sided Laplace Transform

In real-world applications one normally deals with casual systems in which the impulse response is only non-zero for positive values of time. In recognition of this fact, the single-sided Laplace transform (SLT) is defined to allow for only causal signals and systems. That is, the integral within its definition is only evaluated for positive and zero values of time. The SLT is very useful in calculating the response of a system to a causal input, especially when the system is described by a linear constant coefficient differential equation with non-zero initial conditions.

Note that the properties of the DLT are *not* exactly the same as those of SLT. This book will concentrate only on the SLT, which will be referred to hereafter simply as the Laplace Transform (LT). Its definition is:

$$X(s) = \mathcal{L}\{x(t)\} = \int_{0^-}^{\infty} x(t)e^{-st}dt,$$

Note that $0^-$ is used in the above definition rather than 0 to allow for analyzing delta functions $x(t) = \delta(t)$. The inverse transform (ILT) is given by:

$$x(t) = \mathcal{L}^{-1}\{X(s)\} = \frac{1}{2\pi j}\int_{\sigma_1-\infty}^{\sigma_1+\infty} X_d(s)e^{st}ds.$$

## Properties of the LT

The properties of the LT can be found in Tables. Some of the most useful properties are:

$$(1)\frac{dx(t)}{dt}\overset{\mathcal{L}}{\longleftrightarrow} sX(s)-x(0^-)$$

$$(2)\int_{0^-}^{t} x(\lambda)d\lambda \overset{\mathcal{L}}{\longleftrightarrow}\frac{X(s)}{s}$$

These properties effectively transform differentiation and integration into algebraic quantities. Because of these properties the LT can be used to transform differential equations into algebraic equations. This has application in many areas, most notably the analysis of electric circuits.

## Region of Convergence of the LT

The region of convergence (ROC) is the region of the s-plane in which the LT is convergent (i.e., has finite values).

**Example** Find the Laplace transform and its ROC for the signal $x(t) = e^{\alpha}tu(t)$, where $\alpha$ is a constant.

*Solution:*

$$\begin{aligned} X(s) &= \int_{0^-}^{\infty} x(t)e^{-st}dt = \int_{0^-}^{\infty} e^{-\alpha t}e^{-st}dt \\ &= \int_{0^-}^{\infty} e^{-(\alpha+s)t}dt = \left[\frac{e^{-(\alpha+s)t}}{\alpha+s}\right]_0^{\infty} \\ &= -\frac{1}{\alpha+s}\left[\lim_{t\to\infty}\left\{e^{-(\alpha+\sigma)t}\cdot e^{-j\omega t}\right\}-e^0\right] \end{aligned}$$

The term $e^{-j\omega t}$ is always bounded, the ROC therefore depends on only the term $e^{-(\alpha+\sigma)t}$. Now,

$$\lim_{t\to\infty} e^{-(\alpha+\sigma)t} = \begin{cases}\infty, & \text{when} -(\alpha+\sigma)>0 \text{ or } \sigma< -\alpha \\ 0, & \text{when} -(\alpha+\sigma)<0 \text{ or } \sigma> -\alpha\end{cases} \tag{1.23}$$

Hence, the ROC is the region defined by $\sigma = \text{Re}\{s\} > -\alpha$ (see Fig. 1.17). For values of $\alpha$ which do lead to a convergent LT, the LT is given by:

$$X(s) = \frac{1}{s+\alpha} \text{ (note that } \alpha \text{ can be positive or negative).}$$

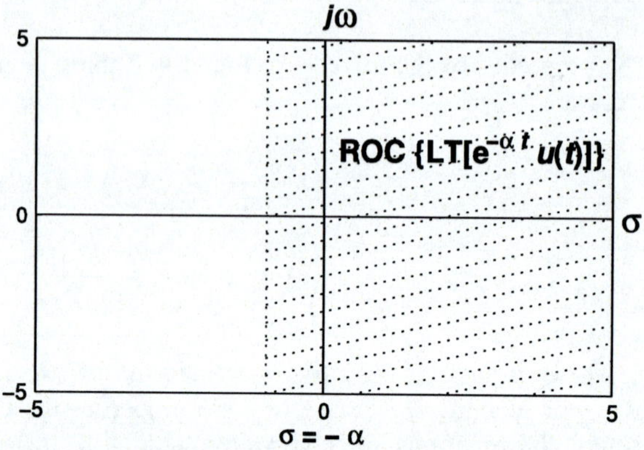


**Fig. 1.17** Region of convergence (*shaded*) of the Laplace transform for $x(t) = e^{-\alpha t}u(t)$

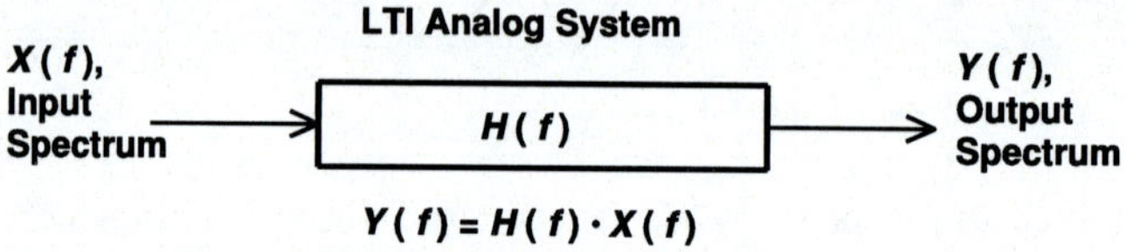


**Fig. 1.18** Frequency-domain representation of an LTI system

#### 1.2.3.4 Mathematical Frequency-Domain Representation

An analog time-domain signal $x(t)$ can be represented equivalently in the frequency domain by its Fourier transform. Similarly, an LTI system with impulse response $h(t)$ can be represented equivalently in the frequency domain by its transfer function, $H(f)$ (see Fig. 1.18).

Recall that the system output can be described in the time domain as the convolution of the input signal and the impulse response of the system:

$$y(t) = h(t) * x(t)$$

Recall also that convolution in the time domain is transformed into multiplication in the frequency domain, and vise versa (see Tables, Laplace Transform Pairs and Theorems). In the frequency domain, therefore, the system output is given by the multiplication of the transfer function by the Fourier transform of the input signal:

$$Y(f) = H(f) \cdot X(f)$$

where $Y(f) = \mathcal{F}\{y(t)\}$ and $X(f) = \mathcal{F}\{x(t)\}$. Similar results are obtained if one uses the Laplace transform:

$$Y(s) = H(s) \cdot X(s)$$

Eigenfunctions of LTI Analog Systems

If the input signal to an LTI analog system is $x(t) = e^{j\omega t}$, then

$$\begin{aligned} y(t) &= h(t) * x(t) = x(t) * h(t) \\ &= \int_{-\infty}^{\infty} h(\lambda)e^{j\omega(t-\lambda)}d\lambda = e^{j\omega t}\int_{-\infty}^{\infty} h(\lambda)e^{-j\omega\lambda}d\lambda \end{aligned} \quad (1.24)$$

If one defines $H(\omega) = \int_{-\infty}^{\infty} h(\lambda)e^{-j\omega\lambda}d\lambda$, then $y(t) = e^{j\omega}tH(\omega)$. That is, the output of the system is just a scaled version of the input. In other words, $e^{j\omega}t$ is an *eigenfunction* of the system, and the associated eigenvalue is $H(\omega)$.

### 1.2.3.5 Stability of Analog LTI Systems-Frequency Domain

In a previous subsection system stability in the time domain was addressed. Here system stability is considered in the framework of the complex frequency domain. In this latter domain an analog system is typically characterized by its transfer function, $H(s)$ (i.e. by the Laplace transform of its impulse response $h(t)$). It can be shown that any practical transfer function $H(s)$, can be re-expressed as the ratio of two polynomial functions of $s$: $H(s) = N(s)/D(s)$. It can also be shown that an analog system is BIBO stable if and only if [3]:

1. The degree of $N(s) <$ degree of $D(s)$.
2. All poles (i.e., zeros of the denominator $D(s)$) are in the *left half* of the s-plane.

**Example** Plot the pole-zero diagram of the system

$$H(s) = \frac{s(s+1)}{(s+2)^2(s+3)}$$

and conclude whether the system is BIBO stable or not.

*Solution:* Zeros are located at $s = 0, -1$; poles are located at $s = -2$ (double) and $s = -3$. The numerator polynomial is of a lower order than the denominator polynomial, All poles are in the left half of the $s$ plane. See the pole-zero diagram in Fig. 1.19. The system is therefore stable.

## 1.2.4 Signal Correlation and Its Applications

The correlation between two deterministic energy signals $s(t)$ and $r(t)$, is defined by the integral:

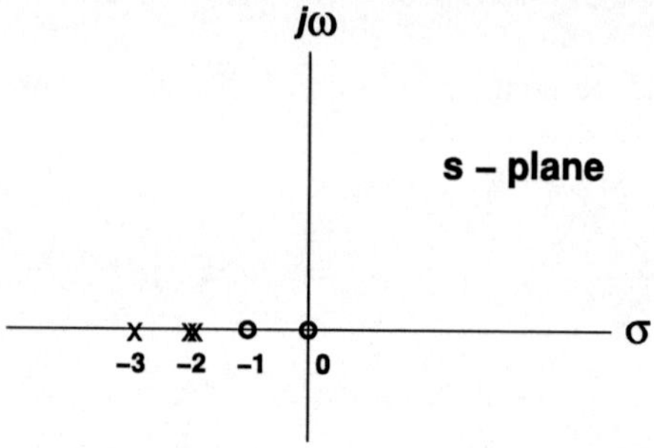


**Fig. 1.19** Pole-zero diagram of the system $H(s) = s(s + 1)/(s + 2)^2 (s + 3)$

$$R_{sr} = \int_{-\infty}^{\infty} s(\lambda)r(\tau + \lambda)d\lambda \quad (1.25)$$

When the two signals being correlated are different it is common to refer to the correlation integral as the *cross-correlation function*, If the two signals are the same, the integral is typically referred to as the *auto-correlation* function.

This cross-correlation function is of the same form as the convolution integral in (1.3), but the argument of the second function has a "+" sign in place of a "−" sign and also incorporates the time-delay $\tau$ instead of the true time variable, $t$. Note also that $R_{sr}(\tau) = R_{rs}(\tau)$. The correlation gives an indication of how the two signals are "related" to each other when the time difference between them is $\tau$. In other words, it is a *measure of similarity* between two signals separated by $\tau$. If $r(t)$ equals $s(t)$, the correlation is referred to as the *auto-correlation* function of the signal $s(t)$, and is denoted by $R_s(\tau)$. For a periodic deterministic signal $x(t)$ with period $T_o$, the auto-correlation function is defined as:

$$R_x(\tau) = \frac{1}{T_o} \int_{0}^{T_o} x(\lambda)x(\tau + \lambda)d\lambda$$

There are also correlation formulas for random signals that will be defined later.

It can be shown that the auto-correlation function has the following properties (see Tutorial 26):

**P1**: $R_x(\tau)$ is even (or symmetric about time-delay axis), $R_x(\tau) = R_x(-\tau)$.

**P2**: $R_x(\tau)$ always has its absolute maximum at the origin, $\tau = 0$.

(see, for example, Fig. 1.20, which shows the autocorrelation function of a non-periodic signal).

The correlation integral between two signals tends to have a large maximum value at $\tau = 0$ if the signals are exactly the same (auto-correlation), and has a finite non-zero value over some range of $\tau$ when the two signal are *somewhat similar*. It tends to be *approximately zero* if the two signals are very dissimilar (as would be the case, say, if one of the signals was a deterministic signal and the other was

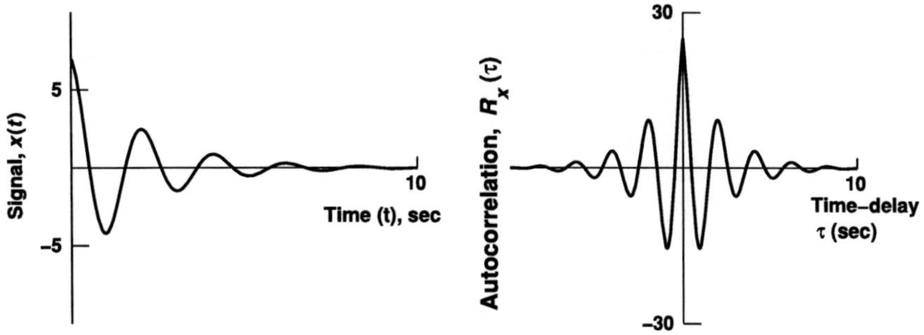


**Fig. 1.20** The signal $x(t) = 7\,e^{-t/2}\cos(3t)$ and its autocorrelation function

random noise). These properties of the correlation function are frequently used as the basis for detecting signals in noise.

### *1.2.5 Signal Power and Energy*

If $x(t)$ is a signal, then its *instantaneous power* at any time $t$, is denoted by $p(t)$, and is defined as the power dissipated in a 1Ω-resistor when a voltage of amplitude $x(t)$ volts is applied. This power is given by the multiplication of the voltage and the current as defined below:

$$p(t) = |x(t)|^2,$$

where absolute value is used to cater for the possibility of complex signals.

Since energy is defined as the integral of power w.r.t. time, the instantaneous energy ($e(t)$), of the signal at any time instant $t$, is given by:

$$e(t) = p(t)dt = |x(t)|^2 dt,$$

and the total energy $E$ of the signal is given by:

$$E = \lim_{T\to\infty} \int_{-T/2}^{T/2} |x(t)|^2 dt.$$

The total average power is the time average of the total Energy, and is given by:

$$P = \lim_{T\to\infty} \frac{E}{T} = \lim_{T\to\infty} \frac{1}{T} \int_{-T/2}^{T/2} |x(t)|^2 dt.$$

Signals are classified as *power signals, energy signals*, or *neither*. Power signals have finite power (hence, infinite energy, according to the above definitions).

Energy signals, on the other hand, have finite energy (hence, zero power). Most signals in practice are either power or energy signals. There are signals, however, that are neither power nor energy signals. The unit ramp $x(t) = r(t) = tu(t)$, is one such signal—it has infinite power and infinite energy.

#### 1.2.5.1 Power in Periodic Signals

If $x(t)$ is a periodic signal with a period $T_o$, then the total energy in this signal is:

$$E = \lim_{T_o \to \infty} \int_{-T_o/2}^{T_o/2} |x(t)|^2 dt \to \infty,$$

but the signal power is *finite*. Periodic signals are therefore power signals, and the power within them can be expressed as:

$$P = \frac{1}{T_o} \int_{-T_o/2}^{T_o/2} |x(t)|^2 dt = \frac{1}{T_o} \int_0^{T_o} |x(t)|^2 dt.$$

**Example** If $x(t) = A\sin(\omega_o t)$, then the signal power is given by:

$$P = \frac{1}{T_o} \int_0^{T_o} A^2 \sin^2(\omega_o t) dt = \frac{A^2}{\omega_o T_o} \int_0^{T_o} \left[\frac{1}{2} - \frac{1}{2}\cos(2\omega_o t)\right] \omega_o dt \text{ (Tables, Formula 3).}$$

Letting $\phi = \omega_0 t$ and $\omega_o T_o = 2\pi$ in the above equation yields

$$P = \frac{A^2}{2\pi} \int_0^{2\pi} \left[\frac{1}{2} - \frac{1}{2}\cos(2\phi)\right] d\phi = \frac{A^2}{2\pi} \left[\frac{1}{2}\phi - \frac{1}{4}\sin(2\phi)\right]_0^{2\pi} = \frac{A^2}{2} \text{ W/Ohm}$$

The same result is obtained for the signal $x(t) = A\cos(\omega_o t)$.

#### 1.2.5.2 Parseval's Theorem

Parseval's Theorem states that the power in periodic signals and the energy in non-periodic signals can equivalently be obtained from the frequency domain according to the following relations:

1. For periodic signals: $P = \sum_{k=-\infty}^{\infty} |X_k|^2$ where the $\{X_k\}$ are the CFS coefficients.

2. For non-periodic signals: $E = \int_{-\infty}^{\infty} |X(f)|^2 df$, where $X(f)$ is the FT of the signal. A plot of $|X_k|^2$ versus $f$ (defined at discrete frequencies $f = kf_o$, $f_o$ being the fundamental frequency) is known as the ***power spectrum***, or the ***power spectral density*** (PSD) of the signal. The function $|X(f)|^2$ versus $f$ is called the ***energy spectrum*** or ***energy spectral density*** (ESD) of the signal.

#### 1.2.5.3 The Wiener–Kinchin Theorem

The Wiener–Kinchin Theorem states that for a periodic signal $x(t)$, its PSD is the FT of its autocorrelation function:

$$R_x(\tau) = \frac{1}{T_o}\int_0^{T_o} x(\lambda)x(\tau+\lambda)d\lambda \overset{\mathcal{F}_{\tau\leftrightarrow f}}{\longleftrightarrow} \underbrace{\sum_{k=-\infty}^{\infty} |X_k|^2 \delta(f - k/T_o)}_{PSD}.$$

For an energy signal $x(t)$, the ESD is the FT of its autocorrelation:

$$R_x(\tau) = \int_{-\infty}^{\infty} x(\lambda)x(\tau+\lambda)d\lambda \overset{\mathcal{F}_{\tau\leftrightarrow f}}{\longleftrightarrow} \underbrace{|X(f)|^2}_{ESD}.$$

A similar relation for random signals will be presented later.

#### 1.2.5.4 Examples

**Example 1:** A periodic current signal $x(t)$ flowing through a 1Ω resistor has the form:

$$x(t) = 10\cos(2t) + 4\sin(6t) \text{ A},$$

1. Determine whether $x(t)$ is a power or energy signal.
2. Find the complex Fourier series of $x(t)$.
3. Find the fundamental period $T_o$ and the fundamental radian frequency, $\omega_o$.
4. Find the Fourier transform of the signal.
5. Plot the magnitude spectrum of the signal.
6. Plot the power spectrum of the signal.
7. Find the signal power from the time domain and from the frequency domain.

*Solution:*

1. Since $x(t)$ is periodic, it is a power signal.
2. Since $x(t)$ is already in the form of a trigonometric FS, one can just use Euler's formula (Tables, Formula 1) to obtain the complex FS:

$$x(t) = 10\left[\frac{1}{2}\left(e^{j2t} + e^{-j2t}\right)\right] + 4\left[\frac{1}{2j}\left(e^{j6t} - e^{-j6t}\right)\right].$$

3. The fundamental frequency $\omega_o$ is determined by the *lowest frequency* in the signal, i.e., $\omega_o = 2$ (rad/s) $= 2\pi f_o = 2\pi/T_o$, hence, the fundamental period is $T_o = \pi$ (s), and $f_o = 1/\pi$ (Hz). Note that if the two frequencies are not *multiples* of each other, one cannot use the above method and the two components are in fact two different periodic signals.
4. Using Fourier transform Tables, the following expression is obtained:

$$X(f) = 5\left[\delta\left(f - \frac{1}{\pi}\right) + \delta\left(f + \frac{1}{\pi}\right)\right] - 2j\left[\delta\left(f - \frac{3}{\pi}\right) + \delta\left(f + \frac{3}{\pi}\right)\right].$$

5. See Fig. 1.21 (left).
6. See Fig. 1.21 (right).
7. In the time domain, power is given by:
   $P = \frac{1}{T_o}\int_0^{T_o} |x(t)|^2 dt$. Using the well-known result for the power in a sinusoidal signal yields:
   $P = 10^2/2 + 4^2/2 = 58$ W. In the frequency domain one can apply Parseval's theorem to get: $P = \sum X_k^2 = 2^2 + 5^2 + 5^2 + 2^2 = 58$ W.

**Example 2** A current signal $x(t)$ flowing through $1\Omega$ resistor has the form $x(t) = e^{-t/10}u(t)$ A.

1. Determine whether $x(t)$ is a power or energy signal.
2. Find the Fourier transform of $x(t)$.
3. Find the ESD of $x(t)$.

*Solution:*

1. $E = \int_{-\infty}^{\infty} x^2(t)dt = \int_0^{\infty} e^{-2t/10}dt = 5$ Joules, hence, $x(t)$ is an energy signal.
2. From **Tables-Fourier Transform Pairs**, $X(f) = \frac{1}{0.1 + j2\pi f}$.
3. $\text{ESD}(f) = |X(f)|^2 = \frac{1}{0.01 + 4\pi^2 f^2}$ (J/Hz).

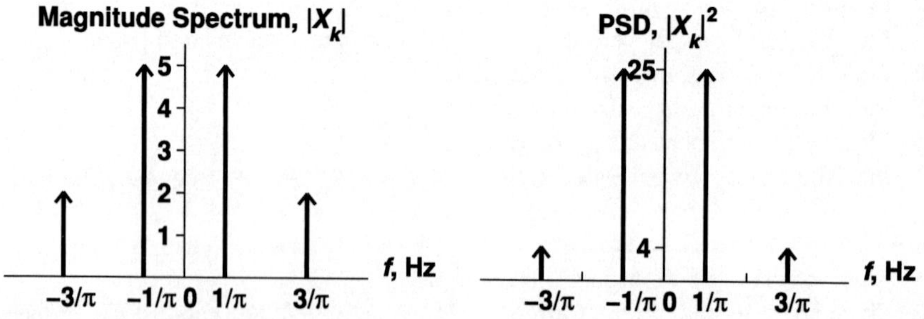


**Fig. 1.21** Magnitude spectrum and PSD of $x(t) = 10\cos(2t) + 4\sin(6t)$

## 1.3 Random Signals

### 1.3.1 *Definition*

A random signal is a signal whose values cannot be perfectly predicted or known a priori. Rather, the signal values need to be characterized by a probability function.

To understand and analyze such signals, it is necessary to use tools which are derived from the theory of probability and statistics.

### 1.3.2 *Overview of Probability and Statistics*

#### 1.3.2.1 Probability and Sample Space

Probability

Probability is a mathematical construct that describes the statistical regularity of the outcomes of a repeating situation or experiment.

Sample space

The sample space is the set of all possible outcomes of a repeating experiment.

**Example 1** In a coin-tossing experiment, outcomes are either "heads" or "tails", hence the sample space is $S = \{h, t\}$. If tossing is repeated a large number of times, then there will be approximately 50% heads and 50% tails. Mathematically this is expressed as: $p(h) = 0.5$, $p(t) = 0.5$.

*Note*: If the sample space of a countable set of events $\{e_k\}$ is $S = \{e_k | k = 1, ..., N\}$, then $\sum_{k=1}^{N} p(e_k) = 1$, i.e., the summation of the probabilities of all possible events (outcomes) should be 1 (100%).

**Example 2** In a die-tossing experiment, the possible outcomes are:

$$S = \{1 - \text{dot}, 2 - \text{dots}, 3 - \text{dots}, 4 - \text{dots}, 5 - \text{dots}, 6 - \text{dots}\}.$$

If a die is tossed a large number of times $N$, then the expected number of each possible outcome is $N/6$. Alternatively, one can say that the probability of each possible outcome is 1/6.

**Example 3** Consider a DC signal which is contaminated by random additive noise. There is a change in amplitude from one sample to the next. The set of all possible changes is given by $S$, where these changes depend on the probability function (distribution) of the noise process. Typically for a random noise signal, $S = \mathbb{R}$ (the set of real numbers).

#### 1.3.2.2 Random Variables

A random variable $X$ is a *real-valued function* whose domain is the set of all possible events (i.e., the sample space $S$) of a repeating experiment.

$$X : S \to M, \text{ where } M \subseteq \mathbb{R} \quad (M \text{ is a subset of } \mathbb{R})$$

**Example** In a coin-tossing experiment, if one defines $X(h) = 1$, $X(t) = -1$, then $X{:}\{h, t\}{:} \to \{1, -1\}$ is a random variable.

*Notes*:

1. Random variables can be discrete (as in the above example) or continuous (as in the amplitude of Gaussian noise).
2. In case of noise where $S = \mathbb{R}$, one can define the random variable $X$ as the noise amplitude itself. That is:

$$X : \mathbb{R} \to \mathbb{R} | X(r) = r \forall r \in \mathbb{R}$$

#### 1.3.2.3 Joint Probability

Assume that one has two experiments, Experiment 1 and Experiment 2. Assume also that $A$ is a possible outcome (or event) for Experiment 1 and that $B$ is a possible outcome for Experiment 2. Then the joint probability of $A$ and $B$ (denoted by $p(A \cap B)$) is the probability that $A$ is the outcome of Experiment 1 and $B$ the outcome of Experiment 2.

**Example** If $n(t)$ is noise, and one defines the events $A = \{n(t_1) > 0.1\}$ and $B = \{n(t_2) > 0.5\}$, then

$$p(A \cap B) = p\{n(t_1) > 0.1 \text{ and } n(t_2) > 0.5\}.$$

that is, $p(A \cap B)$ is the probability that the sample of noise at $t_1$ is greater than 0.1 and the sample of noise at $t_2$ is greater than 0.5.

#### 1.3.2.4 Conditional Probability

Conditional probability (CP) is the probability of an event $A$ given that an event $B$ has already occurred. CP is denoted by the expression $p(A|B)$. The conditional probability and the joint probability are related to one another according to:

$$p(A|B) = \frac{p(A \cap B)}{p(B)}.$$

### 1.3.2.5 Independent Events

The events $A$ and $B$ are independent when $p(A|B) = p(A)$. Hence, using the above formula: $p(A \cap B) = p(A) \cdot p(B)$.

**Example 1** A box contains 20 cubes, 15 of them red and 5 of them blue. Two cubes are to be selected randomly (without replacement). What is the probability that the 1st is red and the 2nd is blue?

*Solution:* Let $R = \{\text{1st cube is red}\}$; $B = \{\text{2nd cube is blue}\}$.

$$p(R) = \frac{15}{20} = \frac{3}{4}; \; p(B|R) = \frac{5}{19}$$

$$p(B \cap R) = p(R) \cdot p(B|R) = \frac{3}{4} \cdot \frac{5}{19} = \frac{15}{76}.$$

**Example 2** Two coins are tossed. What is the probability that the first one is a head and the second is a tail?

*Solution:* $p(H \cap T) = p(H) \cdot p(T) = \frac{1}{2} \cdot \frac{1}{2} = \frac{1}{4}$ (because they are independent events). If one had asked for the probability that a head and a tail was obtained from two consecutive coin tosses (without specifying the order in which they were obtained), the answer would be $2(1/4) = 1/2$.

### 1.3.2.6 Probability Density Function

The pdf of a random variable $X$, $p_X(X)$, is a non-negative function (with a total area of 1) that shows how the values of $X$ would be distributed if a large number of experiments (trials) were conducted. The constraints on the pdf are expressed mathematically as:

$$p_X(X) \geq 0, \quad \int_{-\infty}^{\infty} p_X(x)dx = 1$$

$$p_X(x_1 \leq X \leq x_2) = \int_{x_1}^{x_2} p_X(x)dx.$$

Note that $x$, $x_1$, and $x_2$ are values attained by the random variable $X$.

### 1.3.2.7 Statistical Mean

The statistical mean (or expected value) of a random variable $X$ is defined as

$$m_X = \mathcal{E}(X) = \int_{-\infty}^{\infty} x p_X(x) dx \tag{1.26}$$

and it represents the center around which the values of $X$ are expected to be distributed.

#### 1.3.2.8 The Second Moment

The second moment of a random variable $X$ is defined as

$$m_X^{(2)} = \mathcal{E}(X^2) = \int_{-\infty}^{\infty} x^2 p_X(x) dx \tag{1.27}$$

and it represents the expected value of the square of the deviations of a random variable $X$ from its mean value $m_X$.

#### 1.3.2.9 The Variance

The second central moment (or variance) of a random variable is defined as

$$\sigma^2 = \text{var}(X) = \mathcal{E}\left\{(X - m_X)^2\right\} = \int_{-\infty}^{\infty} (x - m_X)^2 p_X(x) dx.$$

The quantity $\sigma_X = \sqrt{\text{var}(X)}$ is called the standard deviation of $X$. The variance indicates how far the values of $X$ are spread around the mean. Hence, the variance gives a *measure of the randomness* of a random signal. The quantity $\sigma_X = \sqrt{\text{var}(X)}$ is called the standard deviation of $X$.

*Note*:

$$\sigma_X^2 = \mathcal{E}\{X^2 - 2m_X + m_X^2\} = \mathcal{E}(X^2) - 2m_X\mathcal{E}(X) + m_X^2 = \mathcal{E}(X^2) - m_X^2 \tag{1.28}$$

#### 1.3.2.10 The Gaussian pdf

The Gaussian pdf is an important probability density function which is often encountered in real-world applications. A random variable $X$ is said to be Gaussian if its pdf is given by:

$$p(x) = \frac{1}{\sigma\sqrt{2\pi}} e^{\frac{(x-m)^2}{2\sigma^2}}$$

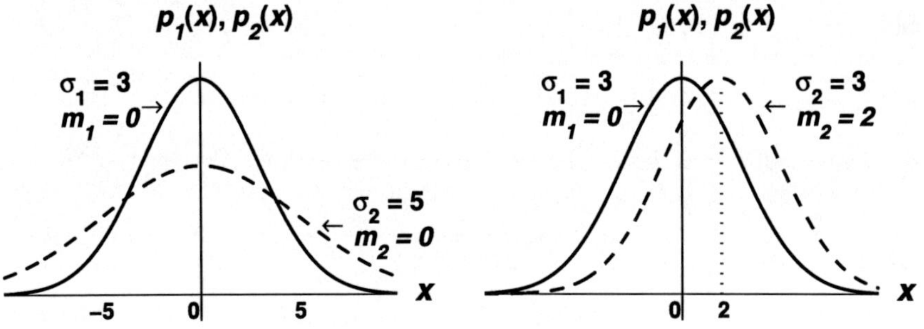


**Fig. 1.22** Gaussian pdf's with different means and variances

where $m$ = the statistical mean, and $\sigma^2$ = the variance. Plots of this pdf for different values of mean and variance are shown in Fig. 1.22.

## *1.3.3 Signals in Noise*

### 1.3.3.1 Gaussian Noise

Noise, $n(t)$, that is encountered in electrical systems frequently has a Gaussian pdf with zero mean, $m_n = 0$. The pdf of this type of signal has the form:

$$p(n) = \frac{1}{\sigma\sqrt{2\pi}} e^{\frac{n^2}{2\sigma^2}}$$

Note that the power of zero-mean Gaussian noise is $\mathcal{E}\left\{(n - m_n)^2\right\} = \mathcal{E}\{n^2\}$ (since $m_n = 0$) = noise variance = $\sigma^2$. (Prove this as an exercise!)

If there are two Gaussian noise signals, $n_1$ and $n_2$, with the variance of $n_2$ being greater than the variance of $n_1$, then the pdf of $n_2$ (pdf-2) has a wider spread around its mean than $n_1$'s pdf (pdf-1) (see Fig. 1.22, left). Furthermore, $n_2$ has more power than the first.

### 1.3.3.2 Signals in Gaussian Noise

If $s(t)$ is a deterministic signal and $n(t)$ is noise, then $z(t) = s(t) + n(t)$ is a random signal. Consider now the case where $s(t) = a$ (a constant). If $n(t)$ is Gaussian noise with zero mean and variance = $\sigma^2$, then the random variable $z(t)$ is also Gaussian with mean and variance given at any time by:

$$\begin{aligned}\bar{z} = m_z &= \mathcal{E}\{z(t)\} = \mathcal{E}\{(s(t) + n(t))\} = \mathcal{E}\{(a + n(t))\} = \mathcal{E}\{a\} + \mathcal{E}\{n(t)\} \\ &= a + 0 = a,\end{aligned}$$

and

$$\mathrm{var}(z) = \mathcal{E}\left\{(z(t) - m_z)^2\right\} = \mathcal{E}\left\{n(t)^2\right\} = \sigma^2.$$

Hence, the pdf of the signal $z(t)$ at any time $t$ is given by:

$$p(z) = \frac{1}{\sigma\sqrt{2\pi}}\, e^{\frac{1}{2}\left(\frac{z-a}{\sigma}\right)^2}$$

While the above result was derived above for the case where $s(t)$ is a constant, its general form is actually valid for any time signal $s(t)$. For example, if $s(t) = \sin(\omega_o t)$, then $\bar{z} = m_z = s(t) = \sin(\omega_o t)$ while the variance is still $\sigma^2$.

### 1.3.3.3 Power Spectral Density of Random Signals

A random signal $n(t)$ can be classified as a *power* signal, and like other power signals, has a PSD. This PSD is often denoted by $G_n(f)$, and is defined as:

$$G_n(f) = \lim_{T\to\infty} \frac{1}{T}\mathcal{E}|\mathcal{F}\{n(t)\Pi_T(t)\}|^2 = \lim_{T\to\infty} \frac{1}{T}\mathcal{E}|N_T(f)|^2,$$

where $\mathcal{E}$ denotes the expected value.

### 1.3.3.4 Stationary Random Signals

Signals whose statistics (i.e., mean, variance and other higher order moments) do not change with time are referred to as stationary.

### 1.3.3.5 The Autocorrelation Function of Random Signals

The autocorrelation function of a random signal $x(t)$ is defined by:

$$R_x(t_1, t_2) = \mathcal{E}\{x(t_1)x(t_2)\}.$$

In the above definition, the $\mathcal{E}$ denotes expected value, and this expected value needs to be obtained by doing many experiments and averaging the results of all those experiments. This kind of average is referred to as an ***ensemble average***.

$R_x\,(t_1,t_2)$ provides an indication of how strongly the signal values at two different time instants are related to one another.

### 1.3.3.6 Wide-Sense Stationary Signals

A random signal is WSS if its mean and autocorrelation are time-invariant, i.e.,

$$\mathcal{E}\{X(t)\} = m_X = \text{constant},$$

and

$$R_X(t_1, t_2) = R_X(t_1 - t_2) = R_X(\tau),$$

where $\tau = t_1 - t_2$.

Every stationary signal is WSS, but the converse is not true.

#### 1.3.3.7 Wiener–Kinchin Theorem for Random Signals

If $x(t)$ is a WSS random signal, then:

$$G_x(f) \overset{\mathcal{F}}{\longleftrightarrow} R_x(\tau),$$

where $G_x(f)$ is the PSD of the signal and $R_x(\tau)$ is its autocorrelation function, i.e., $R_x(\tau) = \mathcal{E}\{x(t)x(t+\tau)\}$.

#### 1.3.3.8 White Noise

A common type of noise encountered in nature is thermal noise, which is typically Gaussian-distributed. If thermal noise has a PSD which is constant over all frequencies (i.e. if the noise is "white"), then its autocorrelation function is a weighted delta function (according to the Wiener–Kinchin Theorem) (see Fig. 1.23); this means that the values of the noise at different instances of time are completely uncorrelated with each other. In practice, noise is rarely white; it is normally band-limited. A more realistic model for the PSD of many practical noise sources is therefore PSD = $(\eta/2)\ \Pi_{2B}(f)$, with $\eta$ being a constant. The use of the arbitrary constant $\eta/2$ rather than $\eta$ is a convention which is meant to signal to a reader that a *double*-sided PSD (with positive and negative frequencies) is being used. Only half of the total power appears in the positive side of the PSD.

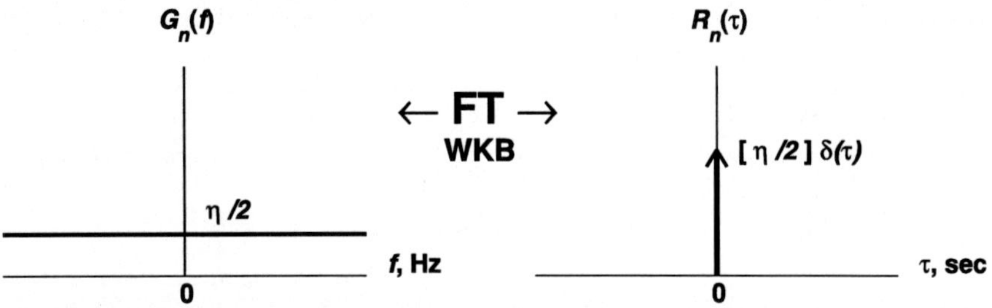


**Fig. 1.23** PSD of white noise with its autocorrelation function

### 1.3.3.9 Effect of Ideal Low-Pass Filter on White Noise

When a random signal enters a system with transfer function *H(f)*, the output signal is also random. The PSD of the output is equal to the input PSD multiplied by the power transfer function of the system. i.e., the PSD of the output is the PSD of the input multiplied by .$|H(f)|^2$). Now assume a white noise input *n(t)* with constant PSD is entering an ideal LPF as shown in Fig. 1.24. The output noise PSD is then given by:

$$G_{no}(f) = |H(f)|^2 G_n(f) = \frac{\eta}{2} \Pi_{2B}(f).$$

Hence, using the WKT, the autocorrelation function of the output noise is:

$$R_{no}(\tau) = \mathcal{F}^{-1}\{G_{no}(f)\} = \frac{\eta}{2} 2B \operatorname{sinc}(2B\tau) \quad \text{(see Fig. 1.25)}.$$

Therefore, the values of the output noise are no longer uncorrelated, except when $\tau = k/2B, k$ being an integer.

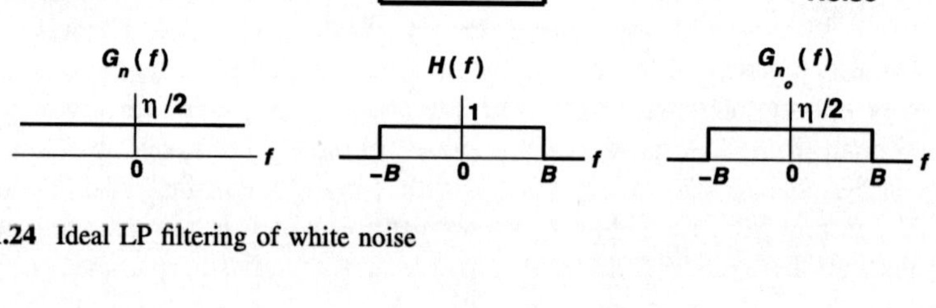


Fig. 1.24 Ideal LP filtering of white noise

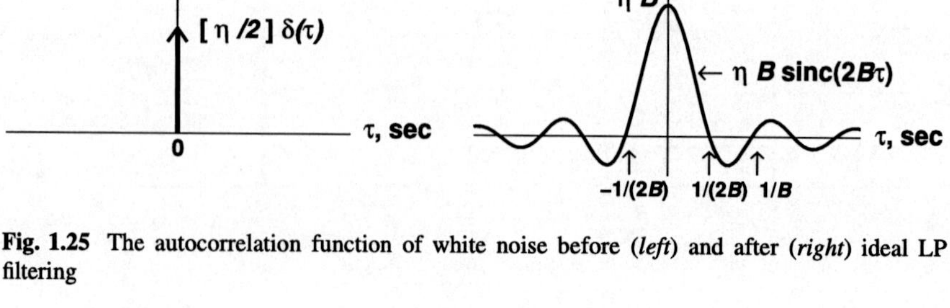


Fig. 1.25 The autocorrelation function of white noise before (*left*) and after (*right*) ideal LP filtering

# 1.4 Applications of Analog Signal Analysis

## 1.4.1 Signal Detection in Noise

In Sect. 1.2.4 the autocorrelation function was defined for *deterministic* periodic and non-periodic signals. The autocorrelation function of *random* signals was defined in Sect. 1.3.3.5; that definition, however, requires ensemble averaging, and in many practical situations, an ensemble of experimental realizations is not available. For this reason it is common in practice to compute the autocorrelation function of random signals using the definition in Sect. 1.2.4—that definition uses a time average in its formulation rather than an ensemble average. This practice is strictly valid only when the random signals are ***ergodic;*** i.e., for signals whose ensemble average equals the time average. There are many signals for which this represents a reasonable approximation to reality. With this approach, the definition for the autocorrelation function of random signals becomes:

$$R_x(\tau) = \frac{1}{T}\int_{-T/2}^{T/2} x(\lambda)x(\tau+\lambda)d\lambda,$$

while the cross-correlation between two random signals is defined as:

$$R_{xy}(\tau) = \frac{1}{T}\int_{-T/2}^{T/2} x(\lambda)y(\tau+\lambda)d\lambda.$$

Naturally occurring noise is often *un-correlated* with deterministic signals. It is also often very nearly uncorrelated with itself (i.e., almost white), so that its autocorrelation function is almost a delta function (see Sect. 1.3.3.8). These properties are exploited in many practical schemes for detecting signals in noise. Consider, for example, the scenario in which a deterministic signal $x(t)$ is transmitted across a communication channel, and a corrupted signal $y(t) = x(t) + n(t)$ is received, One often has to decide whether or not there is a message present inside the received signal or not. One simple way to inform this decision is to correlate $y(t)$ with itself. The autocorrelation of the signal with itself is:

$$\begin{aligned} R_y(\tau) &= \frac{1}{T}\int_{-T/2}^{T/2} [x(\lambda)+n(\lambda)][x(\lambda+\tau)+n(\lambda+\tau)]d\lambda. \\ &= R_x(\tau)+R_n(\tau)+2R_{xn}(\tau) \end{aligned} \tag{1.29}$$

If the noise is white then $R_n(\tau)$ is a spike and $R_{xn}(\tau)$ is approximately zero. Now for most communication signals, $R_y(\tau)$ exhibits a shape which is more than simply a spike and/or a negligibly small background noise. If such a shape is present in the received signal, then, one can infer that a true message is present.

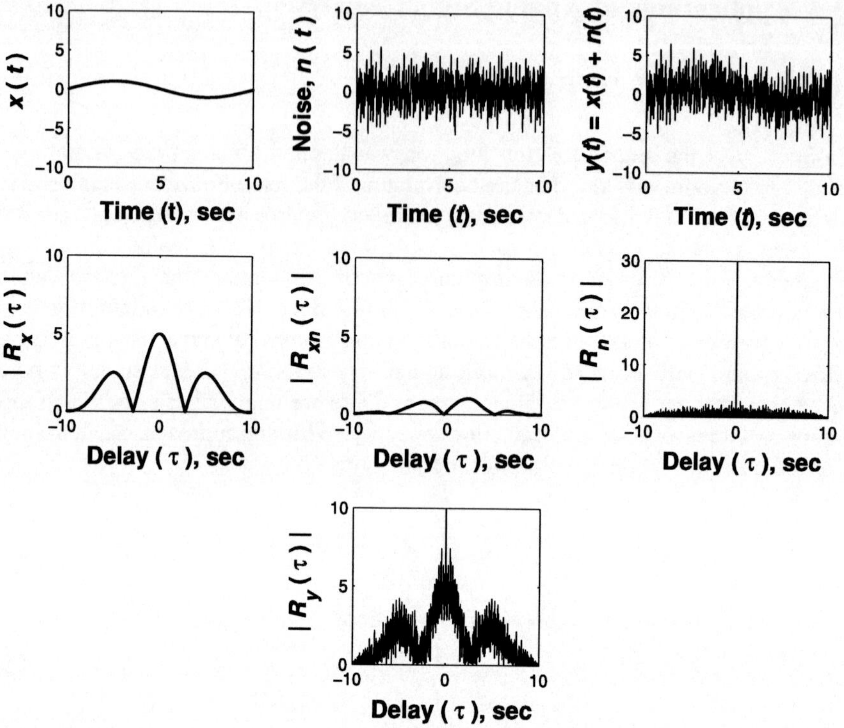


**Fig. 1.26** Signal detection in noise using a correlator

**Example** Consider the sinusoidal signal $x(t) = \sin(\omega_o t)$, with frequency $f_o = 0.1$ Hz. The power of this signal is $1^2/2 = 0.5$ W. The signal and the absolute value of its autocorrelation $R_x(\tau)$ are shown in Fig. 1.26. A random white noise signal $n(t)$ with noise power 5 dB (=3.1623 W) is added to corrupt the sinusoid, giving a noisy signal $y(t) = x(t) + n(t)$. The resulting signal-to-noise-ratio is SNR = 0.1581, or −8.0103 dB. This represents quite a strong level of noise interference. Nonetheless, when the noisy signal is correlated with itself, it is possible to identify a regular non-random pattern with an absolute maximum around the origin. This indicates that there is a deterministic signal imbedded in the noise.

**MATLAB**: The cross-correlation function can be simulated in MATLAB using the command `xcorr(x,y)*Ts`.

## *1.4.2 The Matched Filter*

The matched filter is a linear filter which is often used in detecting signals embedded in random noise. When the matched filter is applied to a received signal,

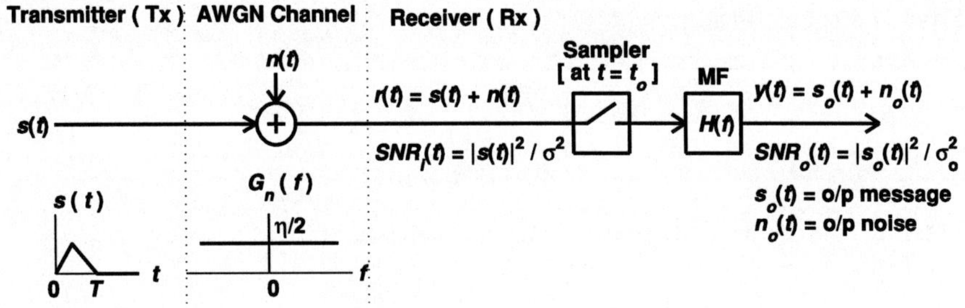


**Fig. 1.27** Matched filter and associated waveforms in a communication system

it produces the *maximum signal-to-noise-ratio* (SNR) enhancement possible for any linear filter. To determine the form the matched filter must take to produce this optimal enhancement, the problem of interest must first be properly defined. Consider that the problem is to detect a particular symbol $s(t)$ of finite duration $T$. The following derivation will seek to determine the shape of the impulse response for the required matched filter. It will actually be found that the optimal impulse response $h(t)$ is dependent on the symbol shape, and hence the name matched filter.

Figure 1.27 shows a communication system with a matched filter. It should be noted that the matched filter is applicable *only when the receiving station knows the symbol library at the transmitter.*

The PSD of the output noise $n_o(t)$ in Fig. 1.27 is:

$$G_{n_o}(f) = G_n(f)|H(f)|^2. \tag{1.30}$$

The output noise power is therefore:

$$\sigma^2 = \int_{-\infty}^{\infty} G_{n_o}(f)df. \tag{1.31}$$

The output signal component is:

$$s_o(t) = \mathcal{F}^{-1}\{H(f)S(f)\} = \int_{-\infty}^{\infty} H(f)S(f)e^{+j2\pi ft}df. \tag{1.32}$$

The output *instantaneous* SNR is:

$$SNR_o(t) = |s_o(t)|^2/\sigma^2 \quad \text{(which is time-dependent).} \tag{1.33}$$

Now it is necessary to find the $H(f)$ that maximizes $\text{SNR}_o(t)$ when $t = t_o$, where $t_o$ is a specific time instant chosen by the sampler. This time instant is either 0 or $T$. From (1.33) it follows that:

$$SNR_o(t_o) = \frac{|s_o(t_o)|^2}{\sigma^2}. \tag{1.34}$$

From (1.32) one obtains:

$$|s_o(t)|^2 = \left| \int_{-\infty}^{\infty} \underbrace{H(f)}_{g_1(f)} \underbrace{S(f) e^{j2\pi f t_o}}_{g_2(f)} df \right|^2. \qquad (1.35)$$

Now the Schwartz inequality (Tables) states that:

$$\left| \int_{-\infty}^{\infty} g_1(f) g_2(f) df \right|^2 \leq \left( \left| \int_{-\infty}^{\infty} g_1(f) df \right|^2 \right) \left( \left| \int_{-\infty}^{\infty} g_2(f) df \right|^2 \right), \qquad (1.36)$$

where equality holds in (1.36) only if:

$$g_1(f) = k\, g_2^*(f) \qquad (1.37)$$

and where $k$ is a constant. * means complex conjugation. Using (1.35) and (1.36), a maximum value for $|s_o(t)|^2$ is obtained as follows:

$$|s_o(t_o)|^2_{\max} = \left( \int_{-\infty}^{\infty} |H(f)|^2 df \right) \left( \int_{-\infty}^{\infty} |S(f)|^2 df \right). \qquad (1.38)$$

Using (1.31), (1.34), and (1.38) it follows that:

$$\mathrm{SNR}_o(t_o) = \frac{\left( \int_{-\infty}^{\infty} |H(f)|^2 df \right) \left( \int_{-\infty}^{\infty} |S(f)|^2 df \right)}{\frac{\eta}{2} \int_{-\infty}^{\infty} |H(f)|^2 df} = \frac{E}{\eta/2}. \qquad (1.39)$$

where $E = \int_{-\infty}^{\infty} |s(t)|^2 dt = \int_{-\infty}^{\infty} |S(f)|^2 df$ is the symbol energy. Using (1.37), this maximum is obtained only when one chooses $H(f)$ to be:

$$\underbrace{H(f)}_{g_1(f)} = k \left[ \underbrace{S(f) e^{j2\pi f t_o}}_{g_2(f)} \right]^* = k S^*(f) e^{-j2\pi f t_o}. \qquad (1.40)$$

Hence, the impulse response of this filter is given by:

$$h(t) = \mathcal{F}^{-1}\{H(f)\} = \mathcal{F}^{-1}\{k S^*(f) e^{-j2\pi f t_o}\} = k p(t - t_o).$$

(using the time-shift property of the Fourier transform (see Tables). Note also that $p(t) = \mathcal{F}^{-1}\{S^*(f)\}$). From the conjugation property of the Fourier transform:

$p(t) = s^*(-t) = s(-t)$ [since $s(t)$ is real]. Hence, the impulse response of the matched filter is:

$$h(t) = k\, s(t - t_o). \qquad (1.41)$$

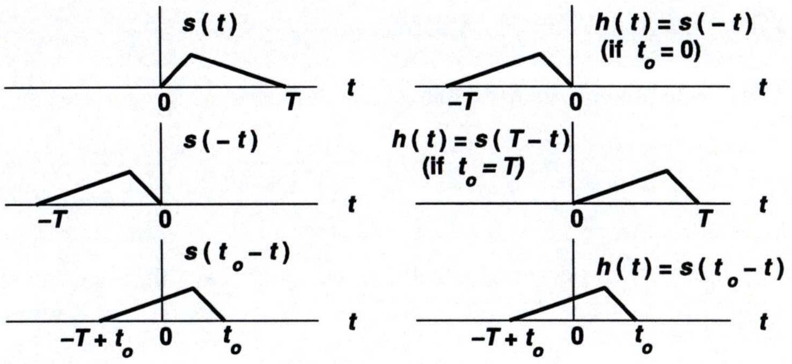


**Fig. 1.28** Matched filter impulse response with various modifications of the original symbol waveform

That is, *the impulse response of a filter matched for a symbol is a scaled time-reversed version of the symbol itself.* Frequently $k$ is chosen to be 1 in practical applications (see Fig. 1.28).

#### 1.4.2.1 Conclusion

If one wants a filter $H(f)$ to give maximum SNR enhancement for a specific symbol $s(t)$ at $t = t_o$, one should *match the filter to the symbol, such that* $h(t) = ks(t_o - t)$.

#### 1.4.2.2 The Output of the Matched Filter at the Time of Optimal SNR

Assuming the system of interest is LTI, the output $y(t)$ of the matched filter in Fig. 1.27 is given by:

$$y(t) = r(t) * h(t) = \int_{t-t_o}^{t+T-t_o} r(\lambda)h(t-\lambda)d\lambda = \int_{t-t_o}^{t+T-t_o} r(\lambda)ks[t-(t-\lambda)]d\lambda, \quad (1.42)$$

At the sampling instant $t = t_o$ one obtains:

$$y(t_o) = k\int_{-\infty}^{\infty} r(\lambda)s(\lambda)d\lambda = k\int_{0}^{T} r(\lambda)s(\lambda)d\lambda. \quad (1.43)$$

That is, the output of the matched filter at the time of optimal SNR, $t = t_o$, is simply *the integration over* $T$ *of the scaled product of the symbol* $s(t)$ *with the received signal* $r(t)$.

#### 1.4.2.3 The Matched Filter is a Correlator

From (1.42) it is possible to write that:

$$y(t) = k \int_{-\infty}^{\infty} r(\lambda)s((t - t_o) + \lambda)d\lambda = kR_{rs}(t - t_o), \quad (1.44)$$

where $R_{rs}(\tau)$ is the cross-correlation between the symbol and the received version of it. Hence the matched filter is essentially a correlator. If $r = s$ (noise-free condition), $kR(\tau)$ will have a symmetric shape with a maximum at $\tau = 0$, i.e., $t - t_o = 0$ or $t = t_o$. This value is $y(t_o)$ as in (1.43).

#### 1.4.2.4 The Optimal Receiver

If there is a finite set of $M$ symbols to be transmitted, $\{s_i(t)|\ i = 1, 2, ..., M\}$ and one wants optimal reception for these symbols, one should use a bank of $M$ filters, each matched to one symbol only (see Fig. 1.29). If the $i$th symbol $s_i(t)$ was transmitted, then the $i$th matched filter will have a maximum at the time of optimal reception, while the other filters will have small values. Hence the receiver will decide that $s_i(t)$ was transmitted. In binary communication systems (such as computer networks), one needs only two filters, matched to the two symbols that represent logic "0" and logic "1".

## 1.5 Analog Filters

Filters play a significant role in signal processing and communication engineering. This section will consider them in some detail.

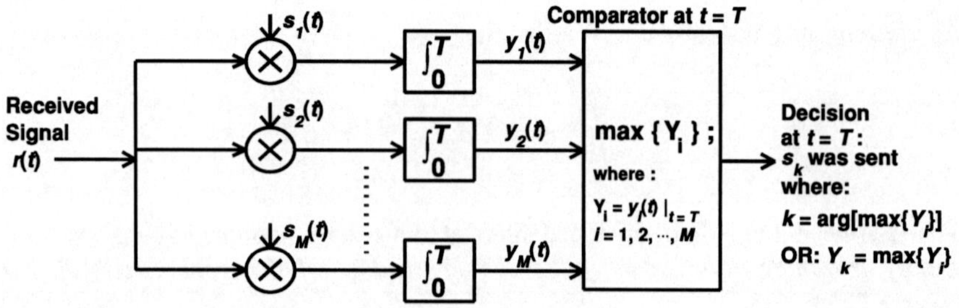


**Fig. 1.29** The optimal receiver for a finite set of symbols

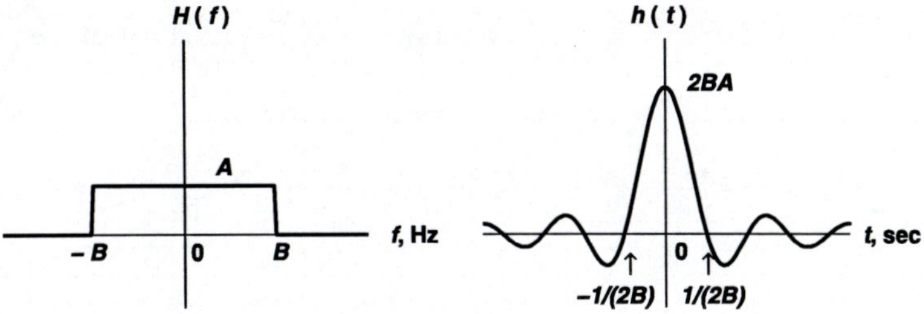


**Fig. 1.30** Transfer function and impulse response of an ideal low-pass filter

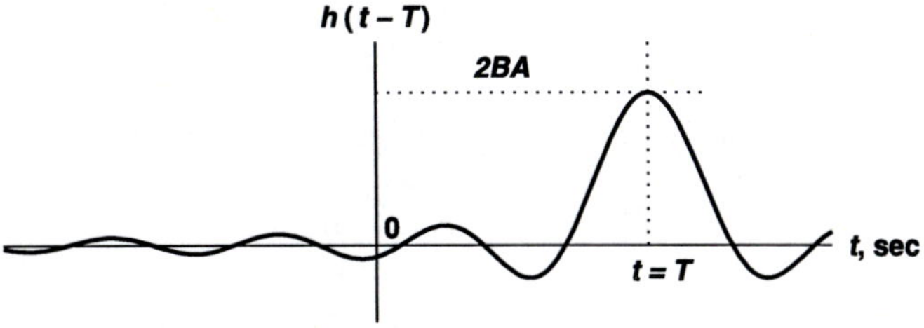


**Fig. 1.31** Quasi-ideal low-pass filter impulse response—it is a time shifted version of the ideal low-pass filter impulse response

## *1.5.1 The Ideal Low-Pass Filter*

The transfer function of an ideal LPF with cutoff frequency $B$ Hz and gain $A$ is a rectangular box function in the frequency domain of width $2B$. Using Tables, its impulse response is found to be a sinc function:

$$H(f) = 2B\Pi_{2B}(f) \xrightarrow{\mathcal{F}} h(t) = 2AB\ \mathrm{sinc}(2Bt)$$

This impulse response is shown in Fig. 1.30.

Since $h(t)$ includes negative values, the ideal LPF is non-causal, and hence ***unrealizable***. If one allows a time-delay $T$ in the system, the delayed impulse response is shifted right by $T$, and one gets $h(t - T)$, as shown in Fig. 1.31. This introduces a linear phase into the transfer function. Since linear phase is a result of only delaying the signal, it corresponds to a harmless phase changes in the transfer function. Note that ***phase distortion*** only occurs when there are ***different delays*** for ***different frequencies***, while ***amplitude distortion*** occurs when there are ***different gains*** for ***different frequencies***. Since $h(t)$ has infinite length, there is still a negative-time portion of $h(t - T)$ despite the delay, but it is of small magnitude, and approximation is possible by truncation.

There are *many approximations* to the ideal LPF that can be implemented in practice. Two of these, the **Butterworth** and **Chebychev-I** filters will be studied in the following subsections.

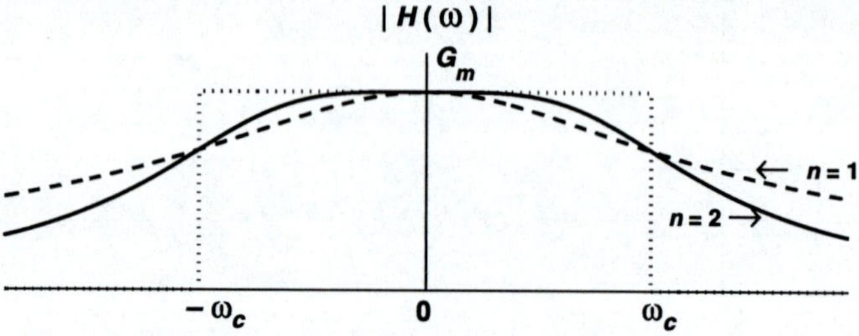


Fig. 1.32 The magnitude frequency response of a Butterworth LPF

## 1.5.2 Butterworth LPF

The magnitude frequency response of a Butterworth LPF (B-LPF) is given by:

$$|H(\omega)| = \frac{G_m}{\sqrt{1+\left(\frac{\omega}{\omega_c}\right)^{2n}}}, \tag{1.45}$$

where $G_m$ is the maximum gain, $n$ is the filter order, and $\omega_c$ is the cutoff frequency (see Fig. 1.32). The Butterworth LPF's magnitude function is *maximally flat* at $\omega = 0$, i.e., all its $2n - 1$ derivatives are equal to 0 at $\omega = 0$. At $\omega = 0$ (DC), the filter gain=$G_m$, and the power gain $GP_o = |H(0)|^2 = G_m^2$—this is the maximum power gain of the filter. At $\omega = \omega_c$ (cutoff), the filter gain $= G_m/\sqrt{2}$, and the power gain $GP_c = |H(\omega_c)|^2 = G_m^2/2 =$ half the maximum power gain $GP_o$. In dB terms:

$$GP_c(\text{dB}) = 10\log_{10}(GP_o/2) = GP_o(\text{dB}) - 10\log_{10} 2 = GP_o(\text{dB}) - 3$$

The cutoff frequency is also called the *3-dB frequency* because the power is 3 *dB* down from the power at DC.

As $n$ increases, the cutoff becomes sharper, but more circuit components are needed to build the filter. The Butterworth LPF has a flat response at $\omega = 0$. Its transfer function for different orders can be obtained from Tables. The $n$th-order B-LPF has $n$ poles and no zeros in its transfer function.

## 1.5.3 Chebychev-I LPF

Chebychev-I filters tend to have sharper roll-off than Butterworth filters but have ripple in the passband, as illustrated in Fig. 1.33. The magnitude frequency response of a Chebychev-I LPF (C-LPF) is given by:

$$|H(\omega)| = \frac{G_m}{\sqrt{1+\varepsilon^2 C_n^2\left(\frac{\omega}{\omega_c}\right)}},$$

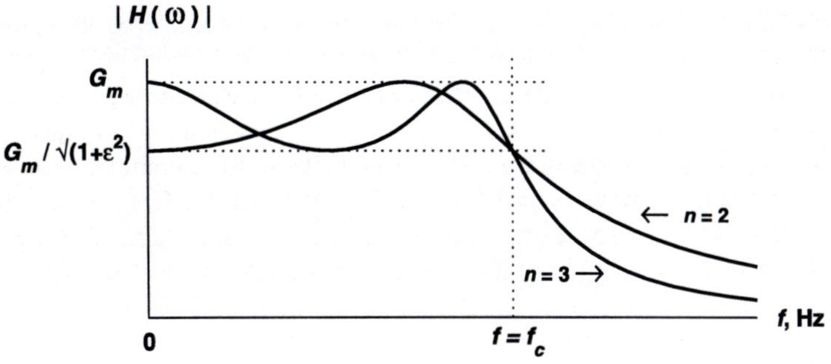


**Fig. 1.33** The magnitude frequency response of a Chebychev-I LPF

where $G_m$ is the maximum gain, $n$ is the filter order, $\omega_c$ is the cutoff frequency, $\varepsilon$ is a constant that determines the ripple in the passband of the magnitude response, and $C_n$ is the $n$th order **Chebychev Polynomial** defined by the iterations:

$$C_o(x) = 1,\ C_1(x) = x,\ C_{n+1}(x) = 2xC_n(x) - C_{n-1}(x)$$

For example, $C_2(x) = 2x^2 - 1$, $C_3(x) = 4x^3 - 3x$, $C_4(x) = 8x^4 - 8x^2 + 1$.

At $\omega = 0$ (DC), the gain $G_o$ is:

$$G_o = \begin{cases} G_m, & n \text{ odd} \\ G_m/\sqrt{1+\varepsilon^2}, & n \text{ even} \end{cases} \tag{1.46}$$

while at $\omega = \omega_c$ (cutoff), the gain is:

$$G_c = G_m/\sqrt{1+\varepsilon^2}\,\forall n. \tag{1.47}$$

For $0 < \omega < \omega_c$, the gain fluctuates between $G_m$ and $G_c = G_m/\sqrt{1+\varepsilon^2}$ (see Fig. 1.33). The maximum power gain variation, called *ripple*, is given by:

$$r\ (\text{dB}) = 10\log_{10}\left(1+\varepsilon^2\right).$$

Like the B-LPF, the $n$th-order C-LPF has $n$ poles and no zeros. However, for the same order $n$, the C-LPF has a sharper cutoff than the B-LPF. Hence, *if the ripple is not disadvantageous in a specific application, Chebychev filters are preferred over Butterworth filters*. This is so because with Chebychev filters one typically needs a lower order and hence, less hardware to achieve the same cutoff sharpness.

### *1.5.4 Design of Butterworth and Chebychev-I LPF's*

This subsection addresses the design of both the filter transfer function $H(s)$ and physical hardware components needed for building Butterworth and Chebychev filters. The physical components are an appropriate combination of resistors (R), inductors (I), and capacitors (C). That is, for the design considered here, the filters are built up as passive RLC networks. Design using active components (op-amp's) is also possible, but will not be considered in this subsection.

Based on the required filter specifications (e.g., cutoff sharpness and stop-band attenuation) the goal is to can find $H(s)$ and a suitable RLC combination. The first step in the design process is to determine the filter order $N$ from standard tables. Since Tables are usually only available for filters with a cutoff frequency of $\omega_c = 1$ rad/s, the filter specifications must first be transformed using frequency denormalization. This is achieved by dividing all the critical frequency points (cutoff frequency, stop edge frequency, etc.) by $\omega_c$. This process is illustrated in the example which follows.

#### 1.5.4.1 Example of a Low-pass Filter Design

**Example 1** Design a LPF with $G_m = 1$, $af_c = 50$ Hz, and stop-band gain $\leq -40$ dB w.r.t $G_m$ for frequencies $\geq 140$ Hz. Find the minimum order $n$ required, if the ripple $\leq 1$ dB and $R_L = 100\Omega$.

*Solution:* $\omega_c = 2\pi(50) = 100\ \pi$, $\omega_1 = (2\pi)140 = 280\pi$, hence $\omega_{1N} = \omega_1/\omega_c = 2.8$. Initially, a butterworth design will be trailed. Use the Gain vs. Normalized frequency plots to determine the minimum order to achieve $-40$ dB for $\omega_N \geq 2.8$ (see Tables and the left hand side of Fig. 1.34). It is seen that $n = 4$ s not adequate to meet the 40 dB attenuation at $\omega_N \geq 2.8$ but $n = 5$ is. If a butterworth filter is used, therefore, one would need a fifth order filter to meet specifications.

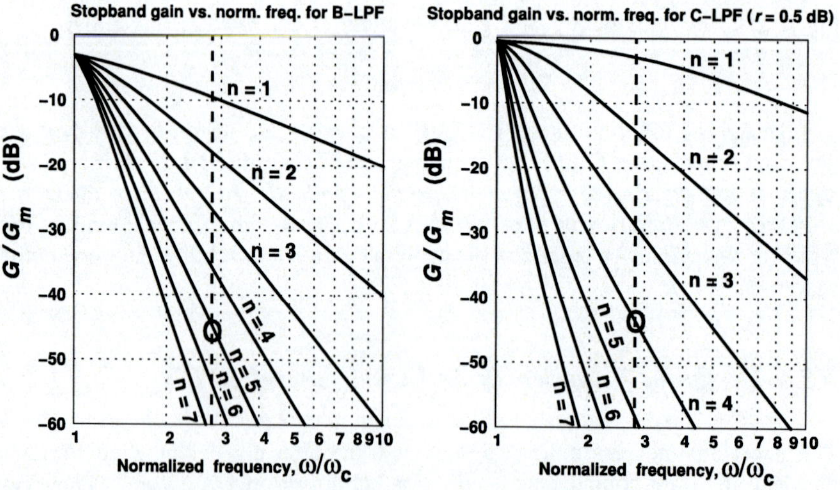


Fig. 1.34 Gain plots versus normalized LPF's for different filter orders

Now a Chebychev filter will be tried. The specifications state the ripple must be less than or equal to 1 dB. Potential designs for both 0.5 and 1 dB ripple will be considered. Checking the C-LPF gain plot for $r = 0.5$ dB in Fig. 1.34 (right), it is found that $n = 4$ is the minimum order which is adequate to meet specifications. Checking the C-LPF gain plot for $r = 1$ dB shows that the minimum order required is also $n = 4$. The $r = 0.5$ dB alternative is chosen as it corresponds to the minimum ripple(hence, from Tables, $\varepsilon^2 = 0.122$). The normalized transfer function is obtained from Tables as follows:

$$H(s_N) = \frac{a_o}{0.3791 + 1.0255 s_N + 1.7169 s_N^2 + 1.1974 s_N^3 + s_N^4}.$$

Since $n = 4$ is even:

$$G_o = G_m/\sqrt{1+\varepsilon^2} = 1/\sqrt{1.122} = 0.9441 = H(s_N = 0) = a_o/0.3791 \to a_o = 0.3579.$$

The denormalized frequency response is given by:

$$H(s) = \frac{a_o}{0.3791 + 1.0255\frac{s}{100\pi} + 1.7169\left(\frac{s}{100\pi}\right)^2 + 1.1974\left(\frac{s}{100\pi}\right)^3 + \left(\frac{s}{100\pi}\right)^4}$$

#### 1.5.4.2 Circuit Design

From Tables one arrives at the circuit in Fig. 1.35 (left). It uses impedance and frequency denormalization given by ISF: $Z = R_L = 100$ and FSF: $W = \omega_c = 100\pi$.

### *1.5.5 Design of Butterworth and Chebychev-I High-Pass, Band-Pass, and Band-Stop Filters*

Standard Tables are also only available for LPF design, but can be adapted via standard transformations to designing high-pass filters (HPFs), band-pass filters (BPFs) and band-stop filters (BSFs),

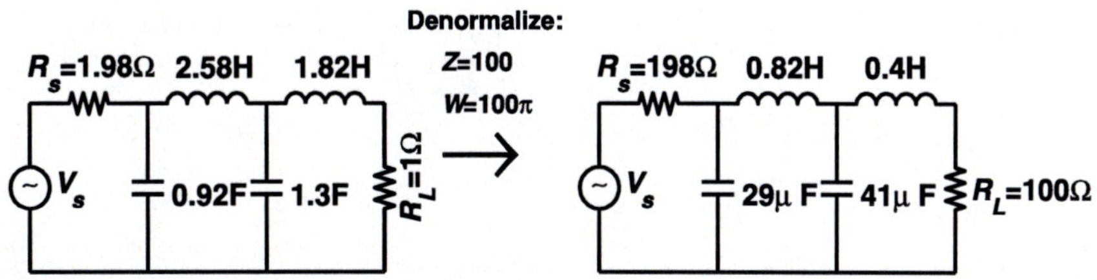


**Fig. 1.35** Circuit Design of a 4th-order C-LPF

The required transformations are:

$$\begin{aligned}
&\text{LP} \to \text{HP}: \omega_{LN} = \frac{\omega_c}{\omega}; \quad s_{LN} = \frac{\omega_c}{s} \quad [\omega_c \text{ is the HP cutoff}] \\
&\text{LP} \to \text{BP}: \omega_{LN} = \frac{\omega^2 - \omega_g^2}{\omega\omega_b}; \; s_{LN} = \frac{s^2 + \omega_g^2}{s\omega_b} \quad [\omega_b = \omega_u - \omega_l; \; \omega_g = \omega_u\omega_l] \\
&\text{LP} \to \text{BS}: \omega_{LN} = \frac{\omega\omega_b}{\omega^2 - \omega_g^2}; \; s_{LN} = \frac{s\omega_b}{s^2 + \omega_g^2}
\end{aligned} \tag{1.48}$$

An example filter design is presented later in this subsection to illustrate the use of these transformations in filter design.

*Note*: Although Butterworth and Chebychev-I LPF's have no zeros, the above transformations indicate that BP, HP, and BS filters do have zeros.

### 1.5.5.1 Circuit Design

Standard Tables provide not only filter transfer functions, but also suitable circuits for implementing those transfer functions. After obtaining the transfer functions and circuit diagrams from the Tables, one needs to perform appropriate transformations to suit the original filter specifications. The standard transformations for physical implementation are shown in Fig. 1.36.

### 1.5.5.2 Impedance Matching

For maximum power transfer in the passband, the load impedance $R_L$ should equal (be matched to) the source impedance. In LPF design Tables, the values of $R$, $L$, and $C$ are given for $R_L = 1$, $\omega = 1$, hence these circuits are *normalized*. For impedance matching, normalized filter circuits should be *denormalized* by scaling all $R$, $L$, and $C$ values as follows:

$$R \to RZ; \quad L \to L\frac{Z}{W}; \quad C \to \frac{C}{ZW},$$

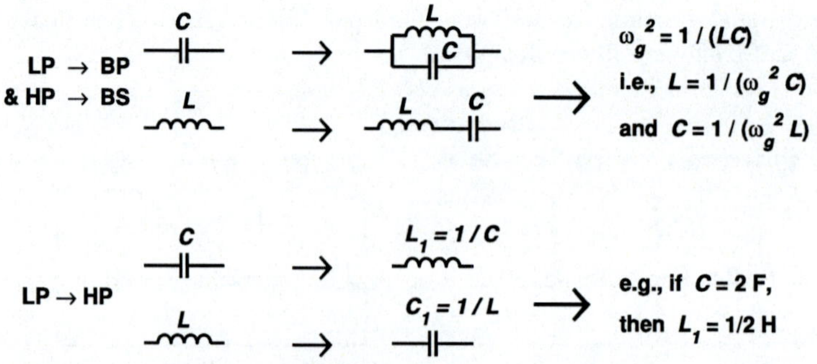


Fig. 1.36 Hardware analog filter transformations

where:

$$Z = \text{Impedance Scaling Factor (ISF)} = R_L$$

$$W = \text{Frequency Scaling Factor (FSF)} = \begin{cases} \omega_o, & \text{For LP and HP} \\ \omega_b, & \text{For BP and BS} \end{cases}$$

*Note*: To denormalize the transfer function $H(s_{LN})$ of a normalized LPF, one simply uses the transformation $s \rightarrow s/\omega_c$ ($\omega_c$ being the required LPF cutoff frequency) to get the true transfer function $H(s)$. There is *no need within this process to find the true transfer function of the HP, BP, and BS filters—if the above transformations from LPF are used with* $s_{LN} \rightarrow s$*, the denormalization is included implicitly.*

#### 1.5.5.3 Hardware Filter Design Rules Using Normalized LPF Standard Circuits

- **HPF Design:**
  1. Transform $LP_N \rightarrow HP_N$ circuit components (Fig. 1.36).
  2. Denormalize $HP_N \rightarrow HP$ (using ISF, FSF).
- **BPF Design:**
  1. Denormalize $LP_N \rightarrow LP$ (using ISF, FSF).
  2. Transform $LP \rightarrow BP$ circuit components (Fig. 1.36).
- **BSF Design:**
  1. Transform $LP_N \rightarrow HP_N$ circuit components (Fig. 1.36).
  2. Denormalize $HP_N \rightarrow HP$(using ISF, FSF).
  3. Transform $HP \rightarrow BS$ circuit components (Fig. 1.36).

#### 1.5.5.4 Example of a High-pass Filter Design

**Example 2** Design a Butterworth HPF with $G_m = 1, f_c = 1$ MHz, and stopband gain $\leq$ 20 dB for $f \leq$ 500 kHz. True load resistance is $R_L = 300\Omega$.

*Solution:*

$\omega_c = 2\pi M$ rad/sec; $\omega_c = 2\pi\frac{1}{2} = \pi$ rad/sec. From LP-HP transformations obtain the normalized LP frequency that corresponds to the HP frequency $\omega_1$, which is $\omega_L = \omega_c/\omega_1 = 2$.

From Butterworth curves in Tables check the order $n$ that gives stop-band gain $\leq -20$ dB for $\omega_{LN} \geq 2$ [note that "$\leq$" for HPF is now "$\geq$" for LPF]. This yields $n = 4$. This same result can also be reached mathematically as follows:

$$n \geq \left[0.5\log\left(A^{-2}-1\right)/\log(\omega_{LN})\right] = [3.314] = 4,$$

where $A = A(\omega_{LN}) = 10^{(A_{dB}/20)}$. From Tables we get the normalized transfer function as follows:

$$H(s_N) = \frac{a_o}{1 + 2.6131s_N + 3.4142s_N^2 + 2.6131s_N^3 + s_N^4}.$$

For B-LPF, $G_o = G_m$. Hence, $G_o = 1 = a_o/1 \rightarrow \therefore\ a_o = 1$.
The true transfer function will be given by:

$$H(s) = \frac{a_o}{1 + 2.6131\left(\frac{2\pi M}{s}\right) + 3.4142\left(\frac{2\pi M}{s}\right)^2 + 2.6131\left(\frac{2\pi M}{s}\right)^3 + \left(\frac{2\pi M}{s}\right)^4}$$
$$= \frac{s^4}{s^4 + 16.4Ms^3 + 134.8M^2s^2 + 648.2M^3s + 1558.5M^4}.$$

The resulting filter has four poles and four zeros.
*Circuit Design*:
$Z = R_L = 300$ and FSF: $W = \omega_c = 2\pi M$ (see Fig. 1.37).

## 1.5.6 Chebychev-II Filters

The Chebychev-II Filter (a.k.a. the inverse Chebychev filter) has a maximally flat pass band (no ripples) and an equiripple stopband. Its magnitude response is given by:

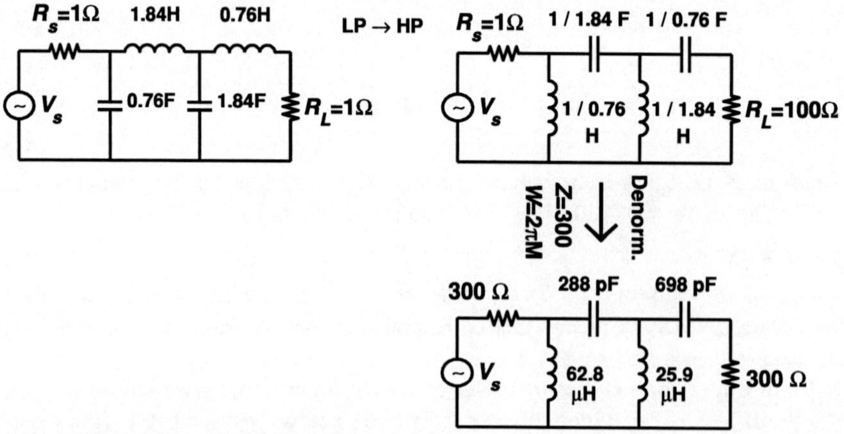


**Fig. 1.37** Design of a Butterworth HPF

$$|H(\omega)| = \frac{G_m}{\sqrt{1 + \frac{1}{\varepsilon^2 C_n^2\left(\frac{\omega}{\omega_c}\right)}}}.$$

## 1.5.7 Elliptic Filters

Elliptic filters have equiripple passband and stopband. As such, they provide the lowest possible filter order for given specifications. However, they have worse phase response (which is highly non-linear). Their magnitude response is given by:

$$|H(\omega)| = \frac{G_m}{\sqrt{1 + \varepsilon^2 E_n\left(\frac{\omega}{\omega_c}\right)}},$$

where $E_n$ is the Jacobian elliptic function of order $n$.

## 1.5.8 MATLAB Analog Filter Design

Butterworth, Chebychev, and Elliptic filters (analog or digital) can be designed in MATLAB using the commands `butter, cheby1, cheby2`, and `ellip`. The maximum gain is always assumed to be 1. The filter order can be determined using `buttord, cheb1ord, cheb2ord`, and `ellipord`—however, these latter commands impose a cutoff frequency rather than allowing for an arbitrary one to be specified.

Example:

1. Design an analog B-LPF with $f_c = 1$ kHz and a stopband gain $\leq -20$ dB for $f \geq 3$ kHz.
2. Design an analog Butterworth BPF with cutoff frequencies $f_l = 1$ kHz and $f_u = 2$ kHz. Stopband gain is $-20$ dB for $f \leq f_1 = 0.5$k and $f_2 \geq 2.5$k Hz.

*Solution:*

1. First, the lowest possible order is found:
   `[n wo] = buttord(w1,w2,RdB,AdB,'s');`
   where `RdB` is the maximum gain loss (below 0 dB) in the passband (defined at the edge radian frequency $\omega =$ `w1`), `w2` is the edge of the stopband with attenuation `AdB` or less, and `wo` is the (imposed by MATLAB) cutoff frequency (rad/s).
   Typically one takes `w1` $= \omega_c$ and `RdB` $=$ `3`. The letter 's' indicates an analog filter (if omitted MATLAB would design a digital filter). In this example the relevant parameters and commands are:

```
fc = 1000;

wc = 2*pi*fc;

f2 = 3000;

w2 = 2*pi*f2;

[n wo]= buttord(wc,w2,3,-20,'s');
```

which yields $n = 3$, $f_o = 1{,}400$ Hz. The filter can now be designed using: `[A B] = butter(n,wo,'s');`

2. Following similar steps:

```
fcbp=[1000 2000];

wcbp=2*pi*fcbp;

f2bp=[500 2500];

w2bp=2*pi*f2bp;

[n wobp] = buttord(wcbp,w2bp,3,-20,'s');

[A B]=butter(n,wobp,'s');

H=freqs(A,B,w);
```

## *1.5.9 Active Filters*

Inductors require more space in circuits than capacitors and resistors and are generally less accurate. Active filters enable filters to be constructed without inductors, simply by incorporating active components such as ***op-amps***. As well as eliminating the need for inductors, active filters have the advantage that they ***can produce a gain***, while passive filters cannot. Active filters are considered briefly in what follows.

### 1.5.9.1 Overview of Active Amplifiers

This section very briefly reviews the key results from the theory of op amps. For a detailed explanation of op-amps, the reader should consult standard Op-Amp references such as [4]. The conventional inverting and non-inverting active amplifier configurations are shown in Fig. 1.38.

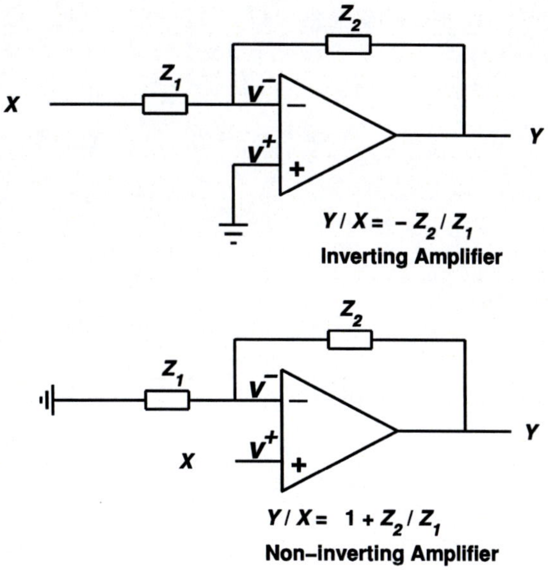


**Fig. 1.38** The inverting and non-inverting active amplifiers

For the inverting amplifier:

$$V^- \simeq 0, \frac{x}{Z_1} \approx -\frac{y}{Z_2}$$

(i.e. the current in $Z_1$ continues almost totally to $Z_2$ without entering the OP-AMP). Hence:

$$\frac{y}{x} = -\frac{Z_2}{Z_1}.$$

For the non-inverting amplifier:

$$\frac{x}{Z_1} = -\frac{y-x}{Z_2} \Rightarrow \frac{y}{x} = 1 + \frac{Z_2}{Z_1}.$$

#### 1.5.9.2 The Active Buffer

The voltage follower shown in Fig. 1.39 has very nearly a unity gain since:

$$y = (V^+ - V^-)A = (x - y)A;$$

where $A$ is the no-load voltage gain of the OP-AMP (normally very large). Hence:

$$\frac{y}{x} = \frac{A}{1+A} \approx 1.$$

The role of this circuit is to *buffer* the load from the signal generator—this buffering is achieved because there is a very high input impedance with a unity

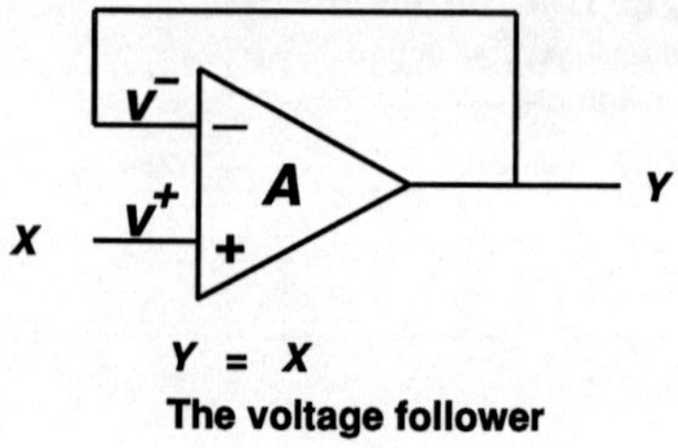


Fig. 1.39 The active buffer

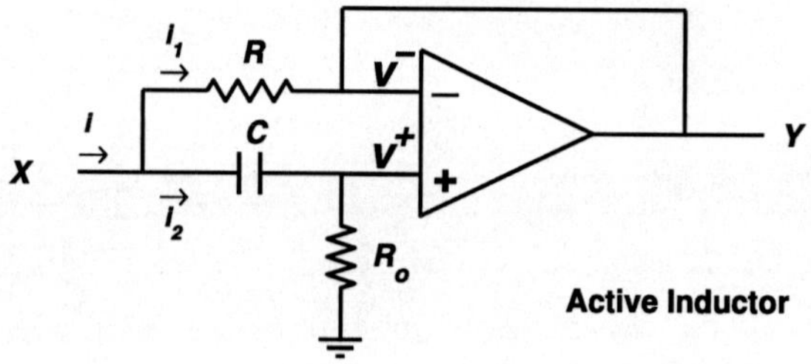


Fig. 1.40 The active inductor

gain. This circuit concept will be used to generate active inductance in the following paragraphs.

### 1.5.9.3 The Active Inductance

Using an OP-AMP as shown in Fig. 1.40, it is possible to produce inductance without the need for an actual coil as follows:

$$y \approx V^{+}(\text{voltage follower}) = x\frac{R_o}{R_o + \frac{1}{j\omega C}}$$

$$i_1 = i_2 + i_3 = \frac{x-y}{R} + \frac{x}{R_o + 1/(j\omega C)} = \frac{x - \frac{xR}{R_o + 1/(j\omega C)}}{R} + \frac{x}{R_o + 1/(j\omega C)}$$

Hence:

$$\text{Input impedance} = z_i = \frac{x}{i_1} = \frac{R + j\omega CRR_o}{1 + j\omega CR} = \frac{R + R_o K^2}{1 + K^2} + j\frac{K(R_o - R)}{1 + K^2}$$

where $K = \omega CR$.

If $R_o \gg R$ and $K \ll 1$, then $K^2 \approx 0$ and:

$$z_i = R + j\omega L \text{ with } L \approx CRR_o.$$

### 1.5.9.4 Butterworth Active Filters

The transfer function of the circuit in Fig. 1.41 corresponds to a first-order Butterworth LPF, as illustrated by the following mathematics:

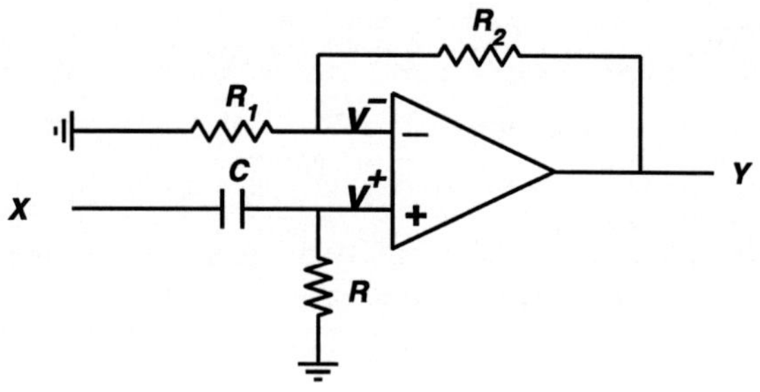


**Fig. 1.41** Active first-order Butterworth LPF

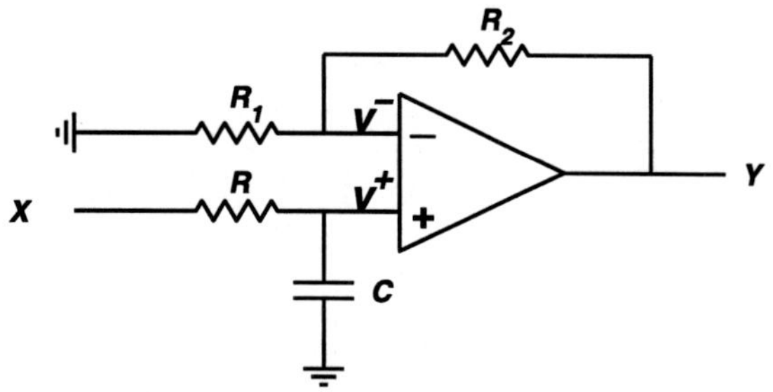


**Fig. 1.42** Active first-order Butterworth HPF

$$\frac{Y}{V^+} = \left(1+\frac{R_2}{R_1}\right).V^+ = X\frac{X_c}{R+X_c} = X\frac{1/(j\omega C)}{R+1/(j\omega C)} = X\frac{1}{1+j\omega RC}$$

$$\therefore \frac{Y}{X} = \frac{Y}{V^+}\frac{V^+}{X} = \left(1+\frac{R_2}{R_1}\right)\frac{1}{1+j\omega RC} = \frac{G_m}{1+j\omega RC} \text{ with } G_m = 1+\frac{R_2}{R_1}.$$

$$\therefore |H(\omega)| = \frac{G_m}{\sqrt{1+\left(\frac{\omega}{\omega_o}\right)^2}} \text{ where } \omega_o = \frac{1}{\sqrt{RC}} \text{ is the cutoff frequency.}$$

Comparing this with the known form for the B-LPF magnitude gain:

$$|H(\omega)| = G_m/\sqrt{1+(\omega/\omega_c)^{2n}},$$

the above is seen to be a first-order B-LPF. Exchanging $R$ and $C$ as in Fig. 1.42 gives the B-HPF with magnitude response:

$$|H(\omega)| = \frac{G_m(\omega/\omega_o)}{\sqrt{1+\left(\frac{\omega}{\omega_o}\right)^2}}$$

This is the standard form for the B-HPF as can be found from Tables:

$$|H_{LP}(s_N)| = \frac{G_m}{1+s_N}, \text{ hence } |H_{HP}(s)| = \frac{G_m}{1+\omega_o/s}$$

or,

$$|H_{HP}(\omega)| = \left|\frac{G_m}{1 + \omega_o/j\omega}\right| = \left|\frac{G_m(\omega/\omega_o)}{\omega/\omega_o + 1/j}\right| = \frac{G_m(\omega/\omega_o)}{\sqrt{1 + (\omega/\omega_o)^2}}$$

Note that the capacitor will be a disconnected element at DC (zero frequency) but a short circuit at high frequency.

## References

1. Bueche, F.J., Hecht, E.: College Physics. Schaum's Outline Series. McGraw-Hill, New York (2000)
2. Carlson, G.: Signal and Linear System Analysis. Wiley, London (1998)
3. Hsu, H.: Signals and Systems. Schaum's Outline Series. McGraw-Hill, New York (1995)
4. Franco, S.: Design with Operational Amplifiers and Analog Integrated Circuits. McGraw-Hill, New York (2001)

# Chapter 2
# Discrete and Digital Signals and Systems

## 2.1 Introduction

Before an analog signal can be processed with a digital computer chip, the signal need to be converted into digital form.

The *first* phase of the digitization process is to sample the analog signal at discrete time instants $t = nT_s$, where $n$ is an integer (1, 2, 3,...) and where $T_s$ is the sampling period (i.e., the time interval between successive samples). Note that this kind of sampling is uniform in the sense that $T_s$ is constant, and the sampling frequency is also constant at $f_s = 1/T_s$. The result of sampling an analog signal in this fashion is a discrete-time signal, also known as a pulse amplitude modulation (PAM) signal.It should be noted that while uniform sampling is simple and intuitively appealing, it is not always the most suitable way to perform sampling. Non-uniform sampling can in some cases actually increase the range of frequencies which can be processed. Nonetheless, for the sake of simplicity, this book will focus only on uniformly sampled signals.

The *second* phase of the digitization process is to quantize the discrete-time signal. The term quantization refers to the approximation of the sampled analog values using a finite set of values. For example, in $M$-bit quantization there are $2^M$ allowable levels, and the signal amplitude must be approximated to one of these levels. If, for example, the expected analog voltage limits are $\pm 3$ volts, the input signal can be quantized in steps of $\Delta = 6/2^M$ volts. Typically the voltage quantization levels are chosen to be symmetric around 0 volts.

In the quantization process approximation via rounding or truncation normally occurs and some information is lost. The quantized signal therefore consists of a true signal plus an error signal which is referred to as quantization noise. The power of this noise is dependent on the quantization step, $\Delta$; with this power being given by $\Delta^2/12$ (see Sect. 3.6.1.1).

The *third* phase of the digitization process involves an encoding of the multibit binary sequence into a practical and convenient form before it is used or transmitted. This process is known as pulse code modulation (PCM), and the resulting

Z. M. Hussain et al., *Digital Signal Processing*,
DOI: 10.1007/978-3-642-15591-8_2, 

signal is known as a PCM signal. In PCM, each sample is represented by one of $2^M$ codewords, each of length $M$ bits. If $f_s$ is the sampling rate, the *data rate* would be $Mf_s$ bps. The binary encoding may correspond to the natural binary code (NBC), which is a straightforward decimal-to-binary mapping of the index of the voltage level (between 0 and $2^M - 1$). Alternatively the encoding may use one of the *Gray Codes*, which, because it reduces bit transmission errors, is particularly relevant for long range transmission.

Practical multibit analog-to-digital converter (ADC) tend to perform all three phases mentioned above within the one device: sampling, quantization, and binary encoding. The accuracy (resolution) of multibit ADCs is strongly dependent on the number of bits $M$, and increasing the number of bits typically improves accuracy.

### 2.1.1 Digital Systems

Digital systems are hardware or software systems that process digital signals. Most digital systems are built up using binary or "on–off" logic, and so the operation of digital processors can be described using binary arithmetic.

In contrast to the resistors, capacitors and inductors which make up analog systems, the common building blocks for DSP systems are shift registers, adders and multipliers. These building blocks may be realized in either hardware or software, and if implemented in hardware, usually incorporate flip-flops, logic gates and synchronously controlled clocks. Like analog systems, discrete-time and digital systems can be analyzed in either the time or frequency domains. Both forms of analysis are considered later in this book.

## 2.2 Ideal Sampling and Reconstruction

Sampling is the process of selecting (and possibly storing) the values of a continuous-time signal $x(t)$ at specific time instants which can be indexed by the ring of integers $Z = \{\ldots, -2, -1, 0, 1, 2, 3, \ldots\}$. As long as the sampling arrangement has been well designed, the discrete-time signal, denoted by $x(n)$, reliably represents the original signal, at least under certain assumptions. In other words, with appropriate design of the sampling strategy, it is possible to process analog signals using digital systems without any loss of information.

### 2.2.1 Ideal Uniform Sampling

Suppose that an analog signal $x(t)$ is sampled uniformly at a rate of $f_s = 1/T_s$, One can represent this kind of sampling mathematically as multiplication of $x(t)$ by the

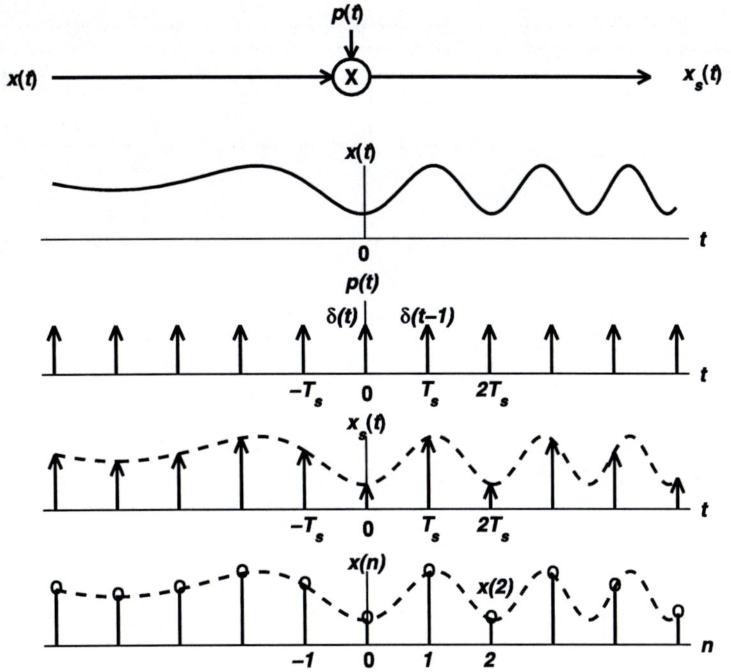


**Fig. 2.1** The sampling process in the time domain

impulse train $p(t) = \sum_{n=-\infty}^{\infty} \delta(t - nT_s)$ (See Fig. (2.1)). The resulting sampled signal in the continuous-time domain is $x_s(t)$, which corresponds to a train of delta functions weighted by the values of $x(t)$ at integer multiples of $T_s$, i.e., weighted by $x(nT_s)$. The *discrete-time representation* of the same sampled signal is $x(n)$, which is the sequence of actual values of $x(t)$ at the discrete-time instants $t = nT_s$. (Note that $x(n)$ is actually $x(nT_s)$, although, $T_s$ is normally omitted from the notation for simplicity). In practical DSP, the discrete-time representation ($x(n)$) is more commonly used than the continuous-time representation of sampled signals ($x_s(t)$).

#### 2.2.1.1 Definitions for Some Important Discrete-Time Signals

Formal definitions for two important discrete-time signals are presented below.

##### The Discrete-Time Unit Pulse

The discrete-time counterpart for the continuous-time unit impulse (delta function) is the discrete-time unit pulse. It is defined as:

**Fig. 2.2** The discrete-time delta and unit-step functions

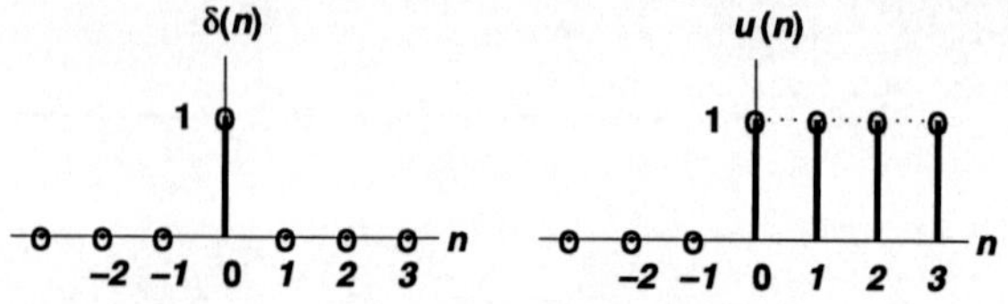


$$\delta(n) = \begin{cases} 1, & n = 0 \\ 0, & n \neq 0. \end{cases}$$

The Discrete-Time Unit-Step Function

The discrete-time counterpart for the continuous-time unit step function is the discrete-time unit step function. Its definition is:

$$u(n) = \begin{cases} 1, & n \geq 0 \\ 0, & n < 0\,. \end{cases}$$

Figure (2.2) shows graphical representations of the above two functions.

Time- and Frequency-Domain Interpretation of Sampling

The sampling process can be illuminated by using various properties from Table 2.1. One of the key results from these *Tables* is that multiplication "·" in the time domain is transformed into convolution "*" in the frequency domain and vice-versa. With these properties the following time domain and frequency domain mathematical descriptions can be obtained:

Time Domain

$$\begin{aligned} x_s(t) &= x(t) \cdot p(t) \\ &= x(t) \cdot \sum_{n=-\infty}^{\infty} \delta(t - nT_s) \\ &= \sum_{n=-\infty}^{\infty} x(nT_s)\delta(t - nT_s) \end{aligned}$$

**Table 2.1** Some common time windows

| | *a* | *b* | *c* | Comment |
|---|---|---|---|---|
| Rectangular | 1 | 0 | 0 | Harsh Gibbs effect (strong oscillations) |
| Hanning | 0.5 | −0.5 | 0 | Mild oscillations |
| Hamming | 0.5 | −0.46 | 0 | Mild oscillations |
| Blackman | 0.42 | −0.5 | 0.08 | Mild oscillations |

Frequency Domain

$$\begin{aligned} X(f) &= X(f) * P(f) \\ &= X(f) * f_s \sum_{kn=-\infty}^{\infty} \delta(f - kf_s) \\ &= f_s \sum_{n=-\infty}^{\infty} X(f - kf_s)\delta(t - nT_s) \end{aligned}$$

where we have been used *Tables* to find that:

$$\sum_{n=-\infty}^{\infty} \delta(t - nT_s) \overset{\mathcal{F}}{\longleftrightarrow} f_s \sum_{k=-\infty}^{\infty} \delta(f - kf_s), \tag{2.1}$$

$$X(f) * \delta(f - kf_s) = X(f - kf_s), \tag{2.2}$$

$$x(t) \cdot \delta(t - nT_s) = x(nT_s)\delta(t - nT_s). \tag{2.3}$$

The sampling process in the frequency domain is illustrated in Fig. (2.3). From the above equations and Fig. (2.3) one can deduce that $X_s(f)$ is essentially a collection of repeated versions of $X(f)$ scaled by $f_s$. These repeated versions are often referred to as "images". Hence, when a signal $x(t)$ is sampled, its spectrum $X(f)$ is:

1. scaled by $f_s$ and
2. repeated every $f_s$.

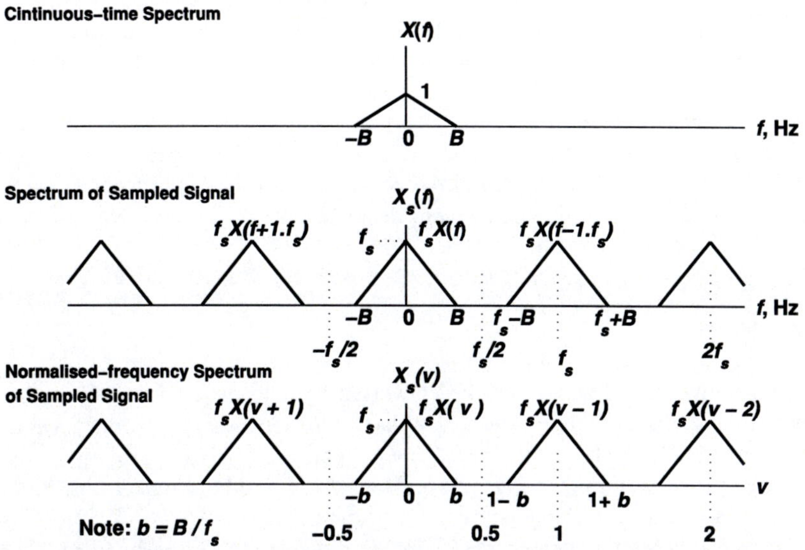


**Fig. 2.3** The discrete-time Fourier transform (DTFT) of a sampled signal

Due to the periodicity of $X_s(f)$, there is redundant frequency information in the spectrum or "discrete-time Fourier transform" (DTFT) of $x_s(t)$. All the non-redundant information in $X_s(f)$ is fully contained in the interval $f \in \{-f_s/2, f_s/2\}$ Hz, or equivalently $f \in \{0, f_s\}$. In the literature $X_s(f)$ is normally plotted as a function of the *normalized frequency* $\nu = f/f_s$, or the *normalized radian frequency*, $\Omega = 2\pi\nu = 2\pi f/f_s = \omega T_s$ [see Fig. (2.3)]. Note that $\nu$ and $\Omega$ have no units, and the period of $X_s(f)$ is 1.

For real signals the frequency content (information) in the interval $[-f_s/2,0)$ is a reflected copy of the frequency content in $[0, +f_s/2)$, and therefore many authors display only the frequency interval $[0, +f_s/2)$ or the normalized frequency interval [0, 1/2] of discrete-time signal spectra.

Note also that $X_s(f)$ is still continuous in the frequency variable. From a practical perspective one cannot compute $X_s(f)$ for a continuous range of frequency values on a digital computer—one can only compute it for a finite number of frequency positions. For this reason, $X_s(f)$ is usually only evaluated in practice at a finite set of discrete frequencies. This discretization of the frequency domain of the DTFT gives rise to the so-called discrete Fourier transform (DFT), which will be studied later in Sect. 2.4.1.

### 2.2.2 *Ideal Reconstruction*

Consider the discrete-time Fourier transform (DTFT) of the sampled signal $x_s(t)$ shown graphically in Fig. (2.3). It is not hard to see that one can reconstruct the original spectrum $X(f)$ from $X_s(f)$ [and hence, the original signal $x(t)$ from $x_s(t)$] by filtering the sampled signal with an ideal low-pass filter whose cutoff frequency is $B$ Hz.

Often in practice reconstruction occurs in a two stage process—the first involves using a digital to analog converter (DAC) which applies a sample and hold function, and the second stage involves applying a low-pass reconstruction filter. Both stages are considered in more detail below.

#### 2.2.2.1 Stage 1

When a digital signal is ready to be converted back to the analog domain, the sequence of digitally stored values is usually fed synchronously into a DAC. Almost all DACs apply a *zero-order hold* (sample-and-hold) function to the sequence of input values [see Fig. (2.4)]. The sample and hold function is effectively a filter with a rectangular impulse response $h(t) = \Pi_{T_s}(t - T_s/2)$. This circuit simply holds the sample $x(nT_s)$ for $T_s$ seconds. The corresponding filter transfer function is a sinc function, as shown in Fig. (2.4).

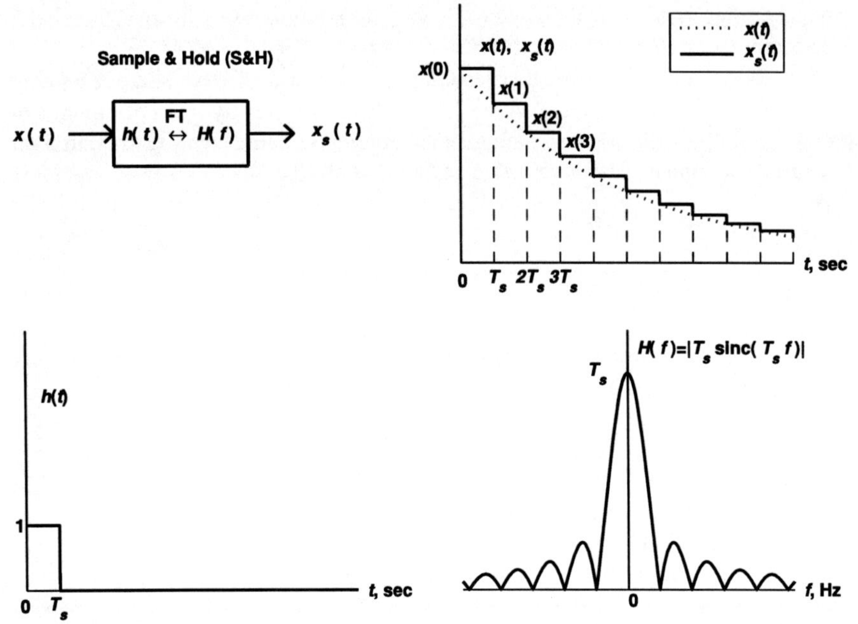


**Fig. 2.4** Sample-and-Hold: operation, impulse response, and transfer function

#### 2.2.2.2 Stage 2

Note that the DAC converts the digital signal into an analog stair-step waveform, which contains many high-frequency artefacts. This is because the sample and hold is effectively a filter with a very slow roll off, and it does not fully eliminate the energy in all the unwanted spectral images. To reconstruct a smooth replica of the original signal, the stair-step function must be passed through a low-pass filter, often called an anti-imaging filter. This filter ideally removes all frequency content above the upper frequency limit, $B$.

For perfect reconstruction of the analog signal to occur within this two stage process, three criteria must be satisfied. The first criteria is that there is no quantization noise in the digital signal. The second criterion is that the sampling frequency should be greater than twice the signal bandwidth $B$, i.e., greater than twice the highest frequency present in the signal:

$$B \leq \frac{f_s}{2}. \tag{2.4}$$

This requirement is necessary so that the repeating spectra shown in Fig. (2.3) do not “run into other”. The minimum sampling rate needed to avoid reconstruction errors ($f_s = 2B$) is called the Nyquist rate. The third criteria is that the filtering provided by the combination of the sample and hold function and the subsequent low-pass filter is ideal. That is, those two filters combine to form a filter with a perfectly flat pass-band and a perfectly sharp cutoff at $f = B$ Hz.

In practice none of the three criteria above are met perfectly. There is usually a very small amount of quantization noise, there is typically a small amount of spectral energy in the original analog signal above $f_s/2$ and the reconstruction filter is usually not ideal. The errors due to these imperfections, however, are often small.

#### 2.2.2.3 Frequency Aliasing

If $B > f_s/2$, replicas of the signal spectrum $X(f)$ within the DTFT overlap, and part of the information content from the input signal is lost [See Fig. (2.5) and compare it with Fig. (2.3)]. This overlap is called frequency aliasing. To avoid aliasing one normally band-limits the signal before sampling to remove the high frequency content that may cause aliasing. This can be achieved using a LPF, which is often called an anti-aliasing filter in this scenario.

One technique which is often used in practice to help reduce errors in reconstruction is to increase the sampling rate of the digital system well above the Nyquist rate. When this is done the spectral images in the DTFT are well separated from each other, and when it is necessary reconstruct the analog signal, it is not as difficult to filter out the unwanted images. In particular, the requirements on the sharpness of the anti-imaging filter are relaxed when one increases the sampling rate. Increasing the sampling rate not only relaxes the requirements on the anti-imaging filter, but also on the front-end anti-aliasing filter—i.e., the increase in the $f_s/2$ value effectively allows the transition band of the anti-aliasing filter to be wider.

Figure (2.6) shows a practical signal processing system which incorporates the anti-aliasing filter, the sampling and ADC, the digital processing, the DAC and the anti-imaging filter.

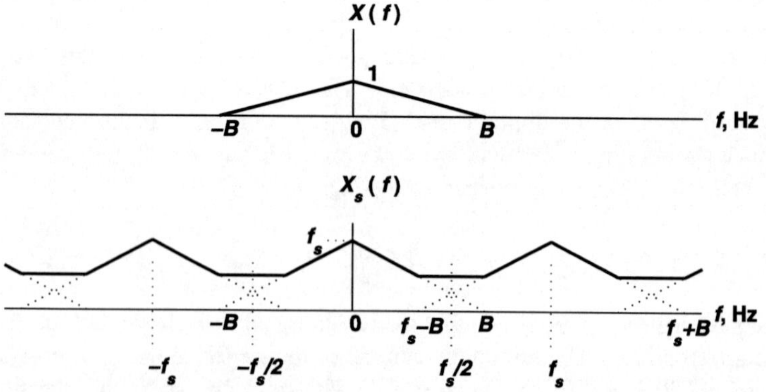


Fig. 2.5 Frequency aliasing $B > f_s/2$

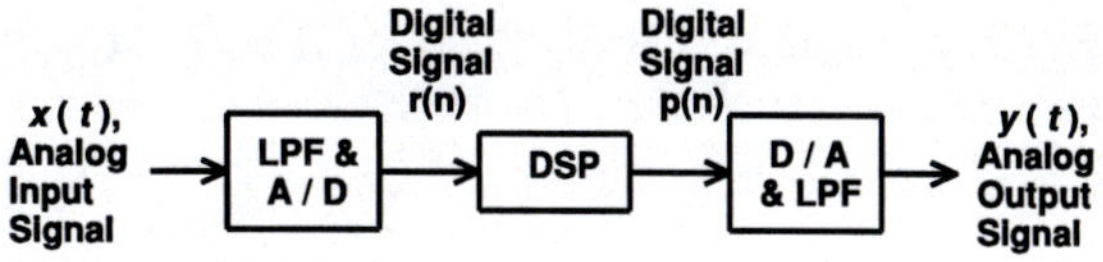


**Fig. 2.6** Block diagram of a practical signal processing system

## 2.3 Time-Domain / Frequency-Domain Representations

Like analog signals and systems, digital signals and systems can be analyzed either in the time-domain or in the frequency domain. The two domains are equivalent, provided that suitable transformations are used. The most common time-to-frequency transformations for continuous-time signals are the Fourier transform and the Laplace transform. The counterparts of these transforms for sampled signals are the discrete-time Fourier transform (DTFT) and the z-transform. The DTFT is useful for analyzing the frequency content of signals, while the z-transform is useful for both analyzing the frequency content of signals and analyzing the stability of systems. The DTFT and z-transforms will be defined and studied in more detail in Sects. 2.3.2.2 and 2.3.3.

### *2.3.1 Time-Domain Representation of Digital Signals and Systems*

This section considers the representation and analysis of digital signals and systems. Fundamental to time domain analysis of discrete-time signals is discrete-time convolution, which is defined in what follows.

#### 2.3.1.1 Discrete Linear Convolution

If $x(n)$ and $y(n)$ are two discrete signals, their discrete linear convolution $w(n)$ is given by:

$$w(n) = x(n) * y(n) = \sum_{k=-\infty}^{\infty} x(k)y(n-k) \; .$$

Note that $k$ is a dummy variable of summation. The above formula is similar to that of continuous-time linear convolution, except that summation is used instead of integration.

If both $x(n)$ and $y(n)$ are finite signals of lengths $N_1$ and $N_2$ samples, respectively, the length of $w(n)$ is finite and Is given by $L = N_1 + N_2 - 1$. Hence, if $x(n)$ and $y(n)$ are of equal length $N$, then the result of the convolution is $L = 2N - 1$ samples long.

*Example (1)* If $x(n) = (0.8)^n\ u(n)$ and $y(n) = (0.4)^n\ u(n)$, then their convolution is:

$$w(n) = x(n) * y(n) = sum_{k=\infty}^{-\infty} x(k)y(n-k) \tag{2.5}$$

Since the variable of summation is $k$, $x$ and $y$ should be re-expressed as functions of $k$:

$$y(k) = \begin{cases} (0.4)^k, & k \geq 0 \\ 0, & k < 0; \end{cases}$$

$$x(k) = \begin{cases} (0.8)^k, & k \geq 0 \\ 0, & k < 0 \end{cases}$$

$$\therefore y(n-k) = \begin{cases} (0.4)^{n-k}, & n-k \geq 0 \Rightarrow \therefore k \leq n \\ 0, & n-k < 0 \Rightarrow \therefore k > n \end{cases}$$

Examination of Fig. (2.7) reveals that the product $x(k)y(n-k)$, is non-zero only when $n \geq 0$. Hence it follows that:

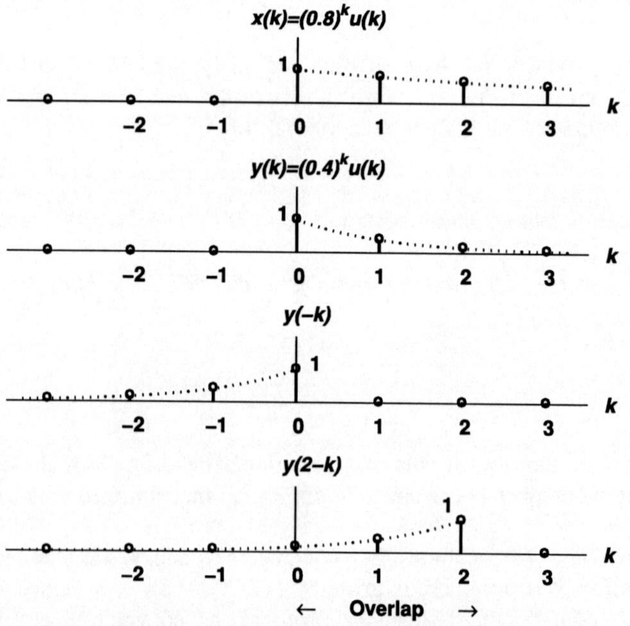


Fig. 2.7 Convolution of $x(n) = (0.8)^n\ u(n)$ with $y(n) = (0.84)^n\ u(n)$

$$
\begin{aligned}
w(n) &= \sum_{k=0}^{n} (0.8)^k (0.4)^{n-k}; \quad n \geq 0 \\
&= (0.4)^n \sum_{k=0}^{n} 2^k = (0.4)^n \frac{1 - 2^{n+1}}{1 - 2} \quad \textit{[Tables, Formula 10]} \\
&= (0.4)^n (2^{n+1} - 1) u(n).
\end{aligned}
$$

*Example (2)* If there are two finite duration signals $x(n) = \{\hat{2}, 3, 4\}$ and $y(n) = \{-\hat{1}, 5\}$, where "$\hat{\ }$" indicates that the sample is taken at $n = 0$, then their convolution can be found directly from the general formula as follows:

$$
\begin{aligned}
w(n) &= \sum_{k=-\infty}^{\infty} x(k)y(n-k) \text{ (General formula)} \\
w(0) &= \sum x(k)y(0-k) = x(0)y(0) = (2)(-1) = -2, \\
w(1) &= \sum x(k)y(1-k) = x(0)y(1-0) + x(1)y(1-1) = (2)(5) + (3)(-1) = 7, \\
w(2) &= \sum x(k)y(2-k) = x(0)y(2-0) + x(1)y(2-1) + x(2)y(2-2) \\
&= (2)(0) + (3)(5) + (4)(-1) = 11, \\
w(3) &= \sum x(k)y(3-k) \\
&= x(0)y(3-0) + x(1)y(3-1) + x(2)y(3-2) + x(3)y(3-3) \\
&= (4)(5) = 20, \\
w(4) &= w(5) = \cdots = 0. \text{ (Note that } L = N_1 + N_2 - 1 = 3 + 2 - 1 = 4)
\end{aligned}
$$

#### 2.3.1.2 Mathematical Representation of Digital Signals and Systems in the Time Domain

While analog signals are normally specified as a function of continuous-time, uniformly sampled digital signals are normally represented in the time domain as a function of the sample number (or the sample count integer). Digital systems are typically characterized in ways that are similar to those used in the continuous-time domain. Rather than being represented by a continuous-time impulse response, a digital system is fully specified by a discrete-time impulse response. This impulse response is the output of the digital system when the input is the discrete delta function, $\delta(n)$. Within this book, the discrete-time impulse response is denoted by $h(n)$.

The system input/output (I/O) relationship is specified by the discrete linear convolution of the impulse response $h(n)$ and the input signal $x(n)$:

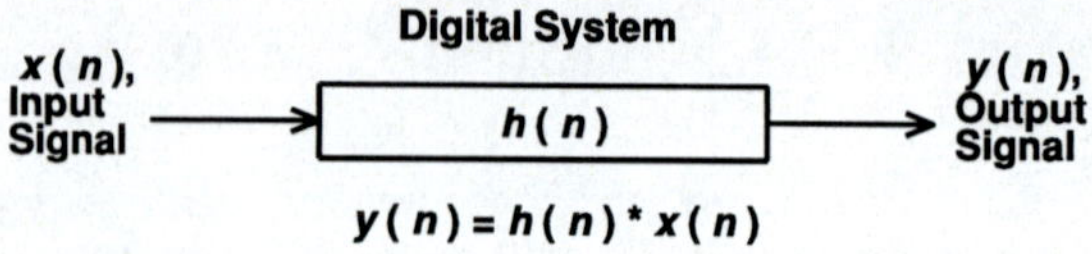


**Fig. 2.8** Discrete-time domain representation of a digital system

$$y(n) = x(n) * h(n) = \sum_{k=-\infty}^{\infty} x(k)h(n-k) \tag{2.6}$$

*Note* For *causal* digital systems $h(n) = 0$ for $n < 0$.

A graphical illustration of a digital system is shown in Fig. (2.8).

### Eigenfunctions of LTI Digital Systems

If the input signal to an LTI digital system is $x(n) = e^{j\omega nT_s} = e^{jn\Omega}$, then the output is given by:

$$y(n) = x(n) * h(n) = h(n) * x(n) = \sum_{k=-\infty}^{\infty} h(k)e^{j(n-k)\Omega} = e^{jn\Omega} \sum_{k=-\infty}^{\infty} h(k)e^{-jk\Omega}.$$

If one defines $H(e^{j\Omega}) = \sum_{k=-\infty}^{\infty} h(k)e^{-jk\Omega}$, then for an input of $e^{jn\Omega}$ one obtains the output $y(n) = e^{jn\Omega}H(e^{j\Omega})$. Hence, $e^{jn\Omega}$ is an eigenfunction, and the associated eigenvalue is $H(e^{j\Omega})$ (Compare with the analogous result obtained for continuous-time systems in Sect. 1.2.3.4).

### Analyzing the Stability of Digital Systems in the Time Domain

For a digital system to be BIBO stable, the output $y(n)$ should be bounded when the input $x(n)$ is bounded, i.e.,

$$|y(n)| < \infty \text{ when } |x(n)| < A(\text{which is a finite constant}), \forall n.$$

Using the inequality $|a + b| \leq |a| + |b|$, it follows that

$$\left|\sum_{k=-\infty}^{\infty} x(k)h(n-k)\right| \leq \sum_{k=-\infty}^{\infty} |x(k)||h(n-k)|.$$

If $|x(k)| < A \forall k$, then

$$|y(n)| = \left|\sum_{k=-\infty}^{\infty} x(k)h(n-k)\right| \leq A \sum_{k=-\infty}^{\infty} |h(n-k)| \leq A \sum_{m=-\infty}^{\infty} |h(m)|,$$

where $m = n - k$. Hence, a digital system is BIBO-stable if $\sum_{k=-\infty}^{\infty} |h(k)| < \infty$. (Compare with the condition for stability of analog systems found in Sect. 1.2.3.5).

## 2.3.2 Frequency-Domain Representation of Digital Signals and Systems

### 2.3.2.1 Discrete-Time Fourier Series for Periodic Digital Signals

In the analog domain the continuous-time signal and the Fourier Series are related according to:

$$x(t) = \sum_{k=-\infty}^{\infty} X_k e^{+j2\pi k f_o t} \Leftrightarrow X_k = \frac{1}{T_o}\int_0^{T_o} x(t)e^{-j2\pi k f_o t}dt \qquad (2.7)$$

If the analog signal $x(t)$ is sampled with a rate of $f_s$ (sampling interval $= T_s = 1/f_s$), a discrete-time signal $x(n)$ is obtained:

$$x_s(t) = x(n) = \sum_{k=-\infty}^{\infty} x(t = nT_s)\delta(t). \qquad (2.8)$$

It will be assumed that the period of the above discrete-time signal is $N$ samples, where $N = T_o/T_s = f_s/f_o$. If $x_s(t)$ is fed into the formula for the Fourier Series in Sect. 1.2.3.1 one obtains:

$$X_k = X(k) = \frac{1}{N}\sum_{n=0}^{N-1} x(n)e^{-j2\pi kn/N} \quad \text{[DFS pair]}$$

The above expression is obtained with the result that $\int_0^{T_o}\delta(t)dt = 1$. The equation linking time domain signals with Fourier series for discrete-time signals is therefore given by:

$$x(n) = \sum_{k=0}^{N-1} X(k)e^{j2\pi kn/N} \Leftrightarrow X(k) = \frac{1}{N}\sum_{n=0}^{N-1} x(n)e^{-j2\pi kn/N}\text{[DFS pair]}.$$

Note that, unlike the continuous-time FS, the summation for $x(n)$ in the DFS does not need to go from $-\infty$ to $\infty$ due to the periodicity of both $X(k)$ and the discrete exponential $e^{j2\pi\, kn/N}$ in $k$. That is, the DFS coefficients are *periodic* (with period $N$).

### 2.3.2.2 The Discrete-Time Fourier Transform for Non-Periodic Digital Signals

Recall from Sect. 2.2 that the formula for the discrete-time Fourier transform (DTFT) for non-periodic signals is:

$$X_s(f) = X(f) * P(f)$$
$$= X(f) \cdot f_s \sum_{kn=-\infty}^{\infty} \delta(f - kf_s) = \sum_{k=-\infty}^{\infty} X(f - kf_s)^{\cdot}$$

The DTFT consists of a sum of scaled and infinitely repeated versions of the spectrum of the original analog signal, $x(t)$. This is depicted in Fig. (2.3).

### 2.3.3 The z-Transform

The Laplace transform (LT) was previously seen to be very useful for stability analysis of continuous-time signals. One can obtain similar stability analysis capabilities for discrete-time signals by using a discrete-time counterpart to the LT. If one substitutes the expression for a discrete-time signal, $x_s(t)$ into the expression for the LT in Sect. 1.2.3.3 and simplifies, one obtains the *discrete-time Laplace transform (DTLT)*:

$$X(s) = X(e^{sT_s}) = \sum_{n=-\infty}^{\infty} x(n)e^{-nsT_s}. \tag{2.9}$$

The DTLT is, like the DTFT, periodic, and can be shown to have the equivalent alternative expression:

$$X_s(s) = f_s \sum_{k=-\infty}^{\infty} X(s - jk\omega_s).$$

where $X(s)$ is the LT of the original analog signal. It is convenient to define a new variable $z = e^{sT_s}$, which can be substituted into (2.9) to obtain the following transform:

$$X(z) = \sum_{n=-\infty}^{\infty} x(n)z^{-n}. \tag{2.10}$$

The above relation is called the z-transform (ZT). It is evident that the z-transform is reduced to the Fourier transform if one substitutes $z = e^{j\Omega}$.

Unlike the DTFT the ZT is *not necessarily periodic* in the frequency variable. This non-periodicity is due to the additional factor $\alpha$ in the exponent ($e^{sT_s}$) which results in a *decaying or expanding factor* ($e^{\alpha}$) in the overall transform.

Under certain conditions (that are normally satisfied in practical signals and systems), the ZT exists as a pair, i.e., there is a z-transform and an inverse z-transform (IZT). Like the LT, the ZT has a region of convergence (ROC).

Computation of the IZT can be quite involved as it requires integration in the complex z-plane. For this reason, it is advisable to use *Tables* to determine the IZT.

#### 2.3.3.1 The Single-Sided ZT

As with the LT, one can define double-sided and single-sided transforms. Mimicking the approach used for the LT, this book will focus only on the single-sided ZT, which is defined as:

$$X(z) = \sum_{n=0}^{\infty} x(n) z^{-n} \,. \tag{2.11}$$

*Example (1)* If $x(n) = a^n\, u(n)$, then its ZT is given by:

$$X(z) = \sum_{n=0}^{\infty} x(n) z^{-n} = \sum_{n=0}^{\infty} a^n z^{-n} = \sum_{n=0}^{\infty} (az^{-1})^n$$

$$= \frac{1-(az^{-1})^{\infty}}{1-az^{-1}} [\textit{from Tables}, \text{ Formula 10}] = \frac{1}{1-az^{-1}} = \frac{z}{z-a} (\text{if } |z| > a).$$

Figure (2.9) shows this function for $a = 3$.

*Example (2)* If $x(n) = u(n)$, then from *Example (1)*, its ZT is given by $z/(z-1)$ for $|z| > 1$, i.e., the ROC is $|z| > 1$).

*Example (3)* If $x(n) = \delta(n)$, then it ZT is given by $X(z) = \sum_{n=0}^{\infty} \delta(n) z^{-n} = 1 + 0 + 0 + \cdots = 1$.

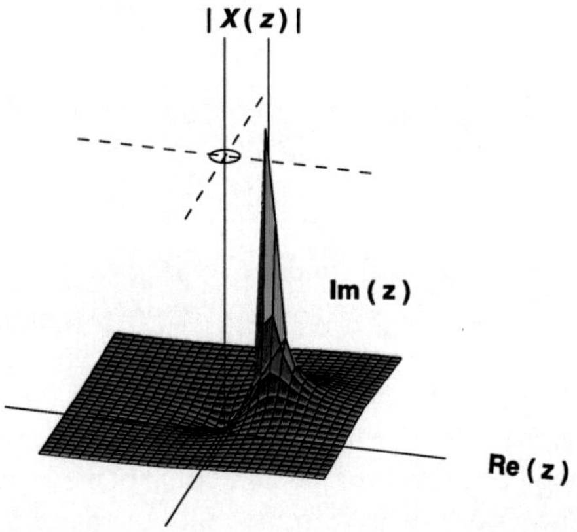


**Fig. 2.9** A plot of the magnitude of the z-transform $X(z) = z/(z-a)$ for $a = 3$

#### 2.3.3.2 The Time-Shift Property of the ZT

Assume that $y(n) = x(n - M)$, i.e., assume that $y(n)$ is a time delayed version of $x(n)$ with the delay being $M$ samples. Then:

$$Y(z) = \sum_{n=0}^{\infty} y(n)z^{-n} = \sum_{n=0}^{\infty} x(n-M)z^{-n}.$$

Letting $k = n - M$ and assuming that $x(k) = 0$ for $k < 0$ gives

$$Y(z) = \sum_{k=-M}^{\infty} x(k)z^{-k-M} = z^{-M}\sum_{n=0}^{\infty} x(k)z^{-k} = z^{-M}X(z)\ .$$

That is, a delay of $M$ samples in the time-domain corresponds to a multiplication by $z^{-M}$ in the z-domain.

#### 2.3.3.3 Relationship Between the FT and ZT of a Discrete-Time Signal

It has been seen previously that to retrieve the FT from the LT, one substitutes $s = j\omega$. In similar fashion one can recover the DTFT from the ZT by putting $z = e^{sT_s} = e^{j\omega T_s}$. Since $|e^{j\omega T_s}| = 1$, the DTFT is effectively the ZT evaluated along the unit circle of the complex $z$-plane, i.e., along $|z| = 1$.

It should be noted that FT plots are in general 2-D, while LT and ZT plots are 3-D. The extra dimension is due to the additional variable, $\alpha$ in the LT and ZT formulations.

#### 2.3.3.4 Relationship Between the LT and the ZT for Discrete-Time Signals

Recall that the ZT is obtained from the LT by making the assignment $z = e^{sT_s}$ $= e^{(\alpha+j\omega)T_s}$. Since $z = e^{sT_s} = e^{(\alpha+j\omega)T_s} = e^{\alpha T_s}e^{j\omega T_s}$, the magnitude of $z$ is $r = |z|$ $= e^{\alpha T_s}$ and the phase of $z$ is:

$$\theta = \tan^{-1}\left[\frac{Im(z)}{Re(z)}\right] = \tan^{-1}\left[\frac{\sin(\omega T_s)}{\cos(\omega T_s)}\right] = \omega T_s$$

(using Euler's formula to expand $e^{j\omega T_s}$).

Hence, the $z = e^{sT_s} = e^{(\alpha+j\omega)T_s}$ assignment inherent in the Laplace to z-transform mapping causes the imaginary axis in the $s$-plane (i.e., $s = 0 + j\omega$) to map to the *circumference of the unit circle* in the $z$-plane (i.e., $|z| = 1$). This is illustrated graphically in Fig. (2.10). The left half of the s-plane (with $\alpha < 0$) is transformed into the interior of the unit circle in the z-plane (i.e., $|z| < 1$), while the right half of the s-plane is transformed into the exterior of the unit circle (i.e., $|z| > 1$).

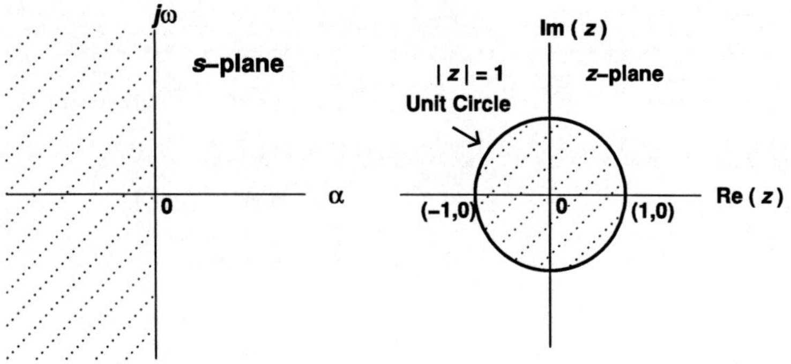


**Fig. 2.10** Mapping the $s$-plane onto the $z$-plane using $z = e^{sT_s}$

Note that this relation is not a general relation between analog and digital representations. It is only applicable to s-plane and z-plane representations of discrete-time signals, since it is based on the relation between the DTLT and ZT.

### 2.3.4 *Mathematical Representation of Signals and Systems in the Frequency Domain*

A discrete-time signal $x(n)$ can be represented by its DTFT in the frequency domain . Similarly, a digital system with impulse response $h(n)$ can be represented in the frequency domain by its transfer function, $H_s(f) = H(e^{j\omega T_s}) = H(e^{j\Omega}) =$ which is the DTFT of its impulse response $h(n)$ [See Fig. (2.11)].

Recall that characterization of the system function is achieved in the time domain via the impulse response. The output of a system can be obtained for a given input via discrete linear convolution of the impulse response with the input:

$$y(n) = h(n) * x(n)$$

In the frequency domain, the *system function* information is contained in the transfer function. The system output for a given input is found by multiplying the transfer $H_s(f)$ by the DTFT of the input signal $X_s(f)$:

$$Y_s(f) = H_s(f) \cdot X_s(f),$$

where $Y_s(f) = \text{DTFT}[y(t)]$ and $X_s(f) = \text{DTFT}[x(t)]$. This relation can also be written with the alternative notation $Y(e^{j\Omega}) = H(e^{j\Omega}) \cdot X(e^{j\Omega})$, where $\Omega = \omega T_s$. A similar relation is obtained in the z-domain:

$$Y(z) = H(z) \cdot X(z).$$

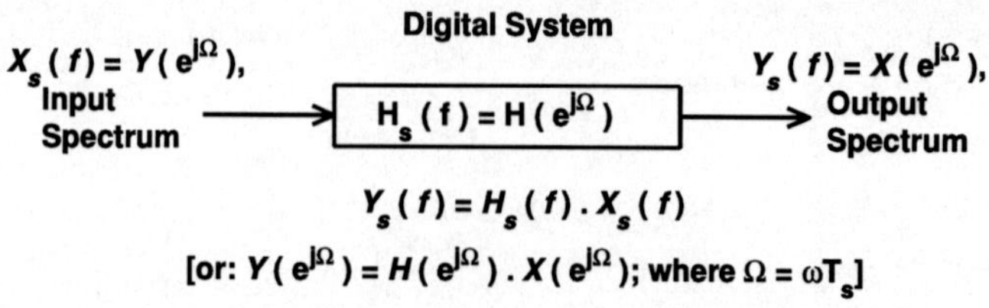


**Fig. 2.11** Frequency-domain representation of a discrete-time system

Again, analogously to the continuous-time case, convolution in the discrete-time domain is transformed into multiplication in both the frequency domain and the z-domain:

$$x(n) * y(n) \overset{Z}{\longleftrightarrow} X(z) \cdot Y(z) \tag{2.12}$$

$$x(n) \cdot y(n) \overset{Z}{\longleftrightarrow} X(z) * Y(z) \tag{2.13}$$

(See also *Tables*, z-Transform Pairs and Theorems).

#### 2.3.4.1 Relationship Between the ZT Transfer Function and the Frequency Response

The (periodic) frequency response $H_s(\omega) = H(e^{j\omega T_s})$ of a digital system can be obtained from its ZT-transfer function *H*(*z*) with the substitution $z = e^{j\omega T_s}$ as follows:

$$H(e^{j\omega T_s}) = H(z)|_{z=e^{j\omega T_s}}.$$

#### 2.3.4.2 Stability of Digital Systems in the z-Domain

It has been shown previously that a digital system is BIBO-stable if its impulse response is absolutely summable, i.e., if $\sum_{k=-\infty}^{\infty} |h(k)| < \infty$. Practically it is often difficult to analyze the system stability in this way. Equivalently, a *causal* digital system is BIBO-stable *if and only if all the poles of its z-transfer function H*(*z*) *are inside the unit circle* (i.e., $|z_p| < 1$, where $z_p$ is the location of the *p*th pole in the *z*-plane). [Note that this condition is for causal digital systems only, otherwise the condition would be that the ROC of *H*(*z*) should contain the unit circle].

*Example (1)* If the system impulse response is $h(n) = a^n\, u(n)$, then its z-transfer function is $H(z) = z/(z - a)$ [from *Tables*. This system has a zero at $z = 0$, and a pole at $z = a$. It is BIBO-stable if $|a| < 1$, i.e., if the pole is within the unit circle

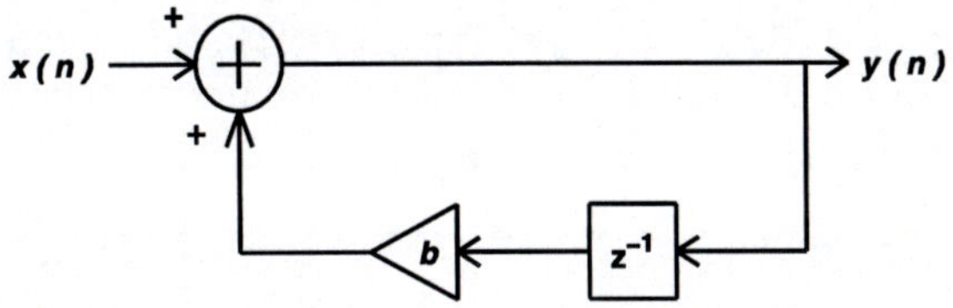


**Fig. 2.12** An example of a recursive digital system (leaky integrator)

*Example (2))* The difference equation that describes the recursive system in Fig. (2.12) is given by:

$$y(n) = x(n) + by(n-1), \tag{2.14}$$

where, $b$ is a multiplicative constant (gain). Taking the ZT of both sides of (2.14) yields:

$$Y(z) = X(z) + bz^{-1}Y(z).$$

Re-arranging terms leads to

$$[1 - bz^{-1}]Y(z) = X(z).$$

$$\therefore H(z) = \frac{Y(z)}{X(z)} = \frac{1}{1 - bz^{-1}} = \frac{z}{z - b}.$$

From *Tables* (z-Transform Pairs), the IZT is $h(n) = b^n u(n)$.

The system frequency response is:

$$\begin{aligned} H(\omega) &= H(z)|_{z=\exp(j\omega T_s)} \\ &= \frac{e^{j\omega T_s}}{e^{j\omega T_s} - b} = \frac{e^{j2\pi f T_s}}{e^{j2\pi f T_s} - b} = \frac{e^{j2\pi f/fs}}{e^{j2\pi f/f_s} - b} \\ &= \frac{e^{j2\pi v}}{e^{j2\pi v} - b} = \frac{e^{j\Omega}}{e^{j\Omega} - b} \end{aligned}$$

where $v = f/f_s$ and $\Omega = 2\pi v$ are the normalized cyclic frequency and normalized angular frequency, respectively. The magnitude and phase responses are:

$$\begin{aligned} H(\omega)| = |H(e^{j\omega T_s})| &= \frac{|e^{j\omega T_s}|}{|e^{j\omega T_s} - b|} = \frac{1}{|[\cos(\omega T_s) - b] - j\sin(\omega T_s)|} \\ &= \frac{1}{\sqrt{[\cos(\omega T_s) - b]^2 - [\sin(\omega T_s)]^2}} \angle H(\omega) \\ &= \angle H(e^{j\omega T_s}) = -\tan^{-1}\left[\frac{b\sin(\omega \cdot T_s)}{1 - b\cos(\omega \cdot T_s)}\right] \end{aligned}$$

As a special case, let $b = 0.5$ and $f_s = 1$ Hz (i.e., $T_s = 1$ s). At $f = 0.1$ Hz (i.e., $\omega = 2\pi(0.1) = 0.6283$rad/s), the magnitude response is $|H(e^{j\omega T_s})| = 1.506$ and

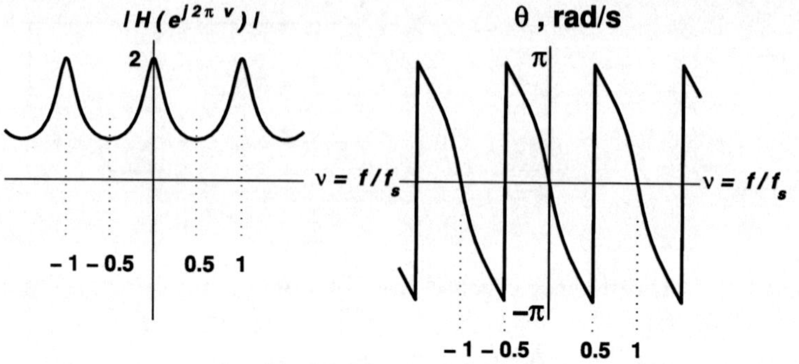


**Fig. 2.13** Magnitude and phase response of a leaky integrator

the phase response is $\angle H(e^{j\omega T_s}) = -0.458$ rad. Figure (2.13) shows the magnitude and phase response of this system as a function of the normalized frequency. If one needs to recover the actual response versus frequency from the normalized frequency response one simply scales the frequency axis by $f_s$.

From Fig. (2.13) one can see that the shape of the transfer function is that of a LPF; the circuit is in fact what is sometimes referred to as a leaky integrator. If $b = 1$, it would be an integrator, as will be seen later.

## 2.4 A Discrete-Time and Discrete-Frequency Representation

The DTFT of a discrete-time signal *x*(*n*) is still continuous in the frequency variable, and hence it is not possible to implement with practical digital technologies such as computers. It is necessary, then, to find a discrete-time and discrete-frequency representation for practical analysis. This doubly discrete representation will be referred to as the *Discrete Fourier Transform (DFT)*.

### 2.4.1 The Discrete Fourier Transform

In practice one has to restrict oneself to computing the Fourier transform at a limited number of representative sample frequencies. It was seen previously that with appropriate precautions no information is lost provided that certain requirements on the sampling rate are met. In particular, it is necessary that the sample rate in the time domain is at least twice as high as the highest frequency component present in the analog signal. It will be seen here that there is an analogous result for frequency domain sampling—one does not lose any information by sampling in the frequency domain, provided that certain conditions are met on the frequency sampling rate.

Consider now the problem of sampling the DTFT in the frequency domain. It is assumed that the frequency sampling is uniform, and that the spacing between samples is $F_s$. Now it is critical that this frequency domain sampling does not lead to information loss. That is, it is necessary for the time domain signal to still be perfectly recoverable, despite this sampling.

It is now required to determine the maximum frequency sampling interval which ensures no loss of information. Assume initially that this rate is $F_s = f_s/N$. This choice implies that the number of samples in the frequency domain is the same as the number of samples in the time domain. Then the resulting DFT is defined by:

$$X_s(k) = \sum_{n=0}^{N-1} x_s(n)e^{-j2\pi kn/N}.$$

Now consider the inverse Fourier Transform of $X_s(k)$. By using an analysis very similar to that done in Sect. 2.3.2.1 it is possible to show that the inverse is given by:

$$x_p(n) = \frac{1}{N}\sum_{k=-\infty}^{\infty} X_s(k)e^{j2\pi kn/N} \Longleftrightarrow X_s(k) = \sum_{n=0}^{N-1} x_p(n)e^{-j2\pi kn/N}$$

Now it is important to realize that both $X_s(k)$ and $e^{-j2\pi}kn/N$ are periodic, and because of this periodicity, $x_p(t)$ is also periodic, with period $N$. This indicates that just as sampling in the time domain causes periodicity in the frequency domain, so sampling in the frequency domain causes periodicity in the time domain. Furthermore, with a frequency sampling interval of $F_s = f_s/N$ the periodicity is seen to be $N$, which is just adequate to prevent the time domain images from "running into each other". That is, the frequency sampling interval of $F_s = f_s/N$ is just enough to prevent time-domain aliasing. If the sampling interval were any greater aliasing in the time domain would occur.

In summary then, sampling in the time and frequency domains causes repetition in both domains. The DFT is normally obtained from the DTFT by sampling at a rate of $F_s = f_s/N$, because this is just enough to avoid time-domain aliasing. With this sample rate the number of samples in both the time and frequency domains is $N$. Although the time and frequency domains both repeat doubly discredited systems, it is common to only display one image. With this in mind the usual definitions for the DFT and inverse DFT (IDFT) are:

$$x(n) = \frac{1}{N}\sum_{k=0}^{N-1} X(k)e^{j2\pi kn/N} \Longleftrightarrow X(k) = \sum_{n=0}^{N-1} x(n)e^{-j2\pi kn/N}. \tag{2.15}$$

Note that

$$IDFT\{DFT[x(n)]\} = 1; \tag{2.16}$$

that is, the DFT and IDFT are reversible transforms.

#### 2.4.1.1 Approximation of the FT Using DFT

The DFT can be used to approximate the original (continuous-time) FT. The quality of the approximation depends largely on the sampling rate used in the time domain. As long as the time domain sampling rate is greater than the highest frequency present in the original analog signal, and the frequency domain sampling rate is $F_s = f_s/N$, the approximation is perfect.

The effectiveness of the DFT for approximating the Fourier Transform derives from the fact that the DFT is simply a sampled version of the DTFT, which is in turn a scaled and periodic version of the FT (see Sect. 2.3.2.2).

#### 2.4.1.2 Relationship Between the DFT and the DFS Coefficients

A simple comparison between the DFS of a periodic sequence (with period $N$) and the DFT of one period of this sequence reveals that the two related by a simple multiplicative constant:

$$X_{k,DFS} = X_{k,DFT}/N, \tag{2.17}$$

where $X_{k,DFS}$ is the pertinent DFS coefficient and $X_{k,DFT}$ is the corresponding DFT sample. This is reminiscent of a similar relation in the analog domain, where the FS coefficients of a periodic signal (of period $=T_o$) are related to the FT of a single period of the same signal according to:

$$X_k = X_{1p}(f)/T_o \,|_{f=kf_o}, \tag{2.18}$$

where $X_k$ is the pertinent FS coefficient DFS and $X_{1p}(f)\,|_{f=kf_o}$ is the corresponding FT value.

#### 2.4.1.3 The Fast Fourier Transform

Calculation of the Discrete-Fourier Transform directly according to the definition in (2.15) requires of the order of $N^2$ computations, i.e, it requires $O(N^2)$ arithmetic operations. These calculations cannot normally be done in real-time, as is required for many practical applications. For this reason many scientists and engineers have sought alternative techniques for rapidly calculating the DFT. The first to devise an efficient algorithm was probably Gauss, but his work went largely un-noticed until about 1990. In 1965, however, the so-called Fast Fourier Transform (FFT) was re-invented and popularized. [see J. W. Cooley and J. W. Tukey, "An algorithm for the machine calculation of complex Fourier series," *Mathematics of Computation*, vol. 19, pp. 297–301, 1965]. In contrast to calculation via the direct definition, the FFT algorithm can be implemented in $O(N \log(N))$ operations. This algorithm provided not only economy of time in run-time operation, but also economy of computational and storage elements in hardware for DFT implementation.

The FFT algorithm can generally support real-time processing (at least on modern computers and DSP chips).

MATLAB the Fast Fourier Transform algorithm is available on MATLAB as `fft`.

#### 2.4.1.4 Circular Convolution and Its Relation to the Linear Convolution

In time and frequency discretized systems the time domain signal and the frequency domain signal are both periodic. Within these doubly discretized systems, it is not only the input signal which is periodic, but also the impulse response and frequency transfer function. To appreciate the implications of this last fact, consider two time-domain signals $x(n)$ and $h(n)$ which are both periodic. Their *linear* convolution

$$x(n) * h(n) = \sum_{k=-\infty}^{\infty} x(k)h(n-k) \tag{2.19}$$

diverges in general (i.e., does not exist). This is because both signals are infinitely long and so the sum within (2.19) can easily be infinite. With periodic signals, therefore, it is more natural to define a modified form of convolution known as circular convolution. As will be seen subsequently, this is the kind of convolution which is implemented implicitly in doubly discretized systems.

Assuming that two signals have *identical periods* of length $N$, then one considers only one period of each signal and defines circular convolution as:

$$x(n) \otimes h(n) = \sum_{k=0}^{N-1} x(k)h[(n-k)_N],$$

where $p_N = p$ modulo $N$ for any integer $p$. The circular convolution $x(n) \otimes h(n)]] >$ is also periodic with the same period $N$.

Recall that the linear convolution of two non-periodic finite-length signals $x(n)*h(n)$ has length $L = N_x + N_y - 1$. If one desires therefore to have $x(n) \otimes y(n) = x(n) * h(n)$ over one period of the periodic signals $x(n)$ and $h(n)$, then one has to artificially extend the length of both signals by zero-padding both $x(n)$ and $h(n)$ to be of length $L = N_x + N_y - 1$. That is, one adds extra samples to $x(n)$ and $h(n)$, with these samples having the value 0.

#### 2.4.1.5 I/O Relations Using Circular Convolution and the DFT

One of the key areas of application for the DFT is digital filtering. In DFT based digital filtering one has an input signal $x(n)$ and an impulse response $h(n)$, and the filtered output $y(n)$ is given by the convolution of $x(n)$ and $h(n)$. Implementation of the filtering in the time domain, however, requires $O(N^2)$ operations, assuming both $x(n)$ and $h(n)$ have $N$ samples. One can take advantage of the efficiency of the

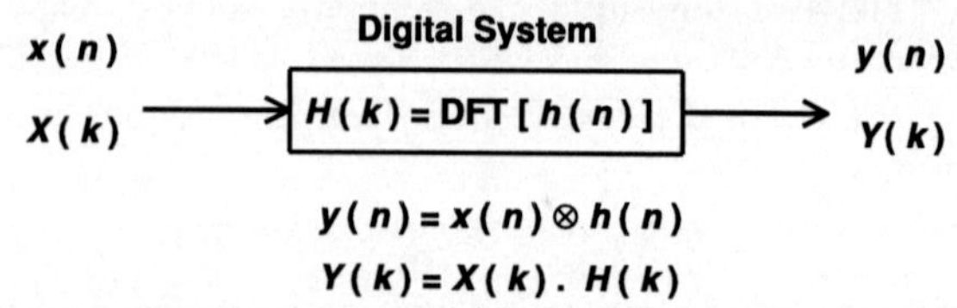


**Fig. 2.14** I/O relations in a digital system

FFT algorithm and implement the filtering in the frequency domain instead. This alternative is discussed in the following paragraphs (Fig. 2.14).

Since convolution in the time domain is equivalent to multiplication in the frequency domain, it is tempting to

1. take the FFTs of $x(n)$, and $h(n)$ to obtain $X(k)$ and $H(k)$, respectively,
2. multiply $X(k)$ and $H(k)$ together to obtain $Y(k)$, and
3. then take the inverse FFT to obtain $y_c(n)$.

This three-step procedure, however, would yield the wrong result in general. To see why the procedure is flawed, imagine that $x(n)$ and $h(n)$ are both $N$ samples long. Then $X(k)$, $H(k)$, $Y(k)$ and $y_c(n)$ would all be $N$ samples long as well. Since $y_c(n)$ is supposed to be the convolution of $x(n)$ and $h(n)$, however, it should be $N + N - 1$ samples long rather than $N$ samples.

The reason for the flaw in the three step procedure above is that multiplication in the DFT (or FFT) domain implicitly corresponds to *circular* convolution in the time domain. For true filtering to occur, one must have *linear* convolution rather than circular convolution, and so one must *zero-pad both* $x(n)$ *and* $h(n)$ to be of at least length $L = N_x + N_h - 1$, where $N_x$ is the length of $x(n)$ and $N_h$ is the length of $h(n)$. If one performs this zero-padding, the circular convolution inherently implemented in the FFT domain becomes equivalent to linear convolution in the time domain.

## 2.5 Signal Correlation, Power, and Energy

### *2.5.1 Definitions*

This subsection presents formulae for correlation, power, and energy of discrete-time signals. These formulae are analogous to those for analog signals, but with integrations changed to summations. The formulae are presented below.

#### 2.5.1.1 Autocorrelation of Non-Periodic Discrete-Time Energy Signals

$$R_x(k) = \sum_{n=-\infty}^{\infty} x(n)x^*(n+k).$$

#### 2.5.1.2 Autocorrelation for Periodic Discrete-Time Power Signals

$$R_x(k) = \frac{1}{N}\sum_{n=0}^{N-1} x(n)x^*(n+k),$$

where $N$ is the signal period.

#### 2.5.1.3 Energy in Non-Periodic Discrete-Time Energy Signals

$$R_x(k) = \sum_{n=-\infty}^{\infty} T_s|x(n)|^2$$

Compare the above formula with the corresponding formula for analog signals: $E = \int_{-\infty}^{\infty} |x(t)|^2 dt$.

#### 2.5.1.4 Power in Periodic Discrete-Time Power Signals

$$P = \frac{1}{N}\sum_{n=0}^{N-1} |x(n)|^2.$$

#### 2.5.1.5 Parseval's Theorem

$$\text{For periodic signals: } P = \frac{1}{N}\sum_{n=0}^{N-1} |x(n)|^2 = \sum_{k=0}^{N-1} \underbrace{|X(k)|^2}_{\text{PSD}}.$$

$$\text{For non-periodic signals: } E = \sum_{n=-\infty}^{\infty} T_s|x(n)|^2 = \int_0^{f_s} \underbrace{T_s^2|X_s(f)|^2}_{\text{ESD}} df,$$

where $X_s(f) = X(e^{j2\pi f T_s})$ is the DTFT.

#### 2.5.1.6 The Wiener-Kinchin Theorem

$$\text{For periodic signals: } R_x(k) \longleftrightarrow \underbrace{|X_k|^2}_{\text{PSD}}.$$

$$\text{For non-periodic signals: } R_x(k) \longleftrightarrow \underbrace{T_s^2 |X_s(f)|^2}_{\text{ESD}},$$

## 2.6 Digital Filters and Their Applications

The advancement of digital computers during the 1960s paved the way for many analog electronic circuits to be emulated in digital computers. The advantages of this kind of digital emulation are many—greater accuracy, greater flexibility, lower cost, lower power requirements, greater reproducibility, etc. This section of the book focuses on a particular type of digital emulation-namely the emulation of conventional filters with digital computer hardware. This type of emulation is commonly referred to as digital filtering.

### *2.6.1 Ideal Digital Filters*

Ideal digital LP, HP, BP, and BS filters can be defined by simple analogy with analog filters. In exploiting the similarities between digital and analog filters, however, it is also important to keep in mind that there are some key differences. One of the key differences is that there is a periodicity in the frequency response of digital filters.

#### 2.6.1.1 Mathematical Formulation

The Digital LPF

The transfer function of a digital LPF over the principal frequency domain $(-f_s/2, f_s/2)$ is given by

$$H_L(f) = H_L(e^{j2\pi f T_s}) = \begin{cases} 1, & 0 \leq |f| \leq f_c \\ 0, & f_c \leq |f| \leq f_s/2. \end{cases}$$

[see Fig. (2.15)].

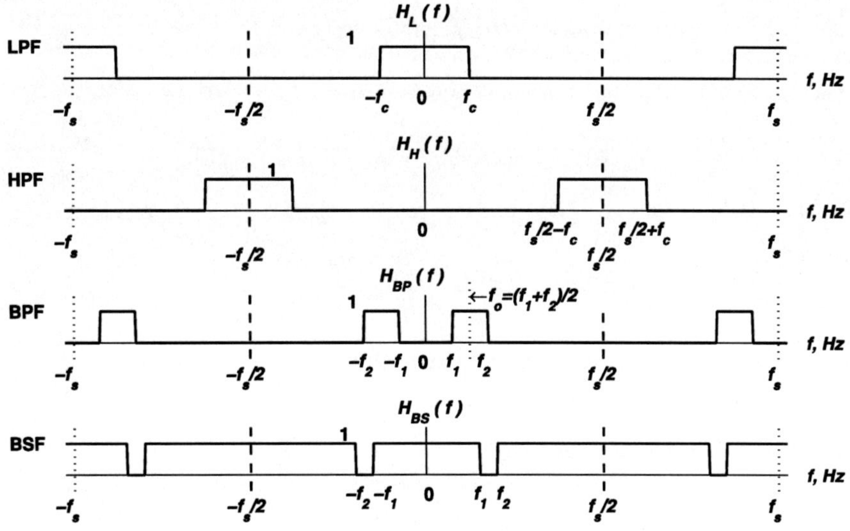


**Fig. 2.15** Transfer functions of ideal digital filters

## Digital Filter Transformations in the Time-Domain

Similar definitions to that of the LPF can be created for HPF, BPF, and BSF filters, These definitions are provided below, and illustrative plots for the corresponding transfer functions are provided in Fig. (2.15).

1. LP → HP : from Fig. (2.15) it is seen that the transfer function of a HPF with cutoff frequency $f_c$ is a spectrally-shifted version of a LPF with the same cutoff frequency, i.e.,

$$\begin{aligned} H_H(f) &= H_L(f \pm f_s/2), \\ \sum_{n=-\infty}^{\infty} h_H(n)e^{-j2\pi nfT_s} &= \sum_{n=-\infty}^{\infty} h_L(n)e^{-j2\pi n(f+f_s/2)T_s} \\ &= \sum_{n=-\infty}^{\infty} h_L(n)(-1)^n e^{-j2\pi nfT_s} \end{aligned} \tag{2.20}$$

noting that $e^{-j2\pi}n\ (f_s/2)T_s = e^{-jn\pi} = (-\ 1)^n$. A simple examination of the above equation indicates that:

$$h_H(n) = (-1)^n h_L(n).$$

i.e., the impulse response of a highpass filter can be obtained from a lowpass one by simply inverting every second sample.

2. LP → BP: from Fig. (2.15) it is apparent that:

$$H_{BP}(f) = H_L(f + f_o) + H_L(f - f_o).$$

where $f_o = (f_1 + f_2)/2$ and the cutoff frequency of the LPF $H_L(f)$ is $f_c = (f_2 - f_1)/2$.

$$\therefore h_{BP}(n) = 2\cos(2\pi n f_o/f_s) h_L(n) [\textit{from Tables}].$$

3. BP → BS: From Fig. (2.15) one can see that $H_{Bs}(f) = 1 - H_{BP}(f)$, hence,

$$h_{BS}(n) = \delta(n) - h_{BP}(n) = \begin{cases} 1 - h_{BP}(0), & n = 0 \\ -h_{BP}(n), & n \neq 0. \end{cases}$$

### 2.6.2 Linear-Phase Systems

Ideally, it would be convenient if a filter had

1. a magnitude response which did not vary with frequency,and so introduced no amplitude distortion,
2. a zero phase response and hence had no phase distortion.

However, these ideals are not achievable in practical filters. A zero phase response implies that the filter's impulse response is symmetric about $n = 0$ and this in turn implies that the filter is non-causal (see Sect. 1.5.1). Practical filters cannot be non-causal and cannot therefore have zero phase response.

A practical impulse response can be obtained from the ideal (non-causal) impulse response by inserting a delay which positions all (or nearly all) of the impulse response after $n = 0$. This delay introduces a linear phase shift to the transfer function according to Fourier and z-transform *Tables*:

$$h(n - P) \overset{\mathcal{F}}{\longleftrightarrow} H(z)z^{-P} \Rightarrow H(e^{j\omega T_s})e^{j\omega PT_s} = H(e^{j2\pi fT_s})e^{j2\pi fPT_s}.$$

The introduced phase shift is linear with negative slope.

Since a linear phase change in the frequency domain corresponds to a simple time-delay in the time domain, it does not cause any change to the overall shape of the filtered signal. This essentially means that no phase distortion occurs. On the other hand, a non-linear phase response can damage the information content of a signal. Therefore, it is desirable in the practical design of a digital filter to aim for a linear phase. Generally, the transfer function of a linear phase digital system is given by:

$$H(e^{j\omega T_s}) = |H(e^{j\omega T_s})|e^{j\omega\tau} = A(e^{j\omega T_s})e^{j\omega\tau}$$

where $A(e^{j\omega T_s})$ is real (i.e., is a zero-phase system), and the phase shift is linear and is given by $\phi = \omega\tau$, where $\tau$ is a constant).

## 2.6.3 Classification of Digital Filters

Digital filters can be classified into two major categories according to their impulse response length:

1. *A finite impulse response (FIR) digital filter*: has an impulse response with a finite number of non-zero samples.
2. *An infinite impulse response (IIR) digital filter*: has an impulse response with an infinite number of non-zero samples.

## 2.6.4 FIR Digital Filters

### 2.6.4.1 Structure and Implementation of FIR Filters

A causal FIR filter can be specified mathematically by the following difference equation [see Fig. (2.16)]:

$$\begin{aligned} y(n) &= h(n) * x(n), \\ &= \sum_{k=0}^{N-1} h(k)x(n-k) \\ &= h_0 x(n) + h_1 x(n-1) + \cdots + h_{N-1} x[n-(N-1)] \end{aligned}$$

where $h_k$ is used in place of $h(k)$ for notational simplicity. Taking the z-transform of both sides yields:

$$Y(z) = h_0 X(z) + h_1 z^{-1} X(z) + \cdots + h_{N-1} z^{-(N-1)} X(z) [\text{Using } \textit{Tables}].$$

Hence, the transfer function is given by:

$$H(z) = Y(z)/X(z) = h_0 + h_1 z^{-1} + \cdots + h_{N-1} z^{-(N-1)} \tag{2.21}$$

Remembering that a $z^{-1}$ corresponds to a single sample delay, one can implement the causal FIR filter using delay elements and digital multipliers as shown in Fig. (2.16).

### 2.6.4.2 Software Implementation of FIR Filters

The I/O relation in (2.6) could easily be implemented using a software program on a DSP chip or a digital computer. However, implementation according to the definition in (2.6) would require $O(N_h\ N_x)$ operations. As suggested in Sect. 2.4.1.5, a more efficient implementation is possible in the frequency domain with the use of FFTs. One uses the relations:

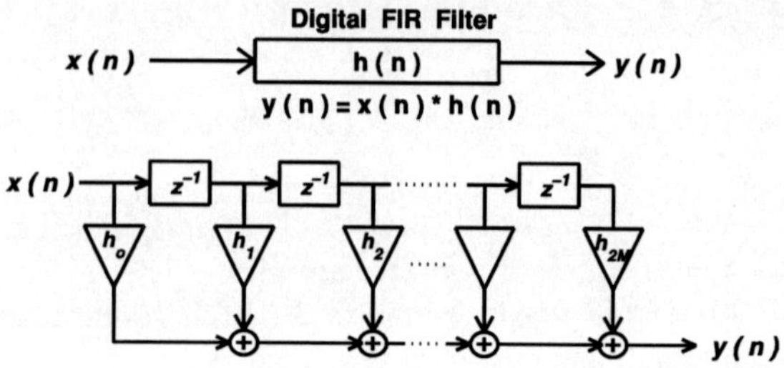


**Fig. 2.16** A finite impulse response (FIR) digital filter and its implementation (with $N = 2M + 1$) being odd

$$Y(k) = X(k) \cdot H(k) \xrightarrow{\text{ifft}} y(n) = Y^{-1}(k).$$

remembering that it is necessary to zero-pad both the input signal and impulse response up to $N_x + N_h - 1$ samples so as to avoid undesirable circular convolution effects.

#### 2.6.4.3 FIR Filtering of Long Data Sequences

In many applications (e.g., speech processing), the signal $x(n)$ can be very long. In this case, one cannot perform FIR filtering using the above technique efficiently since the fft computation time would be very large. Also, saving all of the input data would put a significant strain on the computer memory. Furthermore, real-time processing would be impossible since it would be necessary to wait until the end of the signal to start processing. Instead one can exploit the fact that an FIR filter is a linear system, and obeys the superposition property. Hence, one can partition the data record **x** into a sequence of *blocks*, each of relatively small length $K$, noting that $K >> M$, $M$ being the filter length. Each of these blocks can be filtered separately and the resulting outputs combined to yield an overall output. The process is called the overlap-add method and is illustrated in Fig. (2.17).

MATLAB the above overlap-add method could be implemented on MATLAB using the instruction `fftfilt(x,h)`, where `x` is the signal and `h` is the FIR filter impulse response.

#### 2.6.4.4 Pole-Zero Diagram and Stability of FIR Filters

The FIR transfer function in (2.21) can be written as:

$$H(z) = Y(z)/X(z) = \left(\frac{1}{z^{N-1}}\right)[h_0 z^{N-1} + h_1 z^{N-2} + \cdots + h_{N-1}],$$

which shows that an FIR filter with impulse response of length $N$ has $N - 1$ zeros and a multiple pole of order $N - 1$ at the origin ($z = 0$) (Since this multiple pole is inside the unit circle, it poses *no stability problems*).

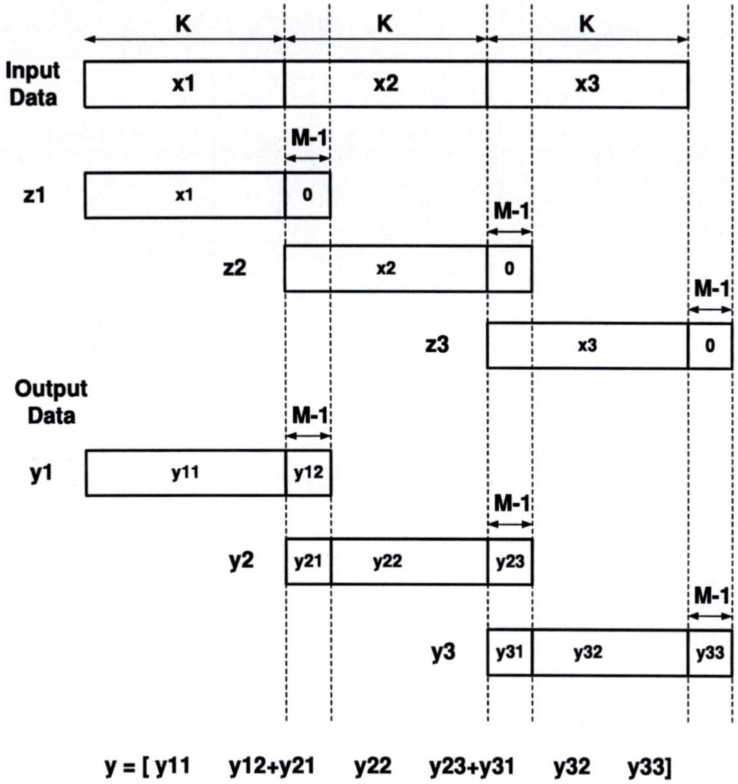


**Fig. 2.17** Overlap-add method for processing long data sequence

MATLAB the pole-zero diagram can be plotted on MATLAB using `zplane(A,B)`, where `A` and `B` are the numerator and the denominator polynomials coefficients (in descending powers of $z$). See also the example pole-zero plot for the Digital Differentiator in Sect. 2.6.6.3.

#### 2.6.4.5 Linear-Phase FIR Filters

A sufficient (but not necessary) condition that a FIR filter of impulse response $h(n)$ with length $N$ is a linear-phase system is that $h(n)$ is either:

1. that the impulse response is symmetric around the midpoint of the filter, i.e., $h(n) = h(N - n - 1)$, or:
2. that the impulse response is anti-symmetric around the mid-point of the filter, i.e., $h(n) = -h(N - n - 1)$.

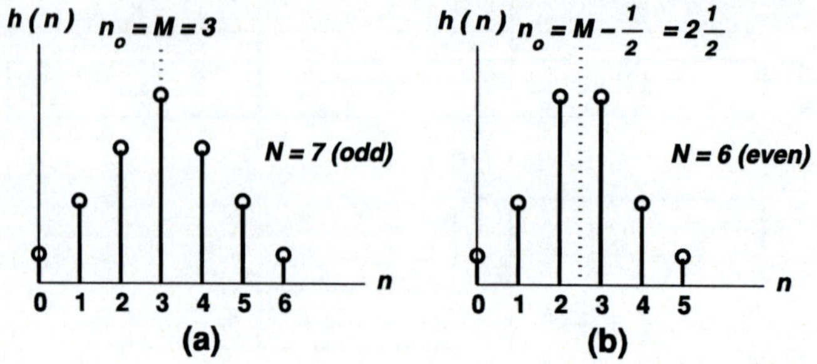


**Fig. 2.18** Symmetric impulse response for a FIR filter. **a** odd length. **b** even length

If $N$ is an odd integer, i.e., $N = 2M + 1$, the midpoint of symmetry is $n_o = M$. If $N$ is an even integer, i.e., $N = 2M$, the midpoint of symmetry is $n_o = M - \frac{1}{2}$ [see Fig. (2.18)].

The frequency response of a linear phase FIR filter (symmetric or anti-symmetric) is given by:

$$H(e^{j\omega T_s}) = H(z)\,|_{z=\exp(j\omega T_s)} = \begin{cases} A_o(\omega)e^{-j\omega T_s M + j\alpha}, & N = 2M + 1(\text{odd}), \alpha = 0 \text{ or } \pi \\ A_e(\omega)e^{-j\omega T_s(M-1/2)+j\beta}, & N = 2M(\text{even}), \beta = 0 \text{ or } \pi \end{cases} \quad (2.22)$$

where $A_o(\omega)$ and $A_e(\omega)$ are arbitrary *real*-valued functions.

*Example* If $h(n) = \delta(n) + \delta(n - 1)$, then:

$$N = 2, M = 1, , n_o = (N - 1)/2 = M/2 = \frac{1}{2}, \text{ and}$$
$$H(e^{j\omega T_s}) = 1 + e^{-j\omega T_s} = 2e^{-j\omega T_s/2}[(e^{j\omega T_s/2} + e^{-j\omega T_s/2})/2]$$
$$= 2e^{-j\omega T_s/2}\cos(\omega T_s/2).$$

Hence, the magnitude response is given by $A_e(e^{j\omega T_s}) = 2|\cos(\omega T_s/2)|$, and the phase response is $\phi = -\omega T_s/2$.

In the z-domain, the system function is $H(z) = 1 + z^{-1} = (z + 1)/z$, so that there is a pole at $z = 0$ and a zero at $z = -1$ [see Fig. (2.19)]. It is essentially a LPF. [As an exercise plot the frequency response vs. true frequency $f$ (Hz) or $\omega$ (rad/s)!]. The pole-zero diagram can be found on MATLAB using `zplane(A,B)`, where `A=[1 1]` and `B=[1 0]`.

#### 2.6.4.6 Efficient Hardware Implementation of Linear Phase FIR Filters

Recall that the transfer function of a general FIR filter of length $N$ is:

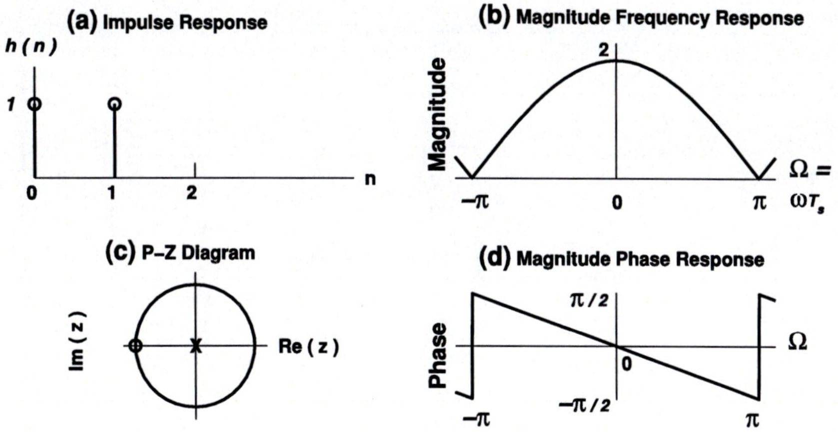


**Fig. 2.19** Impulse and frequency response of the filter $h(n) = \delta(n) + \delta(n-1)$

$$H(z) = Y(z)/X(z) = h_0 + h_1 z^{-1} + \cdots + h_{N-1} z^{-(N-1)}.$$

A straightforward hardware implementation which follows directly from the expression in (2.21) is shown in Fig. (2.16). If the filter is designed to have linear phase (as many FIR filters are), then its impulse response is typically symmetric or anti-symmetric and one can exploit this symmetry to halve the number of coefficient multipliers. These elements are generally costly and consume a lot of hardware resources. An efficient implementation which takes advantage of the coefficient symmetry is shown in Fig. (2.20) for odd and even length impulse responses.

## *2.6.5 Design of FIR Digital Filters*

### 2.6.5.1 Time-Domain Design

The frequency response of an ideal digital LPF over the principal frequency domain $(-f_s/2, f_s/2)$ is given by:

$$H_L(f) = H_L(e^{j2\pi f T_s}) = \begin{cases} 1, & 0 \le |f| \le f_c \\ 0, & f_c \le |f| \le f_s/2. \end{cases}$$

The impulse response of this filter can be obtained from (see *Tables* to be:

$$h_L(n) = \frac{2f_c}{f_s}\,\text{sinc}\left(\frac{2f_c}{f_s} n\right).$$

This impulse response and its Fourier transform are depicted in Fig. (2.21).

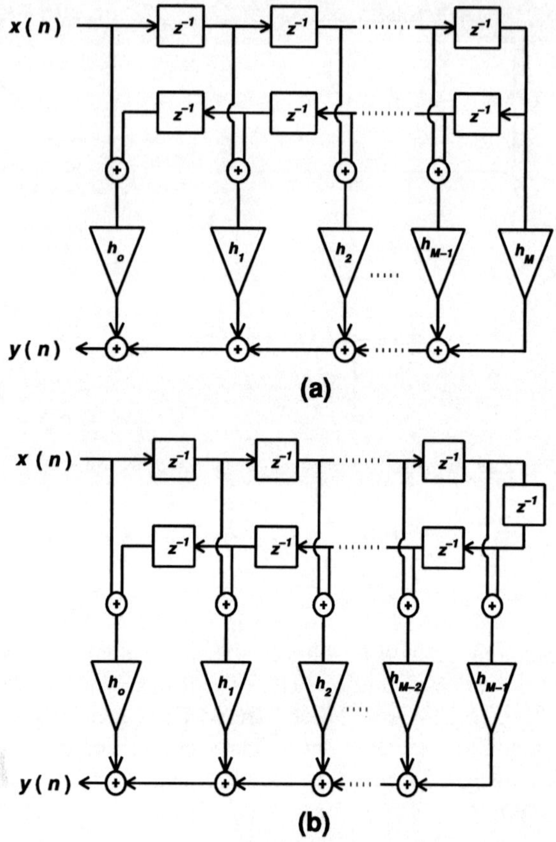


**Fig. 2.20** Efficient implementation of a linear-phase FIR digital filter with symmetric impulse response $h(n)$ **a** with odd length $N = 2M + 1$, **b** with even length $N = 2M$

Since there are non-zero values of $h(n)$ for for $n < 0$, the ideal LPF is non-causal, and is hence physically unrealizable. It cannot therefore be used to process real-time signals. In addition, $\sum |h_L(n)| \to \infty$, and so the filter is also unstable. Since $-\infty < n < \infty$, the ideal LPF is an IIR filter. A practical $2M + 1$ sample FIR approximation to the ideal LPF can be obtained by first shifting $h_L(n)$ right by $M$ samples to get $H_1(n) = h_L(n - M)$. Note that this time-shift causes *only a phase shift* in the frequency domain (see *Tables, Fourier Transform Properties*), and hence $|H_1(f)| = |H_L(f)|$. Second, $h_1(n)$ needs to be truncated (symmetrically around $n = M$) by putting $h_1(n) = 0$ for $n < 0$ or $n > 2M$ (total length is $N = 2M + 1$). This process yields a finite impulse response $h_{Lt}(n)$, which is symmetric about $n = M$, as shown in Fig. (2.22) for the scenario where $M = 5$, $N = 11$, $f_c = 1.25$, and $f_s = 5$ Hz.

As would be expected intuitively, larger values of $M$ tend to give rise to better approximations of the ideal impulse response. It is important to note that when the original infinite length time domain impulse response is truncated, the

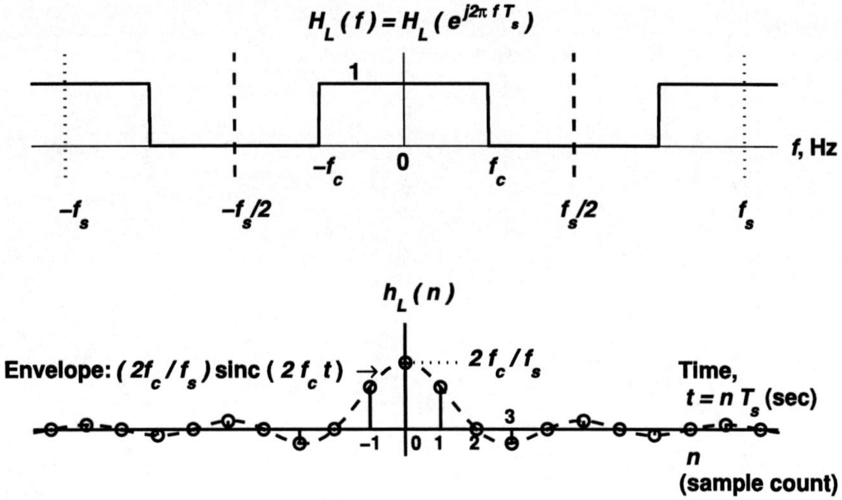


**Fig. 2.21** Transfer function of the ideal digital LPF and its impulse response

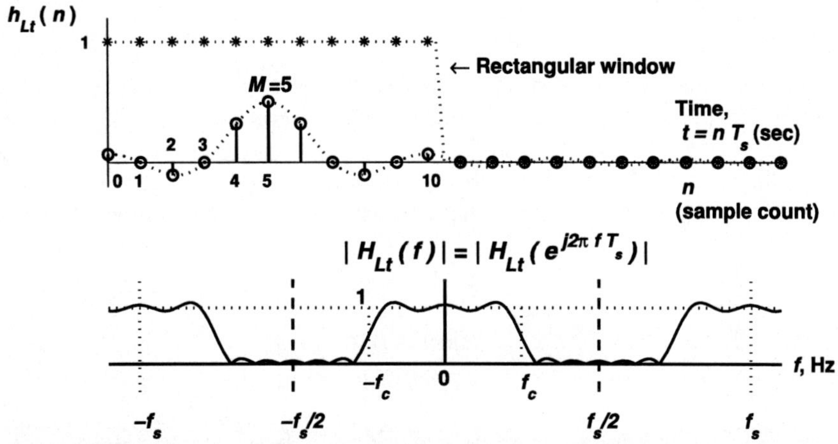


**Fig. 2.22** Magnitude and impulse responses of a practical digital LPF, where the impulse response is obtained by truncating the ideal response using a rectangular time window

effect on the frequency domain is a reduction in the sharpness of the transition band. In particular there is an introduction of ripples into the transfer function $H_{Lt}(f)$, with the appearance of these ripples often being referred to as the *Gibbs Phenomenon* [see (Fig. 2.23)]. This phenomenon is discussed further in the next subsection.

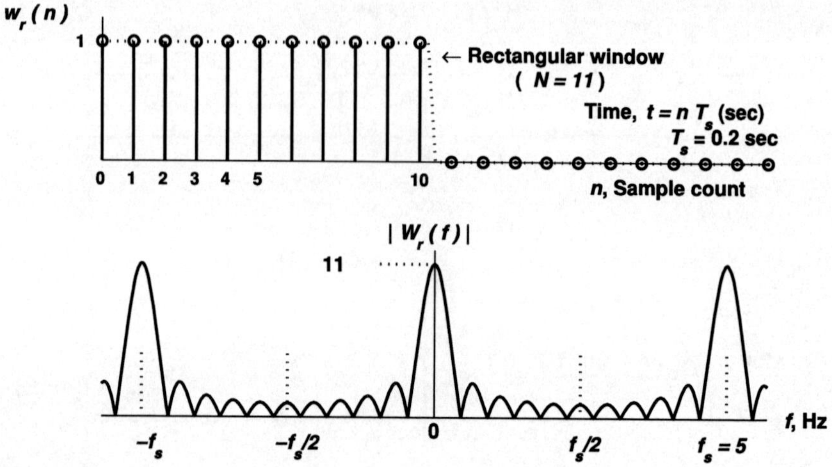


**Fig. 2.23** The rectangular time window and its magnitude spectrum (with total length $N = 11$ points, mid-point $M = (N - 1)/2 = 5$ points, and sampling frequency $f_s = 5$ Hz)

Gibbs Phenomenon

The Gibbs phenomenon pertains to the oscillatory behavior in the frequency response corresponding to a truncated digital impulse response sequence. In the previous subsection the approximated impulse response was obtained by truncating a shifted version of the ideal low-pass impulse response ($h_1(n) = h_L(n - M)$) to get the truncated impulse response $h_{Lt}(n)$.This truncation process is equivalent to multiplying $h_1(n)$ in the time domain by a rectangular time window $w_r(n) = \Pi_N(n - M)$. This time window has a Fourier transform of the form

$$W_r(f) = N\frac{\mathrm{sinc}(\mathrm{N}f/\mathrm{f_s})}{\mathrm{sinc}(f/f_s)}e^{-jM2\pi f/f_s},$$

which is a *sinc-like* frequency function with a major lobe and many side lobes as seen in Fig. (2.23). This multiplication in the time domain is equivalent to convolving $H_1(f)$ with $W_r(f)$ in the frequency domain. This convolution with a function having substantial side-lobes is the reason for the oscillations which appear in the magnitude response.

To reduce the effect of the oscillations in $H_{Lt}(f)$, one can multiply the ideal impulse response with a non-rectangular window which has lower amplitude side-lobes. Some suitable windows include the Hamming window, the Hanning window, the Blackman window, the Gaussian window and the Kaiser window [1] If one does use this kind of *smooth* windowing one typically needs a longer impulse response to have an equivalent quality of approximation.

General formula

A general formula may be used to describe many smooth windows used in practice. This formula is: $w(n) = a + b \cdot \cos(n\pi/M) + c \cdot \cos(2n\pi/M), \quad 0 \leq n \leq 2M$ length $2M + 1$

Some common smooth windows conforming to this formula are listed in Table 2.1 and Fig. (2.24).

#### 2.6.5.2 Frequency-Domain Design

So far the design of FIR filters has focused on approximating the desired impulse response (i.e., the time-domain characterization of the filter). An alternative approach is to design the filter by approximating the desired filter transfer function (or frequency representation). This is the approach that is used in the so-called *frequency sampling method* described next.

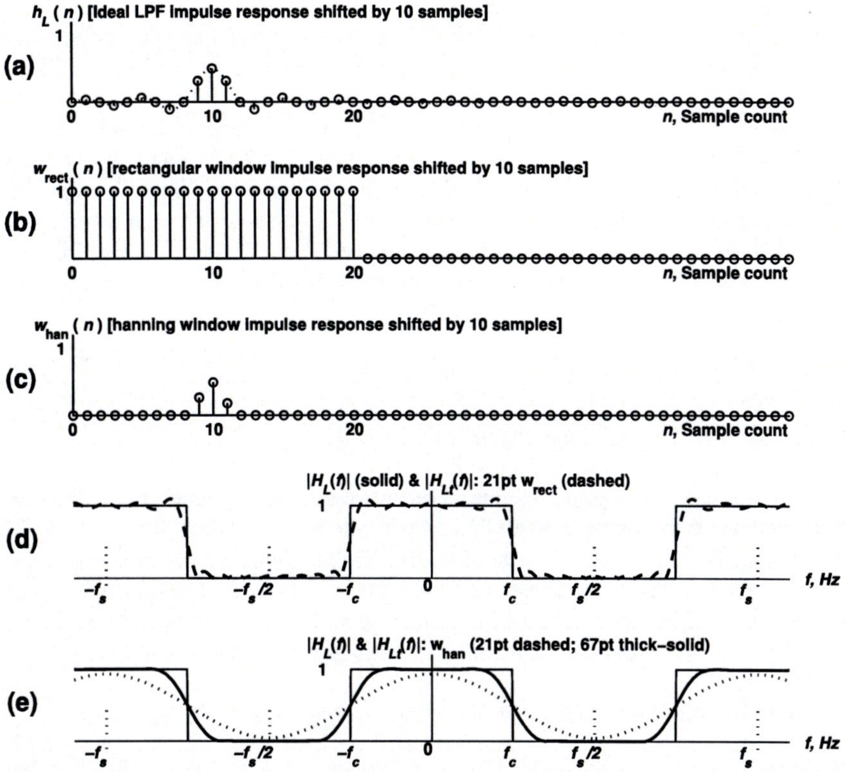


**Fig. 2.24** Rectangular and Hanning windowing of the impulse response of an ideal digital LPF

A. Design of FIR Digital Filters by Frequency-Sampling

A intuitively appealing approach to designing a FIR filter is to simply take the desired frequency transfer function and then inverse Fourier transform to find the filter's impulse response. Within a digital computer, however, one cannot specify the Fourier transform, only the *samples* of the desired Fourier transform. In other words one can define the samples of a desired frequency response $H_d(e^{j\omega T_s}) = H_d(e^{j\Omega})$, then use the IDFT to find the impulse response $h(n)$. Often in digital filters it is the transition band which is particularly critical and so it is necessary to specify this portion of the transfer function most accurately. This implies that one needs to have one or more samples within the transition band if one is to achieve good filter designs. It is also important to note that if one seeks to specify too sharp a transition band one will be confronted by the Gibbs Phenomenon-substantial ripple will then manifest in both the pass-band and the stop-band.

*Example* The goal is to design a 33rd order LPF with cutoff frequency $f_c = f_s/4$ (or, $\Omega_c = 2\pi/4 = \pi/2$). With the frequency sampling method it is necessary to use a 33-point DFT to define the samples of the desired frequency response $H_d(e^{j\omega T_s})$. This desired response is the ideal rectangular response over the principal domain $0 < f < f_s$ which is sampled to yield the set of samples $\{H_d(k) \mid k = 0, 1, \ldots, 32\}$. The frequency sampling method is illustrated in Fig. (2.25) for two scenarios, The first scenario is where no samples are specified in the transition band, and the second is where one sample is specified in the transition band. Once the samples have been specified the IDFT can be taken to recover the filter's impulse response. Figure (2.25) shows the DTFT magnitudes of the impulse response so obtained. It is seen in Fig. (2.25) that when no transition band sample is specified the ripple in both the pass and stop bands is significantly greater [Note that $H_d(9)$ and $H_d(24)$ are the transition samples, chosen to be 0.5].

B. Optimal Frequency-Sampling FIR Filter Design

The frequency-sampling method described above is straightforward but it is an *ad hocl* method. It is possible to design filters which are optimal in the sense of minimizing the maximum error between the desired frequency response and the actual frequency response. These types of filters are referred to as equiripple, and can be designed using the. Parks-McClellan algorithm. The latter utilizes Remez and Chebychev approximation theories to design optimal equiripple linear-phase FIR filters.

MATLAB the Parks-McClellan algorithm is available on MATLAB as:

$$\mathtt{h = remez(N-1, Fd, Ad)}.$$

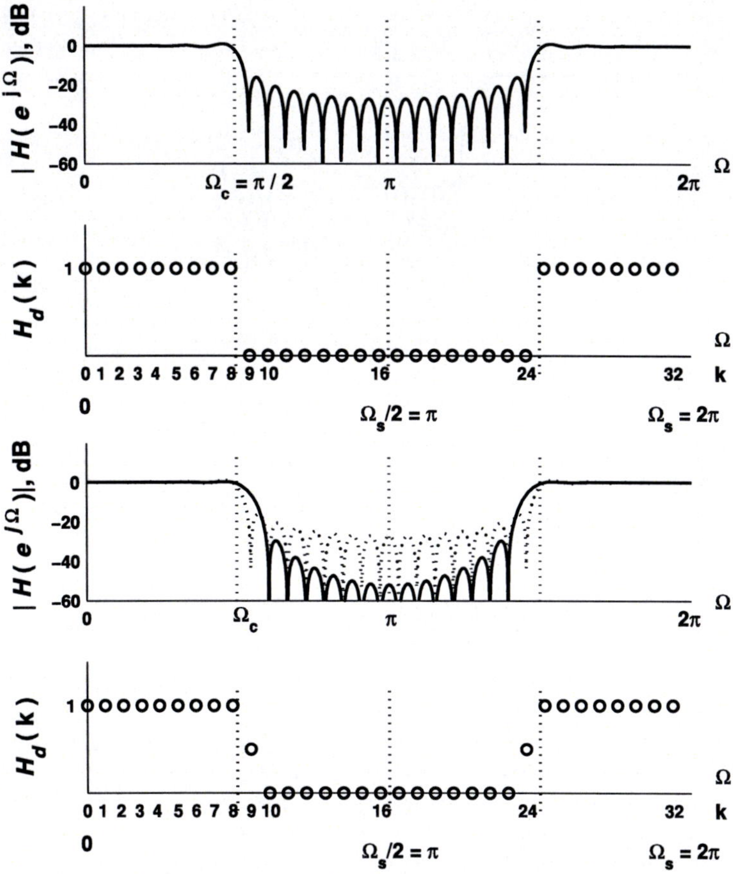


**Fig. 2.25** FIR filter design using frequency-sampling method. *Above* without a transition sample. *Below* with two transition samples (*dotted curve* for the first case without transition samples)

This command yields the impulse response of a length-$N$ (i.e., order $N - 1$) linear phase FIR filter with desired magnitude response, given by the vectors `Fd` and `Ad`. `Fd` is a vector of normalized frequency points, ranging from 0 to 1 (corresponding to the range $0 < f < f_s/2$ Hz), and `Ad` contains the desired magnitudes for points in `Fd`. Example: To design an optimal FIR filter in MATLAB with the specifications in the previous example, one can use:

$$\mathtt{h} = \mathtt{remez}(\mathtt{N} - 1, [0\,0.5\,0.6\,1], [1\,1\,0\,0])$$

Figure (2.26) shows the actual frequency response as compared to the ad-hoc frequency-sampling approach described in Sect. 2.6.5.2-A.

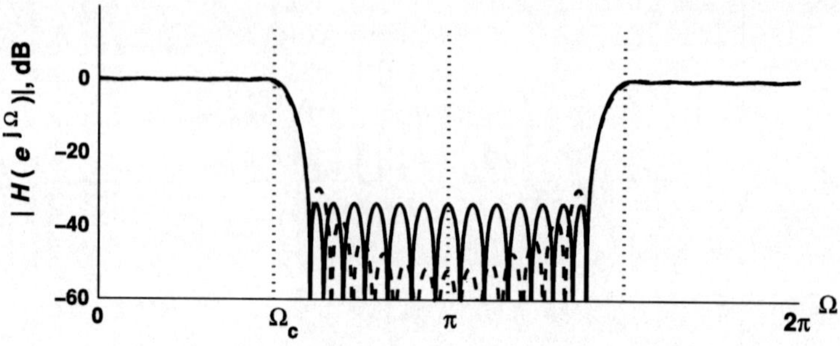


Fig. 2.26 Frequency response of a LPF designed by remez (dotted curve for ad-hoc frequency-sampling design)

## 2.6.6 Applications of FIR Digital Filters

### 2.6.6.1 Communication Channel Equalization

Telephone and wireless channels often behave like band-limited filters with frequency response $H(f)$. This filtering is in general an undesirable frequency dependent distortion which needs to be eliminated - it can be removed via the use of an *equalizer* as explained below.

The transfer function of the channel can be estimated by sending a training signal [e.g., a unit pulse function $\delta(n)$ along the channel and then measuring the impulse response $h(n)$ and the corresponding transfer function $H(f) = \mathcal{F}\{h(n)\}$]. An equalizer (or inverse filter) with a frequency response $H_e(f) = 1/H(f)$ is then applied to counteract the channel distortion. Normally one chooses an FIR filter (also called a transversal filter), and the frequency sampling approach is often used to design $H_e(f)$.

### 2.6.6.2 The Moving Average Filter

Sometimes it is necessary to smooth data before making a decision or interpretation about that data. Averaging is a commonly used smoothing technique and this averaging can be implemented with simple FIR filters. For example, the stock market prices fluctuate from day to day, and even sometimes hour to hour, To observe trends some form of smoothing or averaging is necessary. Typically, one takes an average of the stock price over several days before deciding the actual trend of prices. *Note that the global average is misleading in these cases and cannot be used—rather, local or moving average should be used.* (See Fig. (2.27)).

Since data $x(n)$ is assumed to be *continuously flowing in* for the stock market scenario, local averaging is done at each instant $n$. If, for example, averaging is performed over three consecutive samples, the output becomes:

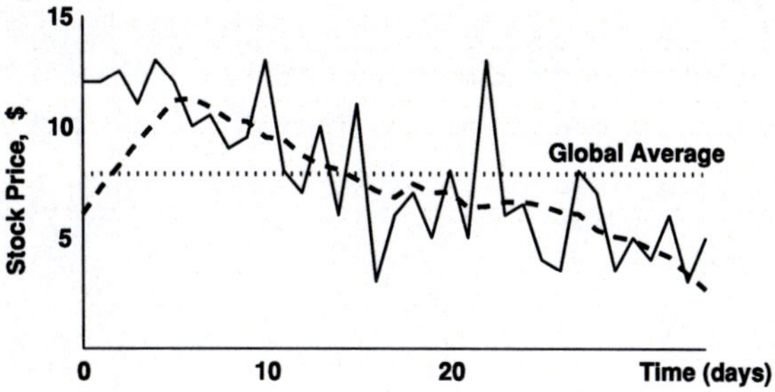


**Fig. 2.27** Moving average of a stock price over 35 days using a 10-tap FIR filter. *Note* that the first $M - 1 = 9$ samples are inaccurate due to insufficient information at start

$$y(n) = \frac{1}{3}[x(n) + x(n-1) + x(n-2)] \tag{2.23}$$

For example, at $n = 2, y(2) = \frac{1}{3}[x(2) + x(1) + x(0)]$, at $n = 3, y(3) = \frac{1}{3}[x(3) + x(2) + x(1)]$. Comparing the above Eq. (2.23) with the general FIR equation:

$$\begin{aligned} y(n) = h(n) * x(n) &= \sum_{k=0}^{N-1} h(k)x(n-k) \\ &= h_0(n) + h_1 x(n-1) + \cdots + h_{N-1} x[n-(N-1)] \end{aligned}$$

it is seen that (2.23) represents an FIR filter with $N = 3$ taps: $h(0) = h(1) = h(2) = \frac{1}{3}$. Although these filter coefficients are all equal in this case they can be different in general. For example, in the stock market studies one may give more weight (importance) to the current sample [i.e., $x(n)$] than to others, e.g.,

$h(0) = 0.5$, 50% weight to the current price (today)
$h(1) = 0.3$, 30% weight to the previous price (yesterday)
$h(2) = 0.2$, 20% weight to the first price (2 days ago)

Figure (2.27) shows an example of a changing price over 35 days along with the overall average and the moving average with $h(n)$ = [.1 .1 .1 .1 .1 .1 .1 .1 .1 .1] ($M$ = 10 taps with equal weight).

#### 2.6.6.3 The Digital Differentiator

Time Domain Approach

The derivative of a signal $x(t)$, $dx(t)/dt$, can be approximated in the digital domain (when the sampling period $T_s$ is small and $x(t)$ is slowly varying) by the relation:

$$y(n) = [x(n) - x(n-1)]/T_s \tag{2.24}$$

as shown in Fig. (2.28). Comparing with the general form of an FIR filter:

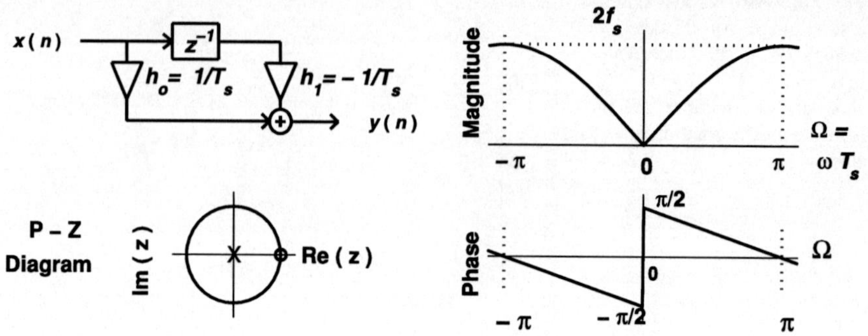


**Fig. 2.28** Digital differentiator-I with frequency response and p-z diagram

$$y(n) = h(n) * x(n) = \sum_{k=0}^{N-1} h(k)x(n-k)$$
$$= h_0(n) + h_1x(n-1) + \cdots + h_{N-1}x[n-(N-1)]$$

It can be seen that (2.32) represents a FIR filter with two non-zero samples impulse response $\{h(n) = (1/T_s)\delta(n) - (1/T_s)\delta(n-1)\}$, i.e., $h(0) = -h(1) = 1/T_s$. From this one can find the system transfer function by taking the z-transform of both sides of (2.32):

$$Y(z) = \frac{1}{T_s}[X(z) - X(z)z^{-1}] = [(1-z^{-1})/T_s]X(z)$$
$$\Rightarrow H(z) = \frac{Y(z)}{X(z)} = \frac{(1-z^{-1})}{T_s}$$

Now consider the magnitude and phase response:

$$H(e^{j\Omega}) = H(z)\,|_{z=exp(j\Omega)} = \frac{1}{T_s}[1 - e^{-j\Omega}]$$
$$= \frac{1}{T_s}e^{-\frac{1}{2}j\Omega}[e^{+\frac{1}{2}j\Omega} - e^{-\frac{1}{2}j\Omega}]$$
$$= \frac{1}{T_s}e^{-\frac{1}{2}j\Omega}[2j\sin(\Omega/2)]$$
$$= \left(\frac{2j}{T_s}\right)e^{-\frac{1}{2}j\Omega}\sin\left(\frac{1}{2}\Omega\right)$$

Hence, the magnitude response is given by:

$$|H(e^{j\Omega})| = \left(\frac{2}{T_s}\right)|\sin\left(\frac{1}{2}\Omega\right)| = 2f_s\left|\sin\left(\frac{1}{2}\omega T_s\right)\right| = 2f_s|\sin(\pi f/f_s)|,$$

and the phase response is:

$$\phi = \left(-\frac{\Omega}{2} + \frac{\pi}{2}\right) + c,$$

where $c = \pi$ is to compensate for the change of sign in $\sin(\frac{1}{2}\Omega)$.

The magnitude and phase responses are plotted in Fig. 2.28. It is seen that the digital differentiator is a high-pass filter.

## Frequency Domain Approach

From Fourier transform *Tables* one can find the transfer function of the continuous-time differentiator as:

$$H_a(\omega) = j\omega \tag{2.25}$$

which can be transformed into the digital domain as follows:

$$H(e^{j\Omega}) = j\Omega/T_s; \quad -\pi \leq \Omega < \pi. \tag{2.26}$$

Figure (2.29) shows the magnitude and phase responses of the above transfer functions (compare with Fig. (2.28)). Note that Fig. (2.29a) represents a theoretical magnitude response, as practical differentiators are band-limited within a cutoff frequency $\omega_c$.

The impulse response of the above filter can be found to be:

$$\begin{aligned} h(n) &= \frac{1}{2\pi}\int_{-\pi}^{\pi} \frac{j\Omega}{T_s} e^{jn\Omega} d\Omega \\ &= \frac{1}{T_s}\frac{\cos(n\pi)}{n} - \frac{\sin(n\pi)}{\pi n^2}; \quad -\infty < n < \infty \end{aligned} \tag{2.27}$$

For a practical design, the above impulse response should be shifted by $2M + 1$ samples and then truncated. The shifting operation is described by:

$$h(n) = \frac{1}{T_s}\frac{\cos[(n-M)\pi]}{(n-M/2)} - \frac{\sin[(n-M)\pi]}{\pi(n-M)^2}; \quad -\infty < n < \infty, \tag{2.28}$$

while the truncation window $w(n)$ is specified by:

$$h_w(n) = w(n) \cdot h(n); \quad 0 < n < 2M \tag{2.29}$$

If a linear phase is required, then the truncation window should be symmetric so that one can get $h(n) = -h(2M - n)$. The filter in (2.29) can be implemented using the approach in Sect. 2.6.4.6.

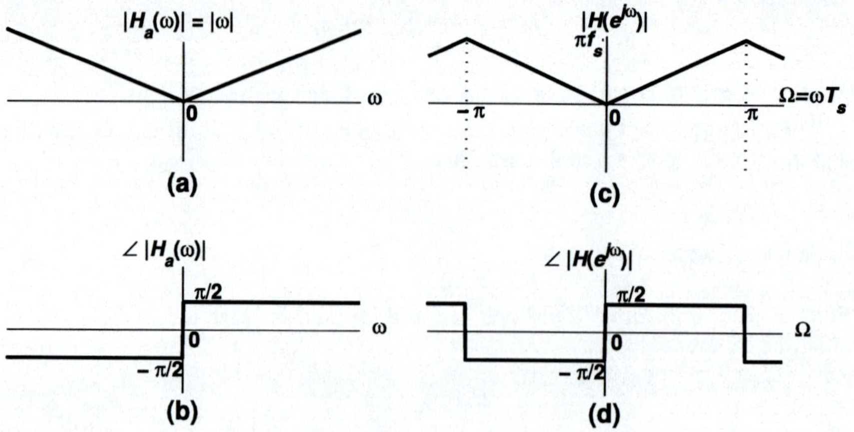


**Fig. 2.29** Frequency response for an analog and digital differentiator-II

#### 2.6.6.4 The Digital Matched Filter

Recall from Sect. 1.4.2 that in the analog domain the impulse response $h(t)$ of the optimum filter for receiving a transmitted symbol $b(t)$ ($0 \leq t \leq T$) is given by the $T$-shifted (delayed) mirror image of the symbol, i.e.,

$$h(t) = b(T - t)\Pi_T(t - T/2).$$

For orthogonal binary transmission, two matched filters are needed, one matched to the "0" and the other matched to the "1". Figure (2.30) shows a digital matched filter for orthogonal binary signalling with the sampling frequency at the receiver equal to *five times the bit rate*. If bipolar signalling is used (i.e., +Vcc represents "1" and −Vcc represents "0"), then one sample per bit is possible, but in practical signalling schemes there are normally at least 4 samples per bit. Increasing the sample rate increases the accuracy of detection, especially in noise, and it does not affect the transmission bandwidth (which depends on the data rate 1/T only). If the sample rate is $N$ samples per bit, then an $N$-stage shift register is needed to store the samples before each symbol is detected for each filter (see Fig. (2.30)).

In digital matched filters it is necessary to have good synchronization between the transmitter and the receiver to decide the start time ($t = 0$). The coefficients of the $i$th matched filter $\{h_i(n) \mid n = 0, 1, \ldots, N - 1\}$, which is matched to the $i$th symbol (bit), are given by:

$$h_i(n) = b_i(N - 1 - n), \quad n = 0, 1, \ldots, N - 1$$

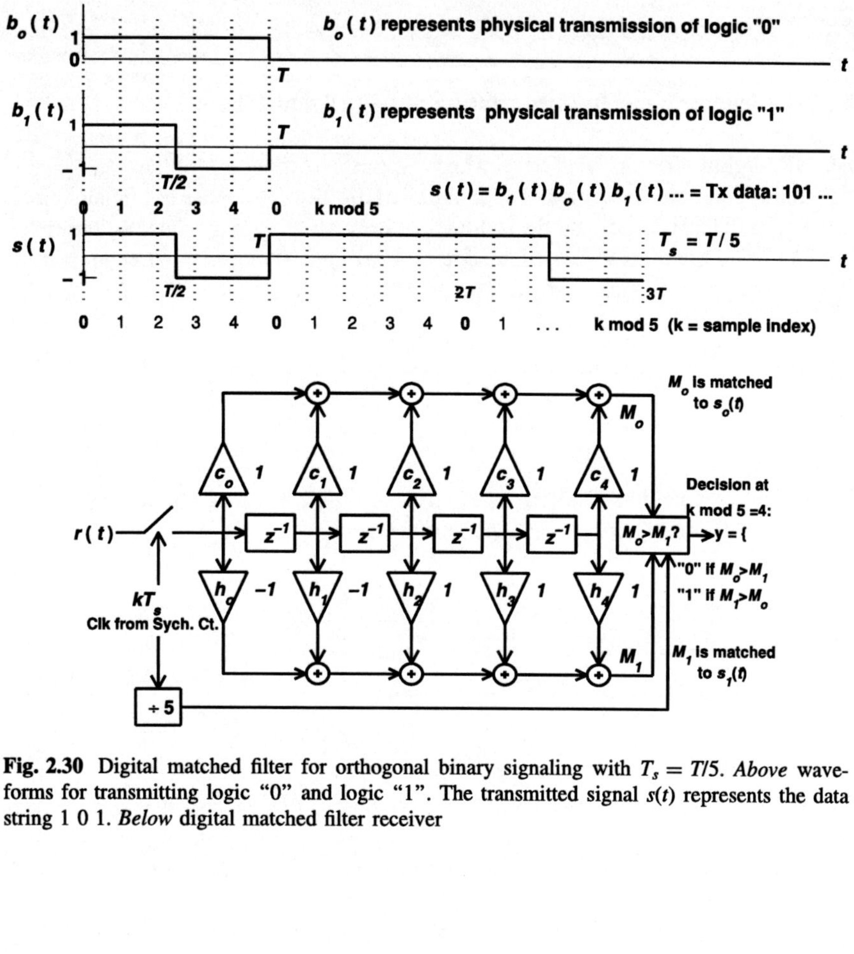


**Fig. 2.30** Digital matched filter for orthogonal binary signaling with $T_s = T/5$. *Above* waveforms for transmitting logic "0" and logic "1". The transmitted signal $s(t)$ represents the data string 1 0 1. *Below* digital matched filter receiver

The output of the *i*th matched filter at the *k*th sampling instant is given by the convolution:

$$M_i(n) = \sum_{n=0}^{N-1} r(k-n)h_i(n)$$

where $r(t)$ is the received signal. Decisions (comparisons) are made only when $kN = N - 1$ [every 5 samples for the scenario represented in Fig. (2.30)]. Note that in Fig. (2.30) the simpler notation $h_n = h_1(n)$ and $c_n = h_0(n)$ is used-under noise-free condition and full synchronization with the transmitter at $k = 4$ (after 5 samples) the shift register has the following contents: $a = 1, b = 1, \; c = 1,$ $d = -1, e = -1$. Hence, $M_o = 1$, $M_1 = 5 > M_o$, and the decision is that "1" was transmitted. As an exercise try to find the decision of the circuit after 10 and 15 samples.

### 2.6.7 IIR Digital Filters

#### 2.6.7.1 Structure and Implementation of IIR Digital Filters

An IIR digital filter is a system whose impulse response $h(n)$ has an infinite number of non-zero samples. To implement IIR filters in practice, recursion is often used. That is, to create the output, not only are previous values of the input used, but also previous values of the output. This kind of recursive implementation is essentially a feedback system. The general formula for the I/O relations of a recursive IIR filter are:

$$\begin{aligned} y(n) = b_o x(n) + b_1 x(n-1) + \cdots + b_M x(n-M) \\ - a_1 y(n-1) - \cdots - a_N y(n-N). \end{aligned} \quad (2.30)$$

The above equation is often referred to as a *difference equation*. Taking the z-transform of both sides of the equation yields:

$$\begin{aligned} Y(n) &= b_o X(z) + b_1 z^{-1} X(z) + \cdots + b_M z^{-M} X(z) \\ &\quad - a_1 z^{-1} Y(z) - \cdots - a_N z^{-N} Y(z). \\ \therefore H(z) &= \frac{Y(z)}{X(z)} = \frac{b_o + b_1 z^{-1} + \cdots + b_M z^{-M}}{1 + a_1 z^{-1} - \cdots - a_N z^{-N}} = \frac{\sum_{i=0}^{M} b_i z^{-i}}{1 + \sum_{k=1}^{N} a_k z^{-k}}. \end{aligned} \quad (2.31)$$

Hence, unlike FIR filters, IIR filters can have poles at locations other than $z = 0$. In fact, the above equations can also represent a FIR filter if one puts $a_1 = a_2 = ... = 0$. Normally, the number of poles $N$ is larger than the number of zeros $M$, and the order of the filter is decided by the number of its poles. For the IIR filter to be stable, *all poles should be inside the unit circle.*

#### 2.6.7.2 IIR versus FIR Filters

IIR filters usually have lower orders than FIR filters with similar performance with respect to sharpness of cutoff, passband ripple, etc. Because of the lower orders IIR filters tend to require fewer delay elements and digital multipliers for hard-ware implementation, and they require fewer computations in software implementations.

One of the key disadvantages of IIR filters is that because of their inherent use of feedback, they can have stability problems. Quantization effects are also much more serious in IIR filters, again due to their use of feedback. Additionally, it is generally not possible to have a linear phase transfer function with IIR filters.

### 2.6.7.3 Direct Form Implementation of IIR Digital Filters

From the general equation for an IIR filter:

$$Y(n) = b_o X(z) + b_1 z^{-1} X(z) + \cdots + b_M z^{-M} X(z) \\ - a_1 z^{-1} Y(z) - \cdots - a_N z^{-N} Y(z),$$

one can readily construct a hardware realization, as shown in Fig. (2.31). The structure shown is called the direct form-I, realization. This form requires $M + N + 1$ multipliers, $M + N$ adders, and $M + N$ memory locations.

Now the transfer function for an IIR filter can also be written as:

$$H(z) = \frac{\sum_{i=0}^{M} b_i z^{-i}}{1 + \sum_{k=1}^{N} a_k z^{-k}} = \left(\sum_{i=0}^{M} b_i z^{-i}\right)\left(\frac{1}{1 + \sum_{k=1}^{N} a_k z^{-k}}\right). \\ = H_1(z) H_2(z) = H_2(z) H_1(z) .$$

The above expression suggests an alternative realization for the filter, and the new realization is called the direct form-II implementation (Fig. 2.32). Importantly the direct form-II requires only max($M$, $N$) memory locations (or delay elements) in contrast to the direct form-I, requires $M + N + 1$. The new realization needs the same number of adders and multipliers as the direct form-I. The coefficients $\{a_i\}$ and $\{b_i\}$ are usually chosen to be *real to avoid complex processing*. It can be shown that with real coefficients the poles and zeros are either real or in complex-conjugate pairs.

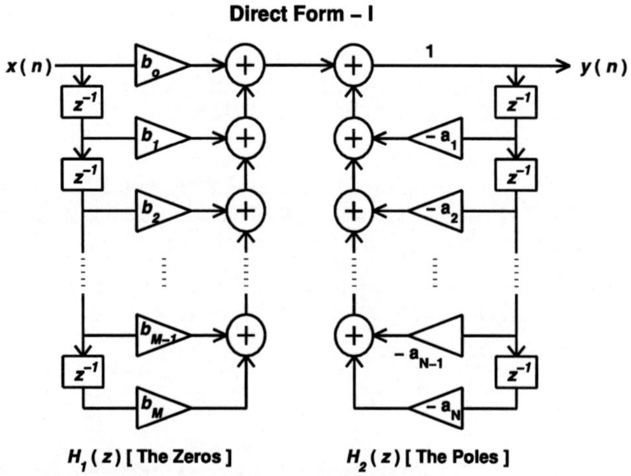


**Fig. 2.31** Direct form-I implementation of IIR digital filters

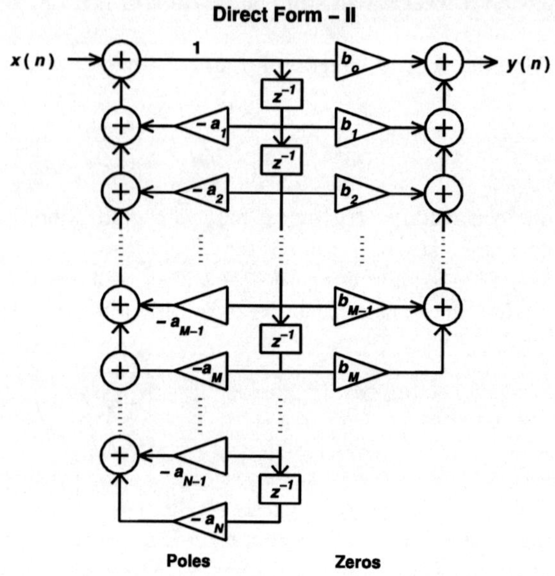


**Fig. 2.32** Direct form-II implementation of IIR digital filters. *Note* to find the poles and zeros of any transfer function *H*(*z*), it is normal to write it as a function of *z*, but for implementation it is normal to write it as a function of $z^{-1}$. The use of $z^{-1}$ terms in the implementation is due to the correspondence of the $z^{-1}$ term to a delay element

#### 2.6.7.4 Practical Implementation of IIR Digital Filters

Since the feedback structure of IIR filters exaggerate quantization errors, these filters are often built up in practice as cascaded first- and second-order units rather than in direct forms - this can be shown to reduce the vulnerability to quantization errors. Ideally one would use only first order sections, but this would necessitate using complex arithmetic. To avoid this latter possibility it is necessary to use second order sections wherever there are complex conjugate poles or zeros.

*Example* Implement the IIR digital filter whose transfer function is given by:

$$H(z) = \frac{2(z-1)(z+2)(z+3)}{z(z+0.5)(z^2+0.3z+0.1)}.$$

It is seen that there are two complex conjugate poles since $z^2 + 0.3z + 0.1$ has two complex conjugate roots. Hence, it is best to use a cascaded form of implementation and write the transfer function as follows:

$$\begin{aligned} H(z) &= \frac{2(z-1)(z+2)(z+3)}{z(z+0.5)(z^2+0.3z+0.1)} \\ &= 2z^{-1}\left(\underbrace{\frac{1-z^{-1}}{1+0.5z^{-1}}}_{\text{1st-order unit}}\right)\left(\underbrace{\frac{1+5z^{-1}+6z^{-2}}{1+0.3z^{-1}+0.1z^{-2}}}_{\text{2nd-order unit}}\right). \end{aligned}$$

The implementation diagram is shown in Fig. (2.33).

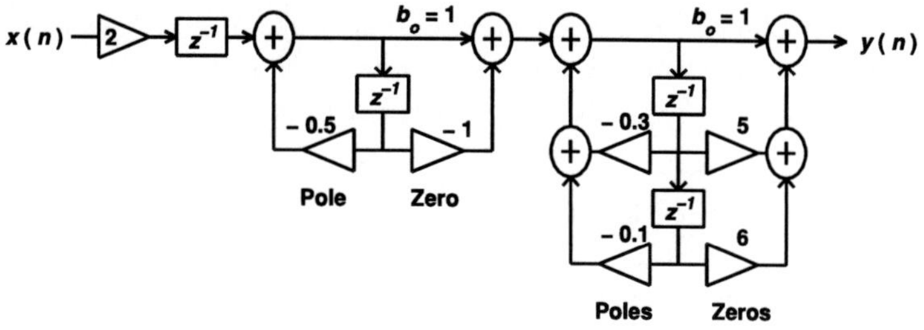


**Fig. 2.33** Implementation of $2(z-1)(z+2)(z+3)/[z(z+0.5)(z^2+0.3z+0.1)]$

## 2.6.8 Design of IIR Digital Filters

Often IIR digital filters are designed by converting analog prototype filters (such as Butterworth, Chebychev, and Elliptic filters) into digital filters via suitable transformation methods. The basic prototype filter type is an analog low-pass filter, and filter transformations are necessary to obtain LP, HPF, BPF, or BSF digital filters.

Typically the required digital filter specifications are initially transformed into their analog counterparts to help in designing the proper analog prototype. For this to be effective the transformation should:

1. preserve stability and
2. map the $j\omega$ axis into the circumference of the unit circle $e^{j\Omega}$.

There are two approaches for transforming digital IIR filters into analog prototypes:

1. Time-domain matching using the impulse response,
2. Frequency-domain matching using the frequency response.

### 2.6.8.1 Time-Domain Design: Impulse Response Matching

In the impulse response matching method (a.k.a. the impulse invariance method), the impulse response of the digital IIR filter is simply taken to be a sampled version of the impulse response of the analog prototype filter. That is,

$$h(n) = h_a(nT_s), \quad n = 0, 1, 2, \ldots.$$

If the transfer function of the analog prototype filter $H_a(f)$ is bandlimited to $(-B, B)$ and $f_s/2 \geq B$, then there would be no aliasing. This in turn would imply that the digital transfer function $H(e^{j\Omega})$ would be a scaled and repeated copy of $H_a(f)$. With this understanding it becomes apparent that the impulse response matching method is not suitable to design a digital HPF, due to aliasing problems.

The analog filter s-domain transfer function is $H_a(s) = LT\{h_a(t)\}$. Now the poles of $H_a(s)$ are transformed to the poles of $H(z)$ through the relation $z = e^{sT_s}$. The method can be summarized as follows (see, for example, A. V. Oppenheim and R. Schafer, *Discrete-Time Signal Processing*, Prentice-Hall, 1989):

1. Expand the analog transfer function by partial fractions as $\sum_{m=1}^{M} \frac{c_m}{s-p_m}$. Note that $p_m$'s can be non-distinct.
2. The required transfer function is

$$H(z) = T_s \sum_{m=1}^{M} \frac{c_m}{1 - e^{p_m T_s} z^{-1}} = T_s \sum_{m=1}^{M} c_m \frac{z}{z - z_m}$$

where $z_m = e^{p_m T_s}$ are the poles in the z-domain.

*Note 1* If there is no aliasing, $H(e^{j\Omega}) = H_a(\omega)$ when $-\pi \le \Omega \le \pi$.

*Note 2* Since the relation $\Omega = \omega T_s$ is applicable, the $j\omega$ axis is transformed into the circumference of the unit circle $e^{j\Omega} = e^{j\omega T_s}$.

*Note 3* Since the poles are transformed according to $z = e^{sT_s}$, stability is preserved.

*Example* Using the impulse invariance method, design a digital Butterworth LPF with the following specifications:

1. $T_s = 0.01$ sec (hence, $f_s = 100$ Hz),
2. $f_c = 10$ Hz (therefore $\omega_c = 20\ \pi$ rad/sec),
3. $G_m = 1$, gain $\le 0.1$ (i.e., $-20$ dB) for $20 \le f \le f_s/2 = 50$ Hz).

*Solution:*

It is necessary to first find an analog Butterworth LPF with the above specifications. The gain should be less than $-20$ dB for a normalized frequency of $f_c > 20/10 = 2$ (normalized w.r.t $f_c$). From the graph in *Tables*—Stopband gain for a B-LPF, the filter order is found to be $n = 4$. From *Tables*—Denominator Polynomial Coefficients for Normalized LPF's, the transfer function of this normalized filter is seen to be:

$$H_N(s) = \frac{a_o}{1 + 2.6131 s_N + 3.4142 s_N^2 + 2.6131 s_N^3 + s_N^4}$$

Now $H(_N(s)|_{s=0} = a_0 = G_{dc} = G_m = 1$, and $s_N = s/\omega_c$. Substituting these values for $a_o$ and $s_N$ into the above equation, yields:

$$H_a(s) = \frac{1.55e7}{1.55e7 + 6.48e5s + 1.34e4s^2 + 164.1s^3 + s^4}.$$

$H_a(s)$ can then be expanded using partial fractions. This can be easily done in MATLAB using `B = [1 164.1 1.34e4 6.48e5 1.55e7],A = [1.55e7], [R,P,K] = residue(A,B)` where `R` gives the coefficients $c_m$, `P` gives the poles

$p_m$, and K is empty if degree(A)<degree(B). For the Butterworth filter above, the poles and coefficients are found to be the following complex conjugate pairs:

$$\begin{aligned} c_1 &= -27.8 + j11.6, \\ p_1 &= -23.6 + j58.3, \\ c_2 &= {c_1}^*, \\ p_2 &= {p_1}^*, \\ c_3 &= 27.8 - j73.7, \\ p_3 &= -58.4 + j22.3, \\ c_4 &= {c_3}^*, \\ p_4 &= {p_3}^*. \end{aligned}$$

The z-domain poles are:

$$\begin{aligned} z_1 &= \exp(p_1 T_s) = 0.65 + j0.43, \\ z_2 &= {z_1}^*[\text{since}(e^x)^* = e^{(x^*)}], \\ z_3 &= 0.54 + j0.12, \end{aligned}$$

and $z_4 = \overset{*}{z_3}$. Therefore,

$$\begin{aligned} H(z) &= T_s\left[\frac{c_1 z}{z - z_1} + \frac{{c_1}^* z}{z - {z_1}^*} + \frac{c_3 z}{z - z_3} + \frac{{c_3}^* z}{z - {z_3}^*}\right] \\ &= T_s\left[\frac{r_1 z^2 - r_2 z}{z^2 - r_3 z + r_4} + \frac{q_1 z^2 - q_2 z}{z^2 - q_3 z + q_4}\right] \end{aligned}$$

where $r_1 = -0.55$, $r_2 = -0.26$, $r_3 = 1.3$, $r_4 = 0.6$, $q_1 = 0.55$, $q_2 = 0.47$, $q_3 = 1.08$, and $q_4 = 0.3$. The above *grouping of terms is to have real coefficients* for easier hardware implementation.

MATLAB the above problem can also be solved using the MATLAB command `[Az Bz] = impinvar(A,B,fs)`.

To check that the impulse responses are similar use:

```
t = 0:Ts:1
sys1 = tf(A,B)
h1 = impulse(sys1,t)
sys2 = tf(Az/Ts,Bz,Ts)
h2 = impulse(sys2,t)
plot(t,h1,t,h2':')
```

*Note* In filter design the following MATLAB commands may prove useful: `B = roots(A)`: Finds the roots of a polynomial whose coefficients are in A [e.g., `B=roots([1 2 3])` finds the roots of (descending order)].

`A = poly(B)`: Converts the roots (in B) to a polynomial.

`C = conv(A,B)`: Multiply two polynomials A and B (Also, find the convolution sum).

`X =fzero('fun',xo)`: Finds a zero of the function "`fun`" (which is either a library function or defined in another file) near `xo` [e.g., `fzero('x/3-sin(x)',pi/2)` gives 2.2789].

`zplane(A,B)`: Plots the pole-zero diagram of a transfer function $H(z) = A(z)/B(z)$.

`h = freqs(A,B,w)`: computes the complex frequency response of the analog filter $H(s) = A(s)/B(s)$ at the angular frequencies in the vector `w`.

`h = freqz(A,B,f,fs)`: computes the complex frequency response of the digital filter $H(z) = A(z)/B(z)$ at the frequencies in the vector `f`, with sampling frequency `fs`.

#### 2.6.8.2 Frequency-Domain Design: Frequency Response Matching

The impulse invariance method enforced a time-domain correspondence between the analog and digital impulse responses. The frequency response matching approach enforces a frequency-domain correspondence between the analog and digital transfer functions. Assume that one starts with an analog filter transfer function $H_a(s)$, which needs to be transformed into a digital transfer function $H(z)$. Unlike the time domain matching approach, it is not possible to achieve an *exact* correspondence between $H_a(s)$ and $H(z)$-this is because the $H(z)$ displays repetition in frequency whereas $H_a(s)$ does not.

As a first attempt at achieving a frequency domain correspondence between the digital and analog filters, one can use the natural mapping:

$$H\left(e^{sT_s}\right) = H_a(s) \quad for - \pi \leq \omega T_s \leq \pi.$$

Hence,

$$H\left(e^{j\omega T_s}\right) = H_a(\omega) \quad for - \pi \leq \omega T_s \leq \pi.$$

The relationship $z = e^{sT_s}$ is enforced in this approach. Thus,

$$s = \frac{\ln(z)}{T_s}.$$

It is apparent that this transformation is non-linear and it is therefore impossible in general to obtain $H(z)$ as a rational function of $z$. However, an approximation to this relation is possible. If $z$ is near 1, the Taylor expansion of $\ln(z)$ is $(z - 1) - (z - 1)^2/2 + O\{(x - 1)^3\}$, while the expansion of $(z - 1)/(z + 1)$ is $[(z - 1) - (z - 1)^2 + O\{(x - 1)^3\}]/2$. Hence, ignoring the smaller terms, the following approximation can be used: $\ln \approx 2(z - 1)/(z + 1)$. Then $s$ can be approximated by:

$$s \approx \frac{2}{T_s}\frac{z - 1}{z + 1}.$$

This relation is called the *bilinear transform*. It should be noted that the approximation in used above is true only when $z$ is *near* 1, which means that the low frequency region maps quite reliably from the analog domain to the digital domain. At higher frequencies, however, there is considerable error in the approximation. It will be seen subsequently that the bilinear transform actually introduces a warping of the analog frequency axis, and the warping is most severe at high frequencies. This warping causes the analog frequency range $0 \to \infty$ to be mapped into the digital frequency range $0 \to \pi$.

To gain further insights into the nature of the bilinear transform one can put $z = re^{j\Omega}$ [recall that the frequency response of any digital system $H(z)$ can be found by substituting $z = re^{j\Omega}$]. Substituting the polar form for $z$ $(= re^{j\Omega})$ in the bilinear transform $s \approx (2/T_s)(z-1)/(z+1)$ gives:

$$\begin{aligned} s &\approx \frac{2}{T_s}\frac{re^{j\Omega}-1}{re^{j\Omega}+1} \\ &= \frac{2}{T_s}\frac{[r\cos(\Omega)-1]+jr\sin(\Omega)}{[r\cos(\Omega)+1]+jr\sin(\Omega)} \\ &= \frac{2}{T_s}\left[\frac{r^2-1}{1+r^2+2r\cos(\Omega)}+j\frac{2r\sin(\Omega)}{1+r^2+2r\cos(\Omega)}\right]. \end{aligned}$$

Since $s = \sigma + j\omega$:

$$\sigma = \frac{2}{T_s}\frac{r^2-1}{1+r^2+2r\cos(\Omega)} \quad \text{and} \quad \omega = \frac{2}{T_s}\frac{2r\sin(\Omega)}{1+r^2+2r\cos(\Omega)}$$

Now for $\sigma = 0$, $|z| = r = 1$ and so $z = e^{j\Omega}$. Hence, the $s = j\omega$ axis in the analog domain is transformed into the $z = e^{j\Omega}$ unit circle, where the new relation between $\Omega$ and $\omega$ is given by:

$$\begin{aligned} \omega &= \frac{2}{T_s}\frac{\sin(\Omega)}{1+\cos(\Omega)} = \frac{2}{T_s}\tan\left(\frac{\Omega}{2}\right) \\ \therefore \Omega &= 2\tan(\omega T_s/2). \end{aligned} \tag{2.32}$$

The above equation is non-linear, as illustrated graphically in Fig. (2.34). The infinite analog frequency domain ($\omega = 0 \to \infty$) is mapped onto the principle digital frequency range $\Omega = 0 \to \pi$.

If $\sigma < 0$, then $r < 1$ (hence the LHS of the s-plane is mapped inside the unit circle), and for $\sigma > 0$, $r > 1$.

It can be shown that filter design using the bilinear transform is independent of $T_s$ [2]; hence, one can put $T_s = 2$ (for convenience) and use the following transform instead:

$$s = \frac{z-1}{z+1} \quad \left[\text{with} \quad \Omega = 2\tan^{-1}(\omega) = 2\tan^{-1}(2\pi f)\right]. \tag{2.33}$$

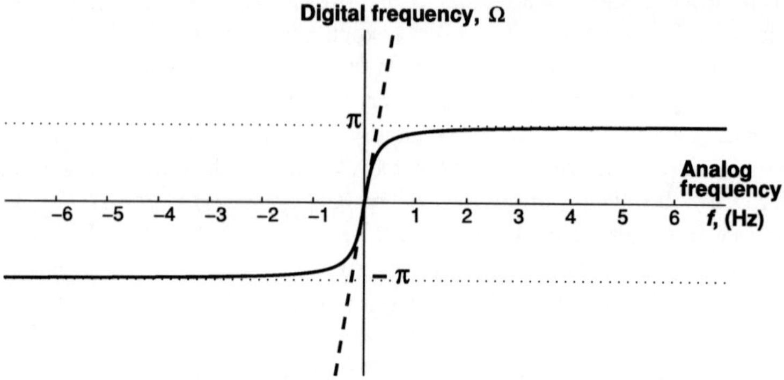


**Fig. 2.34** Frequency domain relationship between analog frequency and digital frequency under the natural transformation $z = e^{sT_s}$, where $\Omega = \omega T_s$ (*dashed curve*, $Ts = 2$) and the bilinear approximation $s = (2/T_s)(z-1)/(z+1)$, where $\Omega = 2\tan^{-1}(\omega)$ (*solid curve*)

The bilinear transformation can be applied to transform all analog Filters, while the impulse invariance method cannot be applied to HPFs. This is so because there is no problem with aliasing, as this transform maps the whole $j\omega$ axis to the unit circle. That is, it is a *1-to-1 mapping*. The impulse invariance transform $z = e^{sT_s}$, on the other hand maps every sector of length $2\pi$ on the $j\omega$ axis to the unit circle - it is a multi-to-1. The price paid for the resilience to aliasing problems is the deformation or warping of the original analog frequency response. It should be noted that this warping is for the frequency response of the transformed analog filter only; it does not imply that the input signal spectrum is also warped.

*Example* Design a low-pass digital IIR filter with cutoff frequency $\Omega_c = 0.6$ using a third order Butterworth analog filter.

*Solution:* From *Tables*-Denominator Polynomial Coefficients for Normalized LPF's, the transfer function of the analog B-LPF is given by:

$$H_a(s_N) = \frac{1}{1 + 2s_N + 2s_N^2 + s_N^3}$$

De-normalizing yields,

$$H_a(s) = \frac{1}{1 + 2(s/\omega_c) + 2(s/\omega_c)^2 + (s/\omega_c)^3}$$

The cutoff frequency is obtained as follows: $\omega_c = \tan(\Omega_c/2) = \tan(0.3) = 0.3093 \approx 0.3$ rad/s. Hence,

$$H(z) = \frac{1}{1 + 6.6\frac{z-1}{z+1} + 22.2\left(\frac{z-1}{z+1}\right)^2 + 37\left(\frac{z-1}{z+1}\right)^3}.$$

#### 2.6.8.3 MATLAB IIR Filter Design Using the Bilinear Transformation

one can design IIR digital filters in MATLAB using analog prototypes selected according to our requirements. The most important MATLAB filters are `butter`, `cheby1`, `cheby2`, and `ellip`.

*Example* Using MATLAB, design the following digital IIR filters. Use the bilinear transform and start with either a Butterworth, Chebychev or Elliptic prototype filter:

1. LPF at $f_s = 10$ kHz with cutoff $f_c = 2{,}000$ Hz, maximum ripple allowed in the passband is $r = 1$ dB, and minimum stopband attenuation = 60 dB starting at 3,000 Hz. Sharpest transition with lowest order is required.
2. HPF with $f_c = 2{,}000$ Hz, minimum stopband attenuation= 60 dB for frequencies below 1,000 Hz, and other specifications as above.
3. BPF with passband 2,000–3,000 Hz, minimum stopband attenuation = 60 dB for frequencies below $f_1 = 1{,}000$ Hz or above $f_2 = 4{,}000$ Hz, and other specifications as above.
4. BSF with stopband 1,000–4,000 Hz, minimum stopband attenuation= 60 dB for frequencies between 2,000 and 3,000 and other specifications as above.

*Solution:*

1. Since the specification requires the sharpest transition and lowest possible order, an elliptic filter is the appropriate choice for the prototype. The corresponding digital IIR digital elliptic filter has $F_c = f_c/(f_s/2) = 2000/(10k/2) = 0.4$, $F_2 = f_2/(f_s/2) = 0.6$, $r_d = 1$, and $A_{dB} = 60$. The order is obtained as follows:
   `[n fo] = ellipord(Fc,F2,3,AdB)`
   which gives $n = 5$ and `fo = 0.4`. Having found the order one can determine the filter transfer function:
   `[b a]= ellip(n,1,60,fo)`
   where b and a are vectors representing the numerator and the denominator coefficients, respectively.
2. $F_2 = f_2/(f_s/2) = 0.2$, $F_c = 0.4$, Since the other specifications are as for the previous example, an elliptic prototype filter must again be used. Its order is found with the command:
   `[n fon] = ellipord(fcn,f2n,3,AdB)`
   which gives $n = 4, f_o = 0.4$. The filter transfer function is determined with the command:
   `[b a] = ellip(n,rdB,AdB,[F1 F2],'high')`

3. The solution to this and the following examples proceed similarly to the previous ones. $F_{c1} = 0.4$, $F_{c2} = 0.6$, $F_c = [F_{c1} \quad F_{c2}]$, $F_1 = 0.2$, $F_2 = 0.8$, $F = [F_1\ F_2]$,
   `[n fo] = ellipord(Fc,F,3,AdB)`
   which gives $n = 3$ and $f_o = [0.4\ 0.6]$.
   `[b a] = ellip(n,rdB,AdB,fo)`
4. $F_{c1} = 0.2$, $F_{c2} = 0.8$, $F_c = [F_{c1}\ F_{c2}]$, $F_1 = 0.4$, $F_2 = 0.6$, $F = [F1\ F2]$,
   `[n fo] = ellipord(Fc,F,3,AdB)`
   which gives $n = 3$ and $fo = [0.4\ 0.6]$
   `[b a] = ellip(n,rdB,AdB,fo,'stop')`

Figure (2.35) shows the magnitude response of the above filters plotted against the normalized frequency $\nu = f/f_s$ [note that normalization in MATLAB is different from many other parts of the literature; in its plots MATLAB typically uses the ratio $f/(f_s/2)$ for normalized frequency instead of $f/f_s$.

#### 2.6.8.4 MATLAB FIR/ IIR Filter Design and Analysis Toolbox

If the following statement is entered on the MATLAB command line:

≫ fdatool

a filter design toolbox will pop up. This toolbox supports many different digital filter design techniques and provides a user-friendly filter design interface. It also enables the user to export the designed filter coefficients to the MATLAB workplace.

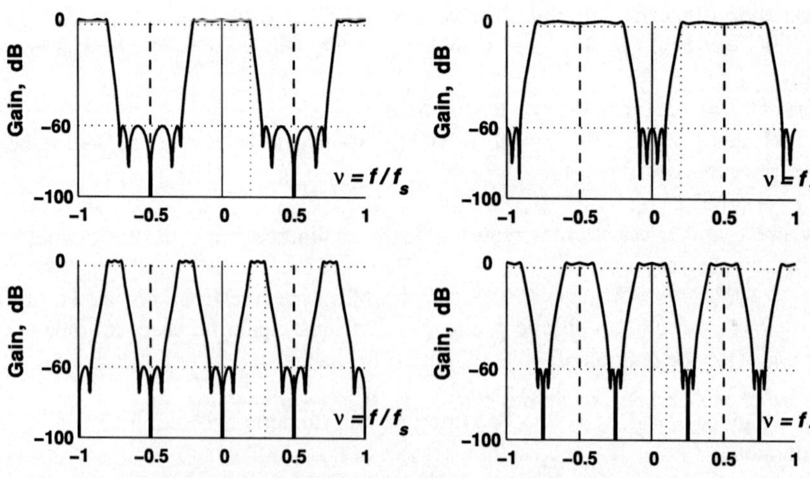


**Fig. 2.35** Magnitude response of elliptic IIR filters: LPF, HPF, BPF, and BSF

## 2.6.9 Applications of IIR Digital Filters

### 2.6.9.1 The Digital Integrator

If the sampling interval $T_s$ is small, then a digital summer provides a good approximation to an integrator. That is:

$$\int_0^T x(t)dt \approx \sum_{n=0}^{N} x(n)T_s = T_s \sum_{n=0}^{N} x(n) \quad (\text{where } N = T/T_s). \tag{2.34}$$

Now

$$y(k) = \sum_{n=0}^{k} x(n) = x(k) + \sum_{n=0}^{k-1} x(n) = x(k) + y(k-1). \tag{2.35}$$

Hence,

$$\begin{aligned} Y(z) &= X(z) + z^{-1}Y(z) \\ \therefore H(z) &= \frac{1}{1 - z^{-1}}. \end{aligned} \tag{2.36}$$

Therefore, the integrator transfer function is $H_i(z) = T_s H(z)$. The transfer function magnitude and phase is plotted in Fig. (2.36) and it is apparent from the magnitude plot that the digital integrator is low-pass in nature. It is the inverse of the digital differentiator, which is a high-pass filter.

MATLAB the pole-zero diagram can be plotted in MATLAB using `zplane(A,B)`, where A is the vector of numerator coefficients and B is the vector of denominator coefficients. In the above example `A = [1 0]` and `B = [1 – 1]` since $H(z) = z/(z - 1)$.

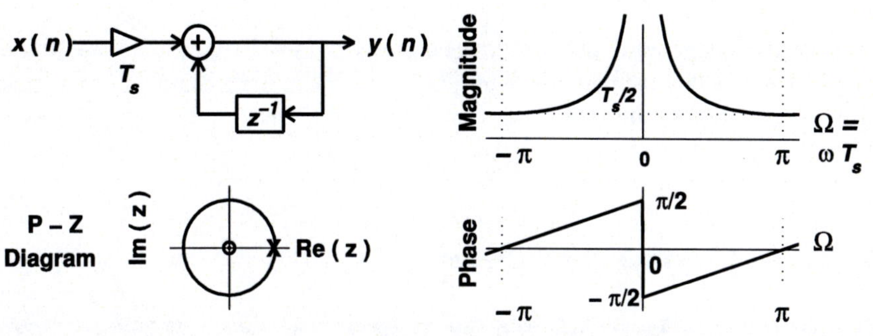


Fig. 2.36 The digital integrator with its frequency response and p-z diagram

#### 2.6.9.2 The Alpha Filter

It was seen in Sect. 2.6.9.1 that the integrator difference equation is $y(n) = T_s[x(n) + y(n-1)]$. If properly scaled, *the integrator can be adapted to become an averager*. Modifying (2.35) to introduce this scaling yields:

$$y(n) = (1-\alpha)x(n) + \alpha y(n-1). \tag{2.37}$$

where $\alpha$ is a real positive integer which is less than 1. The choice of $\alpha$ depends on the importance of the current sample $x(n)$ as compared to the previous average of data. The transfer function is given by:

$$H(z) = \frac{1-\alpha}{1-\alpha z^{-1}} = \frac{(1-\alpha)z}{z-\alpha},$$

with one pole at $z = \alpha$ and one zero at $z = 0$. The implementation of the alpha filter is depicted in Fig. (2.37). Using *Tables*-z Transform Pairs and Theorems, the impulse response of this filter is given by

$$h(n) = (1-\alpha)\alpha^n u(n),$$

hence, the output of the system in the time-domain is specified by:

$$\begin{aligned} y(n) = x(n) * h(n) &= \sum_{k=-\infty}^{\infty} x(n-k)h(k) \\ &= (1-\alpha)\left[x(n) + \alpha x(n-1) + \alpha^2 x(n-2) + \cdots\right] \end{aligned}$$

Noting that $(1-\alpha)[1 + \alpha + \alpha^2 + \cdots] = 1$, the above I/O formula is reminiscent of the moving average FIR filter, except that the number of "coefficients" is now infinite. This IIR averager is much easier to build than the $N$-tap FIR moving average filter, with almost *similar performance*. See Fig. (2.37) for a comparison using the same financial data as was used in Fig. (2.27). As in the FIR example of moving average, a few early samples in the plot should be ignored.

Like the integrator, this system behaves as a LPF. *Since averaging is not sensitive to whether the phase response is linear or not*, the alpha filter has a great deal of appeal for many applications (e.g., in SNR estimation for communication systems).

#### 2.6.9.3 The Sinusoidal Digital Oscillator

An oscillator is a system that generates a specific waveform without an input, except perhaps for an initial impulse or a particular setup of the initial conditions.

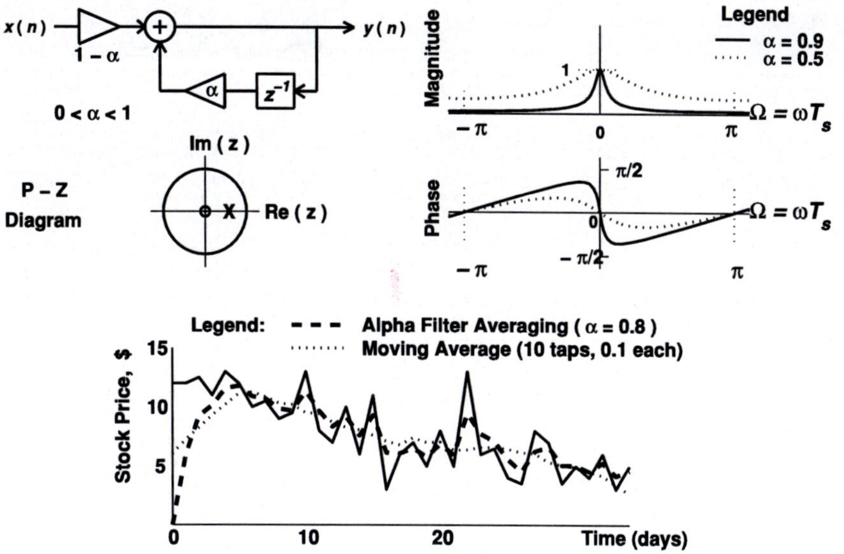


**Fig. 2.37** Alpha filter. *Above* Circuit and its frequency response. *Below* data averaging

To design a digital sinusoidal oscillator, one essentially needs a filter whose impulse response is sinusoidal. From *Tables*- z Transform Pairs and Theorems, such a filter transfer function is specified by:

$$h(n) = \sin(bn)u(n) \overset{ZT}{\longleftrightarrow} H(z) = \frac{\sin(b)z}{z^2 - 2\cos(b)z + 1} \tag{2.38}$$

To build *H*(*z*), one can write it as $H(z) = \sin(b)z^{-1}/[1 - 2\cos(b)z^{-1} + z^{-2}]$, the implementation of which is shown in Fig. (2.38). In this expression *b* corresponds to the normalized radian frequency $\Omega_o = \omega_o T_s$, and hence the frequency of oscillation is:

$$f_o = \omega_o/2\pi = (b/T_s)/2\pi = (b/2\pi)f_s \,\text{Hz},$$

provided that $|b| < \pi$. The two poles of this system are the roots of the equation $z^2 - 2\cos(b)z + 1 = 0$, which are given by:

$$p_{1,2} = \cos(b) \pm \sqrt{\cos^2(b) - 1} = \cos(b) \pm j\sin(b) = e^{\pm jb}. \tag{2.39}$$

Hence, the poles are *exactly on the circumference* of the unit circle, as shown in Fig. (2.39), and the system is strictly speaking, *neither stable nor unstable*.

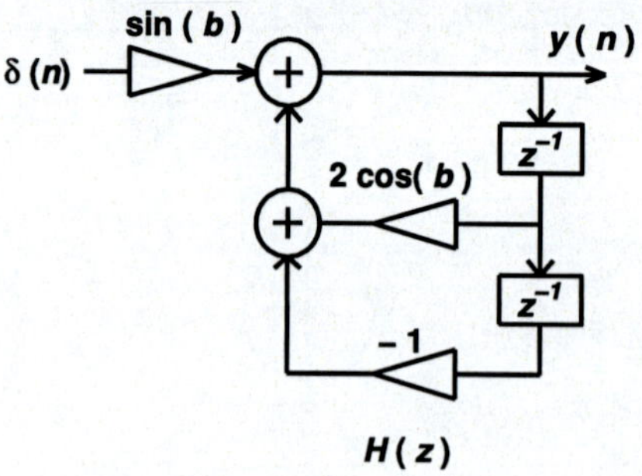


Fig. 2.38 Implementation schemes for a sinusoidal digital oscillator

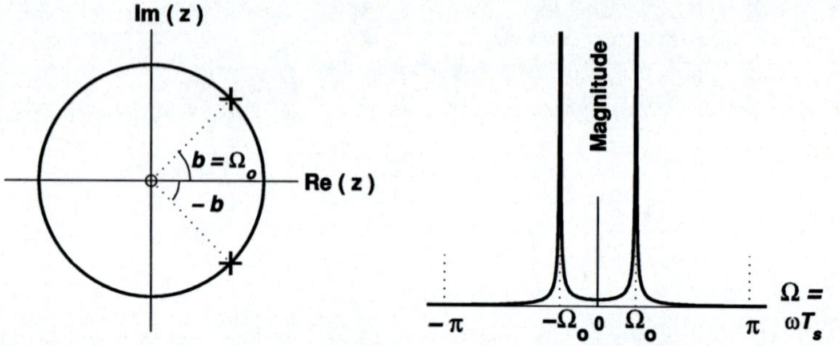


Fig. 2.39 Magnitude response and pole-zero diagram of the digital oscillator

#### 2.6.9.4 The Digital Resonator

The magnitude response of the digital oscillator was seen to consist of two very sharp two spikes. Analogously, one can design a digital resonator, which is a narrowband BPF centered around a resonant frequency $f_o$. Such a resonator could be useful for extracting a fundamental sinusoid (with frequency $f_o$) corrupted by harmonics or other sinusoids. Note that if one attempts to design a resonator using classical filter design techniques, one needs a prohibitively large filter order. To ensure stability, the poles should be inside the unit circle, i.e., $p_{1,2} = re^{\pm\Omega_o}$, with $r$ very near to 1. A possible transfer function for the resonator can then be:

$$H_1(z) = \frac{1}{(z-p_1)(z-p_2)} = \frac{1}{z^2 - 2r\cos(\Omega_o)z + r^2}. \tag{2.40}$$

The magnitude response for this resonator is shown in Fig. (2.40) as the dotted curve. Note that this curve has non-zero values well beyond the frequency of interest, $\Omega_o$.

A transfer function which also satisfies the above criteria with better stopband attenuation is:

$$H(z) = \frac{(z - z_1)(z - z_2)}{(z - p_1)(z - p_2)} = \frac{(z - z_1)(z - z_2)}{z^2 - 2r\cos(\Omega_o)z + r^2}. \quad (2.41)$$

Note that he transfer function in (2.41) has in addition to a pair of poles, two zeros. These zeros are chosen to null the frequency response $H(e^{j\Omega})$ at the digital frequency borders $f = 0$ and $f = \pm f_s/2$. That is:

1. $H(e^{j0}) = H(1) = 0$ can be enforced by putting $z_1 = 1$.
2. $H(e^{\pm j\, Omega_s/2}) = H(e^{\pm j\pi}) = H(-1) = 0$ can be enforced by putting $z_2 = -1$. Having the two zeros at these two positions, the frequency response is "pulled down" more effectively away from the positions of frequency resonance. That is, the stopband attenuation improves.

To facilitate implementation of the resonator, it is helpful to write its transfer function as a function of $z^{-1}$:

$$H(z) = \frac{z^2 - 1}{z^2 - 2r\cos(\Omega_o)z + r^2} = \frac{z^2 - 1}{1 - 2r\cos(\Omega_o)z^{-1} + r^2z^{-2}}.$$

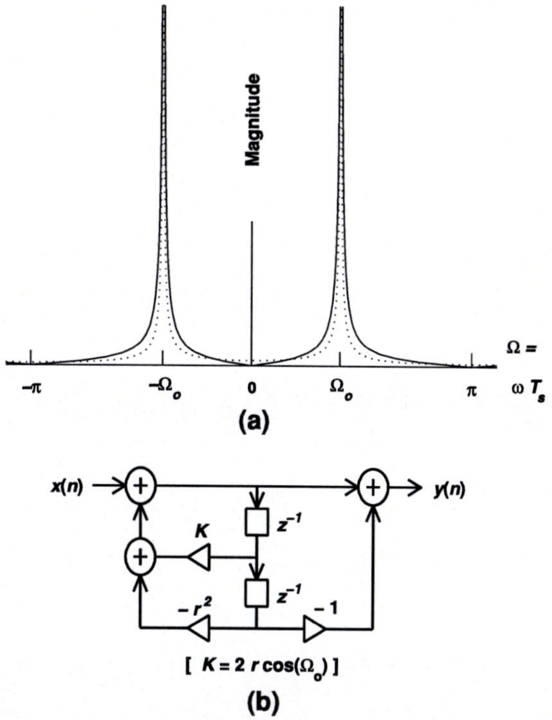


**Fig. 2.40** A digital resonator. **a** Magnitude response with $f_o = 2$ Hz, $r = 0.9$, $f_s = 10$ Hz (*Solid* with 2 zeros; *dotted*: without zeros). **b** Implementation using Direct Form-II

This transfer function can be implemented as shown in Fig (2.40). If $r$ is very close to 1, then the 3-dB (half-power) bandwidth $B_f$ and the maximum gain $G_m$ of this system are approximated, respectively, by:

$$B_f \approx \frac{1-r}{\pi} f_s(\text{Hz}) \quad \text{and} \quad G_m \approx \frac{1}{(1-r^2)\sin(\Omega_o)}$$

Equivalently, the bandwidth can be written as:

$B_\omega = 2\pi . B_f = 2(1-r) f_s$ (rad/s), (where $\omega = 2\pi f$, radian frequency), $B_v = B_f/f_s = \frac{1-r}{\pi}$, (where $v = f/f_s$, normalized frequency), and $B_\Omega = B_\omega/f_s = 2(1-r)$, (where $\Omega = \omega/f_s = 2\pi f/f_s = 2\pi v$, normalized frequency).

### 2.6.9.5 A Digital DC Blocker

In some applications an undesirable DC voltage can appear in the information signal. For example, in some audio applications a DC offset can be added to the recoded sound from the microphone. The ADC may also add some unwanted DC to the digitized signal. This DC component carries no information, and in audio applications cannot even be heard. Nonetheless, in some cases it can hinder processing and cause instabilities. For example, the DC component may drive the signal outside the dynamic range of the processing system which will cause signal clipping. Hence, DC removal is often desirable before other forms of processing are pursued.

For DC removal, it is necessary for the magnitude response to be zero at $f = 0$ Hz and unity elsewhere. This kind of frequency response can only be approximated in practice. To block DC one can place a zero at $z = 1$, and to ensure that there is approximately unity gain for non-zero frequencies, one additionally needs a pole very near to the zero at $z = 1$, and inside the unit circle. The following transfer function has the necessary form:

$$H(z) = \frac{1-z^{-1}}{1-\alpha z^{-1}}, \quad \alpha \text{ close to } 1. \tag{2.42}$$

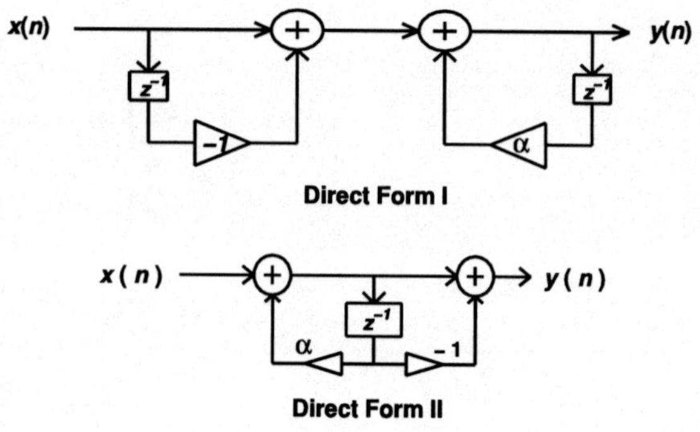


Fig. 2.41 A digital DC Blocker

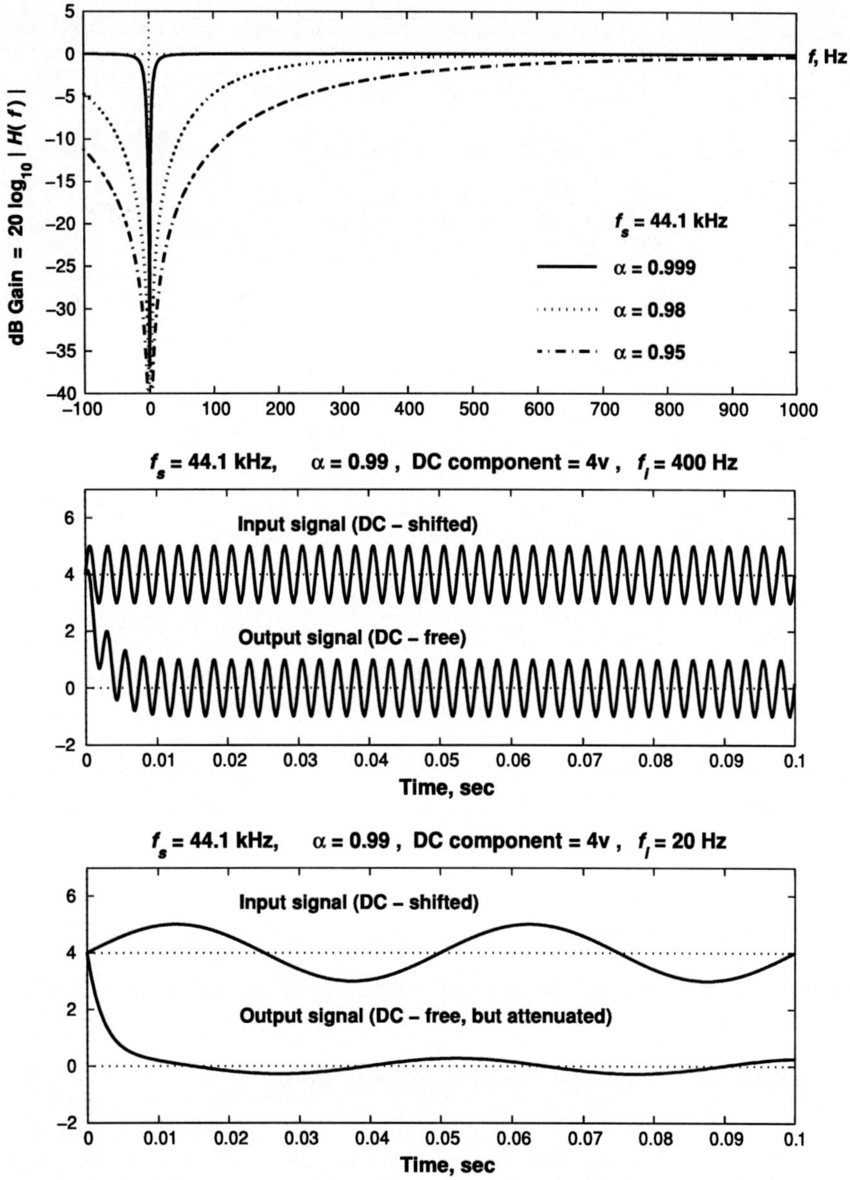


**Fig. 2.42** Frequency and time-domain performance of the DC Blocker

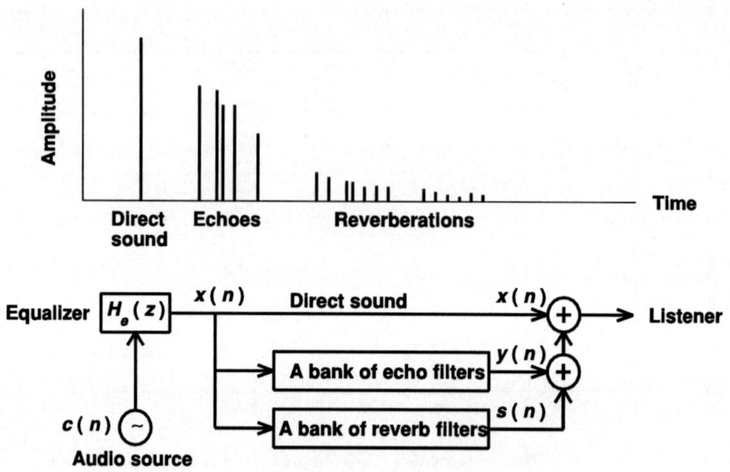


**Fig. 2.43** Audio effects. *Above* an audio signal and its reflections in a listening venue. *Below* simulation of specific acoustics in another venue

Figure (2.41) shows a block diagram of a DC blocker, while Fig. (2.42) shows its practical performance with sinusoidal dc-shifted signals. It is evident that this performance is dependent on how close $\alpha$ is to 1. When $\alpha = 0.99$, the filter removes the DC from the (low-frequency) 20 Hz sinusoid, but it also introduces a phase change and an attenuation.

#### 2.6.9.6 An Application of FIR / IIR Digital Filters: Simulation of Acoustic Effects

A piece of music played in a concert hall does not sound the same as if it is played in a living room. This is due to the echoes (early reflections) and the reverberations (late reflections) which vary from setting to setting. One can actually simulate a concert hall in a living room using the following steps:

1. Equalize the audio transfer function of the room, $H_r(z)$. That is, eliminate any special effects that the living room is inherently creating. This is done by first sending an audio impulse $\delta(n)$, and then measuring the impulse response $h_r(n)$. The transfer function is then given by $H_r(z) = \text{fft}\{h_r(n)\}$. Then one designs an equalizing filter $H_e(z) = 1/H_r(z)$, with say the Remez algorithm.
2. Simulate the echoes and the reverberations as follows:

- For echoes: $y(n) = x(n) + \alpha x(n - N)$. This is so because an echo is a direct reflection of the delayed input signal. Hence, $H(z) = 1 + \alpha z^{-N}$. This is an FIR filter. In practice, choose $NT_s \geq 0.05$

- For reverberations: $s(n) = x(n) + \beta \cdot s(n - M)$. This is so because a reverberation is an accumulation of previous reflections of the signal. Hence, $H(z) = 1/(1 - \beta z^{-M})$. This is an IIR filter with $M$ poles.

Figure (2.43) shows a block diagram of the above procedure.

## References

1. Harris, F.J.: On the use of windows for harmonic analysis with the discrete Fourier transform. Proc. IEEE **66**, 51–83 (1978)
2. Oppenheim, A.V., Schafer R.: Discrete-Time Signal Processing, Prentice-Hall, USA (1989)

# Part II
# Applied Signal Processing

# Chapter 3
# Selected Topics in Applied Signal Processing

## 3.1 Introduction

Signal processing has found application in hugely diverse areas. It is used, for example, to

- decode and receive signals in mobile telephony,
- monitor and guide airborne vehicles,
- detect geological structures which might signal the presence of oil and other minerals,
- automatically tune televisions,
- create special audio effects such as surround sound,
- construct intelligent voice mail systems,
- facilitate computer-based person identification,
- automate ploughing and irrigation in agriculture,
- create special effects in robotic toys,
- provide location information for GPS,
- etc.

This chapter will explore a number of sample applications.

## 3.2 Binary Signal Transmission

In many applications signals are transmitted from one location to another in baseband mode (i.e., as signals whose spectra have only low frequency content). Baseband transmission cannot be used, however, if the signal has to be transmitted through the air via antennas. In such cases modulation is needed so that the signal can be converted to a form which can be effectively radiated from an antenna. This section will consider baseband binary signal transmission, which is important in many applications, especially in communications between computers and digital signal processors operating in the same neighborhood.

Z. M. Hussain et al., *Digital Signal Processing*,

DOI: 10.1007/978-3-642-15591-8_3, 

### *3.2.1 Binary Transmission Using Orthogonal Signals*

In binary communication systems, binary data (which is a sequence of 'on' or 'off' signals) is transmitted through a channel. The signals which are used to convey the on or off information are normally *orthogonal* (or uncorrelated) signals, and will be denoted by $s_0(t)$ and $s_1(t)$. One possible pair of orthogonal signals for $s_0(t)$ and $s_1(t)$ is shown in Fig. 3.1.

If the data rate is $R$ (bps), then the time duration of $s_0(t)$ and $s_1(t)$ is $T = 1/R$. If the number of data bits transmitted over the channel is large, then the on's and off's are equally probable and $p(\text{on}) = p(\text{off}) = 1/2$. As the data moves through the channel it almost invariably acquires noise. For many channels encountered in practice, this added noise can be well modeled as additive white Gaussian noise (AWGN). As the name suggests, this noise is Gaussian noise and is wideband (i.e. approximately white) to the data. It will be assumed here that the power spectral density (PSD) of the added noise is $= \eta/2$ (Watts/Hz). Hence, the received signal has the following form:

$$r(t) = s_i(t) + n(t), \quad i \in \{0,1\}, \quad 0 \le t \le T.$$

In binary transmission systems, the receiving station knows the shape of the two possible signals being transmitted, and hence optimal detection in noise can be achieved through a bank of two matched filters. It will be assumed that $s_0$ and $s_1$ are as shown in Fig. 3.1, and that $s_0$ is transmitted, starting at time $t = 0$. Then the outputs of the two matched filters (correlators) at $t = T$ are:

$$r_0 = \int_0^T r(t)s_0(t)dt = \int_0^T [s_0(t) + n(t)]s_0(t)dt$$

$$= \int_0^T s_0^2(t)dt + \int_0^T n(t)s_0(t)dt = E + n_0$$

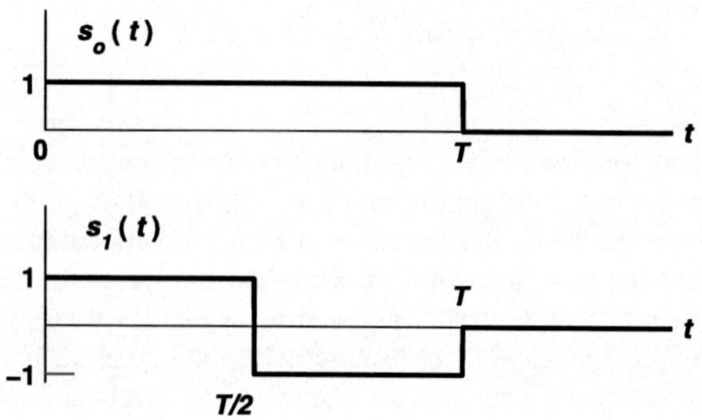


Fig. 3.1 Two signals which are orthogonal (uncorrelated) over the period $T$

$$r_1 = \int_0^T r(t)s_1(t)dt = \int_0^T [s_0(t) + n(t)]s_1(t)dt$$
$$= \int_0^T s_0(t)s_1(t)dt + \int_0^T n(t)s_1(t)dt = 0 + \int_0^T n(t)s_1(t)dt = n_1$$

It is seen from the above that output of the filter matched to $s_1(t)$ is simply noise, whereas the output of the filter matched to $s_0(t)$ is equal to the autocorrelation of $s_0(t)$ plus noise.

The discrepancy in outputs of the two different matched filters provides the basis for deciding whether the on or off signal has been transmitted. Further analysis is provided below.

Since at every time instant $t$ the noise $n(t)$ is a *sample function* of a AWGN process with PSD $= \eta/2$, both $n_0$ and $n_1$ are Gaussian. (Note that integration is essentially a summation, and the sum of weighted Gaussian random variables is also Gaussian (see [1]). Also, both $n_0$ and $n_1$ have zero means since:

$$\mathcal{E}\{n_0(t)\} = \mathcal{E}\left\{\int_0^T s_0(t)n(t)dt\right\} = \int_0^T s_0(t)\mathcal{E}\{n(t)\}dt = 0 = \mathcal{E}\{n_1(t)\} \quad \forall t,$$

where $\mathcal{E}\{\cdot\}$ is the *expectation functional*, i.e., $\mathcal{E}\{x\} = \int xp(x)dx$. The variances of $n_0$ and $n_1$, $\sigma_i(i \in \{1, 2\})$, are given by:

$$\sigma_i = \mathcal{E}\{n_i^2\} = \mathcal{E}\left\{\int_0^T s_i(t)n(t)dt \cdot \int_0^T s_i(\tau)n(\tau)d\tau\right\}$$
$$= \int_0^T\int_0^T s_i(t)s_i(\tau)\mathcal{E}\{n(t)n(\tau)\}dtd\tau$$
$$= \int_0^T\int_0^T s_i(t)s_i(\tau)R_n(t-\tau)dtd\tau \quad \{\text{since } n(t) \text{ is WSS}\}$$
$$= \frac{\eta}{2}\int_0^T\int_0^T s_i(t)s_i(\tau)\delta(t-\tau)dtd\tau$$
$$= \frac{\eta}{2}\int_0^T s_i^2(t)dt = \frac{\eta}{2}E$$

where $E$ is the energy of the signals $s_0$ and $s_1$ [remembering that $s_1^2(t) = s_0^2(t)$].

The above statistics can be used to estimate the reliability of detecting the correct symbol in a binary transmission system. The derivation of this reliability measure is done below.

#### 3.2.1.1 Probability of Error

The matched filter based receiver compares $r_0$ and $r_1$. It will decide that an "off" was transmitted if $r_0 > r_1$, and that an "on" was transmitted if $r_1 > r_0$. If an "off" was transmitted, then an error will occur *only if* $r_1 > r_0$. The probability of such an error is given by:

$$P_e = Pr(r_1 > r_0) = Pr(n_1 > E + n_0) = Pr(n_1 - n_0 > E).$$

Let $x = n_1 - n_0$. Since $n_0$ and $n_1$ are zero-mean Gaussian variables, $x$ is also a zero-mean Gaussian random variable, with variance given by:

$$\sigma_x = \mathcal{E}\{(n_1 - n_0)^2\} = \mathcal{E}\{n_1^2\} + \mathcal{E}\{n_0^2\} - 2\mathcal{E}\{n_1 n_0\} = 2(E\eta/2) = E\eta. \quad (3.1)$$

Note that to obtain the result in (3.1) the following result was used:

$$\mathcal{E}\{n_1 n_0\} = \mathcal{E}\left\{\int_0^T \int_0^T s_0(t)s_1(\tau)n(t)n(\tau)dtd\tau\right\}$$

$$= \int_0^T \int_0^T s_0(t)s_1(\tau)\mathcal{E}\{n(t)n(\tau)\}dtd\tau = 0.$$

Hence, the pdf of the random variable $x$ is given by:

$$p(x) = \frac{1}{\sqrt{2\pi\sigma_x}} e^{-x^2/2\sigma_x^2}.$$

From (3.2) it is possible to find the probability of error. More detail in this process is provided in the following [see also Fig. 3.2].

$$P_e = Pr(x > E) = \int_E^\infty p(x)dx = \frac{1}{\sqrt{2\pi\sigma_x}} \int_E^\infty e^{-x^2/2\sigma_x^2} dx.$$

Now let $r = x/\sigma_x$, then

$$P_e = Pr(x > E) = \int_E^\infty p(x)dx = \frac{1}{\sqrt{2\pi}} \int_{\sqrt{\mathrm{SNR}}}^\infty e^{-r^2/2} dr$$
$$= \frac{1}{2} - \frac{1}{2}\mathrm{erf}\left(\sqrt{\frac{\mathrm{SNR}}{2}}\right) \quad (3.2)$$

where the error function $\mathrm{erf}(y) = \frac{2}{\pi}\int_0^y e^{-u^2} du$ is a reference function in MATLAB, and SNR is the signal-to-noise ratio defined by:

**Fig. 3.2** Gaussian pdf. Note that the *shaded area* represents $Pr(x > E)$

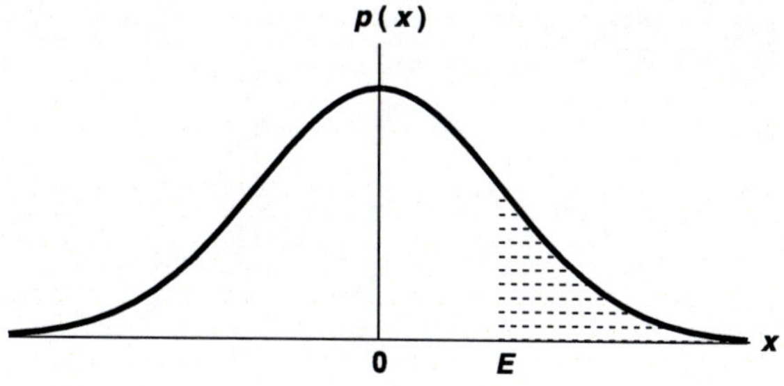


$$\text{SNR} = \frac{E}{\eta},$$

The probability of error $P_e$ in correctly detecting a data symbol is often used as *the basis for performance evaluation of communication systems*. Although the probability of error $P_e$ in (3.2) was obtained under the assumption that an "off" was transmitted, the same kind of result would have obtained if an "on" was transmitted. Hence, the average probability of error is given by:

$$P_{e_{av}} = P_e.$$

Figure 3.3 shows the general shape of $P_e$ versus SNR.

### 3.2.2 Binary Transmission Using Antipodal Signals

Two signals $s_0(t)$ and $s_1(t)$ are said to be *antipodal* if $s_0(t) = -s_1(t) = s(t)$. One possible antipodal configuration is the use of $\pm V$ as the two different signals. An alternative configuration is depicted in Fig. 3.4.

If one uses orthogonal signals for transmission, normally a bank of two matched filters are needed for optimum reception. However, if one uses two *antipodal* signals, *only one* matched filter is needed. The received signal is $r(t) = \pm s(t) + n(t)$, and the matched filter is matched to $s(t)$.

Assume that two antipodal signals are used for transmission in a binary baseband communication system. Following the same kind of analysis as was used for orthogonal signals, the following result can be obtained (see [2]):

$$\text{Output of the matched filter} = \pm E + n_a,$$

where $n_a = \int_0^T n(t) \pm s(t)dt$, and the variance of $n_a = \frac{\eta}{2}E$ (i.e., the same as that of $n_0$ and $n_1$ for orthogonal transmission). The decision process used for antipodal transmission is as follows: if $r > 0$, then $s(t)$ was assumed to be transmitted (which represents the "on" symbol), otherwise $-s(t)$ is assumed to have been transmitted.

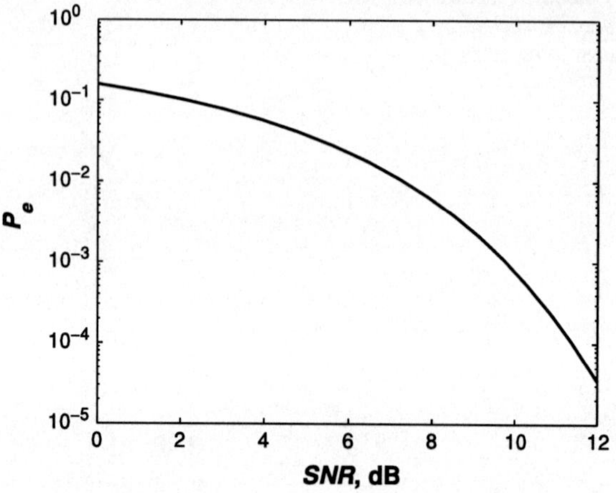


**Fig. 3.3** Probability of error $P_e$ versus SNR in a baseband binary communications system

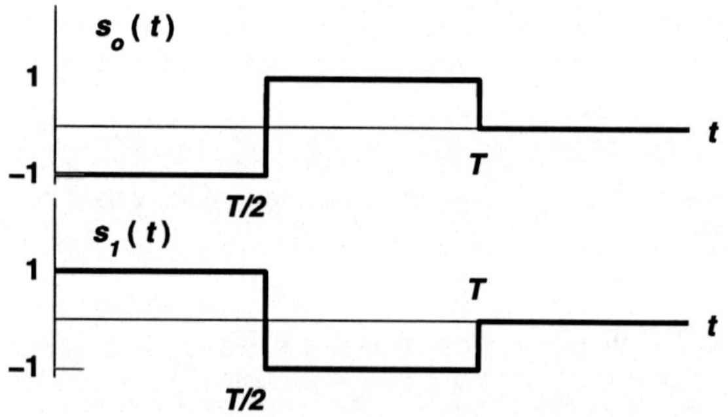


**Fig. 3.4** Two antipodal signals

The probability of error in reception for the antipodal signals can be shown to be (see [2]):

$$P_e = \frac{1}{2} - \frac{1}{2}\operatorname{erf}\left(\sqrt{\mathrm{SNR}}\right). \tag{3.3}$$

Comparison of (3.3) with (3.2) reveals that $P_e$ for antipodal signal transmission is less than $P_e$ for orthogonal signal transmission at the same SNR. Therefore, antipodal signal transmission is more efficient for baseband binary communications. However, orthogonal signals are more efficient (and in fact, optimal) in non-baseband modulation schemes where random phase is introduced to the signal during the transmission process. In such schemes antipodal transmission is inappropriate because the phase scrambling makes it very difficult to reliably discriminate symbols at the receiver (see, for example, [3, Sect. 2.5]).

# 3.3 The Hilbert Transform and the Analytic Signal

## *3.3.1 The Analog and Digital Hilbert Transform*

### 3.3.1.1 The Analog Hilbert Transform

The ideal Hilbert transform filter (HT) is an *all-pass filter* which changes the phase of the input signal by −90°. The transfer function of the analog HT is given by:

$$\mathcal{H}(f) = \left\{ \begin{array}{ll} -j, & f \geq 0 \\ +j, & f < 0. \end{array} \right\} = -j\,\mathrm{sgn}(f).$$

Using *Tables* (Fourier Transform Pairs), the transfer function can be inverted to obtain the impulse response:

$$h(t) = \mathcal{F}^{-1}\{\mathcal{H}(f)\} = \frac{1}{\pi t} \qquad (3.4)$$

Figure 3.5 shows the magnitude and phase responses of an ideal analog HT.

For a Hilbert Transform filter, the input signal is often referred to as the *in phase* signal and the output is known as the *quadrature signal*. The quadrature signal is often used in communications applications and can be obtained for any arbitrary signal $x(t)$ according to the relation:

$$y(t) = HT\{x(t)\} = \hat{x}(t) = \frac{1}{\pi t} * x(t). \qquad (3.5)$$

The Fourier transform of $\hat{x}(t)$ is given by:

$$Y(f) = \hat{X}(f) = -j\,\mathrm{sgn}(f)X(f). \qquad (3.6)$$

from which it is evident that $|\hat{X}(f)| = |X(f)|$.

The Inverse HT

Since $\mathcal{H}^2(f) = [-j\,\mathrm{sgn}(f)]^2 = -1$, a succession of two HT's will be equivalent to a scalar multiplication by (−1). That is,

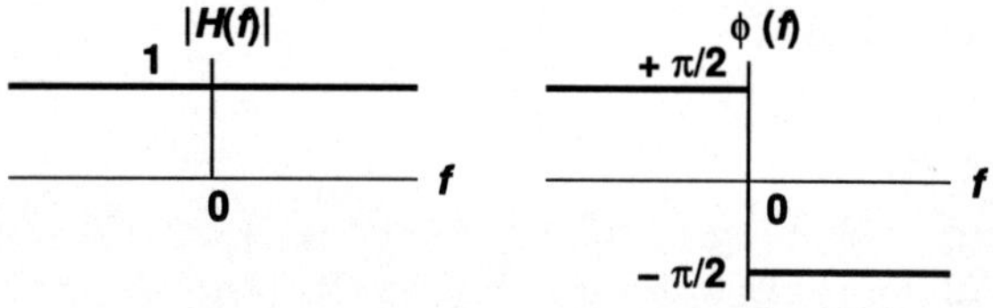


**Fig. 3.5** The magnitude and phase responses of an ideal analog HT

$$\begin{aligned} &HT[HT\{x(t)\}] = -x(t) \\ &\therefore HT\{x(t)\} = HT^{-1}\{-x(t)\} = -HT^{-1}\{x(t)\} \\ &\therefore HT^{-1}\{x(t)\} = -HT\{x(t)\}. \end{aligned} \tag{3.7}$$

In other words, the inverse Hilbert Transform is simply the negative HT.

Q : As an exercise show that the HTs of $\cos(\omega_o t), \sin(\omega_o t)$, and $\delta(t)$ are given respectively by $\sin(\omega_o t), -\cos(\omega_o t)$, and $1/(\pi t)$.

#### 3.3.1.2 The Digital Hilbert Transform

In the digital domain the HT is defined as:

$$\mathcal{H}_d(e^{j\Omega}) = \left\{ \begin{array}{ll} -j, & 0 \leq f \leq \pi \\ +j, & -\pi < f < 0. \end{array} \right\}$$

with:

$$\therefore h_d(n) = \frac{1}{2\pi} \int_{-\pi}^{\pi} \mathcal{H}_d(e^{j\Omega}) e^{jn\Omega} d\Omega = \left\{ \begin{array}{ll} \frac{2}{\pi} \frac{\sin^2(n\pi/2)}{n}, & n \neq 0 \\ 0, & n = 0. \end{array} \right\} \tag{3.8}$$

MATLAB

The HT can be implemented in the digital domain either as an *IIR filter* or as a *FIR filter*. For IIR based implementation the following instruction can be invoked: `[num den]=hilbiir(Ts, Delay, BW,TOL)`. IIR based implementation has the advantage of being lower order, and hence giving less delay, but it is vulnerable to instability. FIR based implementation results in greater delay but has no instability problems. It can be implemented using the following statements:

```
F=[.01 .99];.
g=[1 1];
h=remez(M-1,F,g,'hilbert');
num=h;den=[1 zeros(1,M-1)];
H=freqz(num,den,f,fs);
```

Alternatively one can perform an FIR based implementation by first designing the impulse response with a truncation and widowing of the ideal response $h_d(n)$. This kind of implementation can be realized in MATLAB with the following commands:

```
M=1999;n=0:M-1;
b=(M-1)/2;
hd=(2/pi)*((sin((pi/2)*(n-b)).^2)./(n-b));
hd(b+1)=0;w_han=hanning(M).';
```

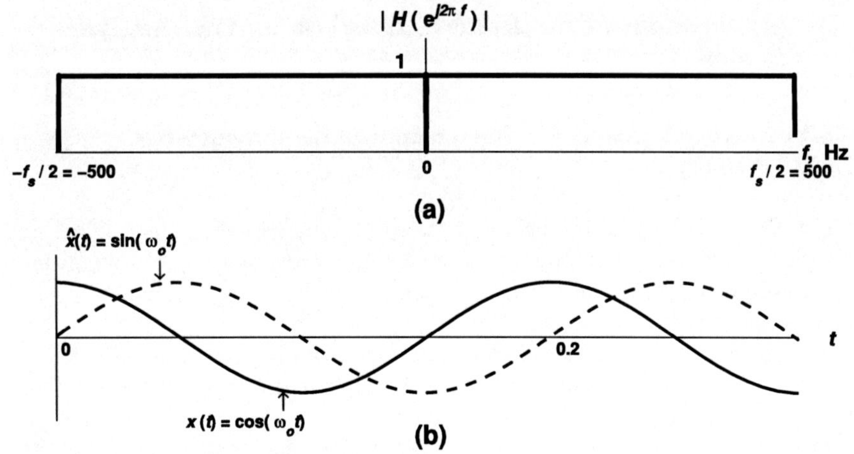


**Fig. 3.6** FIR implementation of the digital HT. **a** The HT magnitude response. **b** HT applied to the sinusoid $x(t) = \cos(\omega_o t)$, which results in a 90° phase shift. Note that the first few samples of $\hat{x}(t)$ are incorrect, hence ignored in the plot

```
h=hd.*w_han;
num=h;
den=[1 zeros(1,M-1)];
H=freqz(num,den,f,fs);
```

Figure 3.6 shows the results of this implementation and its application to a sinusoid.

### 3.3.2 The Analytic Signal

Most practical signals are real and have a positive frequency spectrum as well as a mirror image negative frequency spectrum. It is possible to synthesize a so-called *analytic* signal which has a *no* negative frequency content. The analytic signal $z(t)$ associated with a real signal $x(t)$ is defined as:

$$z(t) = x(t) + HT\{x(t)\} = x(t) + j \cdot \hat{x}(t) \tag{3.9}$$

$$\therefore Z(f) = X(f) + j[-j\text{sgn}(f)]X(f) = X(f)[1 + \text{sgn}(f)]$$

$$\therefore Z(f) = \begin{Bmatrix} 2X(f), & f \geq 0 \\ 0, & f < 0. \end{Bmatrix} \tag{3.10}$$

The analytic signal $z(t)$ associated with the original signal $z(t)$ can be generated in MATLAB using the command:

```
z=x+j*hilbert(x)
```

### 3.3.3 *Applications of the Hilbert Transform and the Analytic Signal*

#### 3.3.3.1 Spectral Economy and Computation of the Instantaneous Frequency

Since the analytic signal has only half the spectral bandwidth of the original real signal, it is spectrally economical and is therefore often used in communications applications. It is used, for example, in single sideband suppressed carrier modulation schemes (see Sect. 3.3.3.2).

The analytic signal also plays an important role in giving a meaningful general definition for the *instantaneous frequency* of a signal $x(t)$ with a continuously time-varying frequency. Assume that $z(t) = x(t) + j\hat{x}(t) = a(t)\exp[j\phi(t)]$ is the analytic signal associated with a general amplitude and frequency modulated signal $x(t)$. The instantaneous frequency (IF) of $z(t)$ is given by:

$$\omega_i(t) = \frac{d\phi(t)}{dt} \tag{3.11}$$

#### 3.3.3.2 Single Side-Band Amplitude Modulation

Natural information-bearing signals normally have relatively low frequency content. For example, the frequency content of human speech signal ranges from about 200 to 3k Hz, while audio signals in general can reach up to approximately 20k Hz. For long-range transmission of speech and audio from one antenna to another, it is often necessary to do a transformation to the radio-frequency portion of the spectrum.

Therefore, one needs to increase the frequency of the signal (without changing the information content) before radio transmission. This translation of the signal information from the low frequency region to the high frequency region is called modulation. Practically important examples of modulation are found in mobile phone systems, which translate baseband speech signals up to around 900 MHz before transmitting these signals through the air. Similarly, TVs which operate in the range 300–800 MHz region, and satellite links operate in the frequency range from a few hundred MHz to more than 40 GHz.

*Amplitude modulation (AM)* is widely used in communication systems. The AM modulator multiplies the message $x(t)$ by a high-frequency sinusoid called the carrier, $c(t) = \cos(\omega_c t)$, to get the AM signal $y(t) = x(t)\cos(\omega_c t)$. The spectrum of this signal is (*Tables*–Fourier Transform Pairs):

$$Y(f) = 0.5\,X(f - f_c) + 0.5X(f + f_c)$$

This type of modulation is called double-sideband suppressed-carrier (DSBSC) AM, since the negative part of the original spectrum appears now in the effective positive frequency part, hence consuming a place in the active bandwidth (BW). Since $x(t)$ is

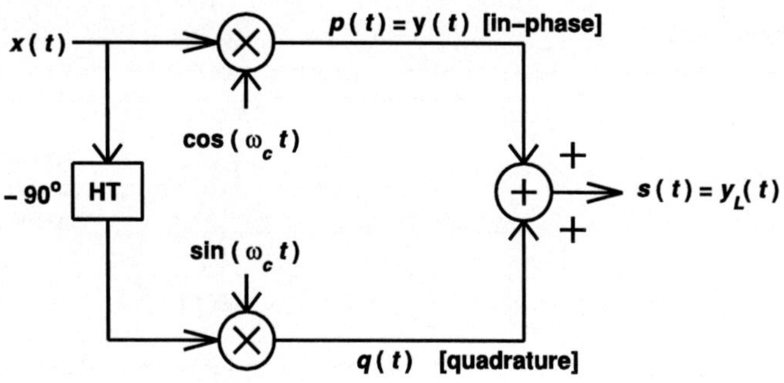


**Fig. 3.7** SSBSC AM generation

real, its magnitude spectrum is symmetric, hence the negative frequency part in the original spectrum is redundant. For spectral economy it is better to remove the redundant part of the spectrum [which is called the lower sideband (LSB), while the original positive part is called the upper sideband (USB)]. This is accomplished by using a Hilbert transformer and forming an analytic sign The resulting system is called single-sideband suppressed-carrier (SSBSC) amplitude modulation. Figure 3.7 shows a circuit diagram for implementing SSBSC AM generation.

#### 3.3.3.3 Spectrum of the SSBSC AM Signal

From Fig. 3.7 one can write that:

$$p(t) = x(t)\cos(\omega_c t) \tag{3.12}$$

and

$$q(t) = \hat{x}(t)\sin(\omega_c t). \tag{3.13}$$

The USB is given by $y_L(t) = p(t) + j * q(t)$ and the LSB is given by $y_U(t) = p(t) - jq(t)$. Figure (3.8) shows the spectra associated with SSBSC AM.

#### 3.3.3.4 Demodulation of SSBSC AM Signals

Figure 3.9 shows the SSBSC AM signal demodulator. Assuming noise-free and Doppler-free conditions, the multiplier produces the following mathematical transformations:

$$\begin{aligned} r(t) &= y_L(t)\cos(\omega_c t) \\ &= x(t)\cos^2(\omega_c t) + \hat{x}(t)\sin(\omega_c t)\cos(\omega_c t) \\ &= 0.5x(t) + 0.5x(t)\cos(2\omega_c t) + 0.5\hat{x}(t)\sin(2\omega_c t) \\ &\quad \text{Using } \textit{Tables} \text{ (Useful Formulas-3)} \end{aligned} \tag{3.14}$$

Fig. 3.8 Spectra associated with SSBSC AM generator

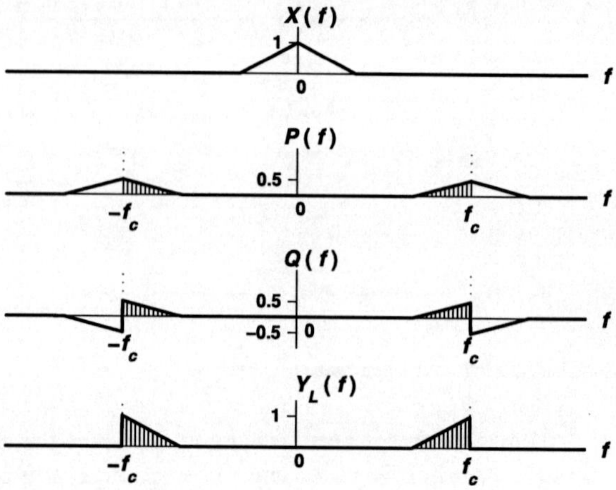


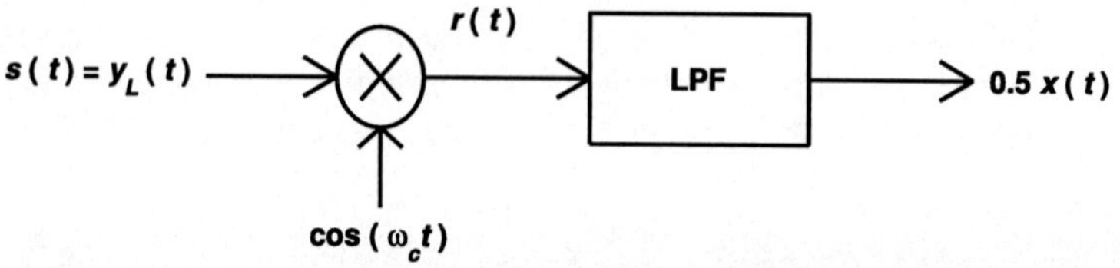


Fig. 3.9 SSBSC AM signal demodulation

The LPF in Fig. 3.9 will then filter out all the signal terms given in (3.14) except the $0.5x(t)$ term. This remaining term is simply a scaled version of the original signal.

## 3.4 Phase-Locked Loops

The phase-locked loop (PLL) represents a very important application in signal processing and communications. The initial idea for the PLL came as early as 1919 in the context of synchronization of oscillators. The theory of PLL was based initially on the theory of feedback amplifiers, and found early application in communications and motor servo systems. Due to the rapid development of integrated circuits (ICs) subsequent to the 1970s, PLLs gained a strong foothold in many areas of applications. Some of these applications include filtering, frequency synthesis, motor-speed control, frequency modulation, demodulation, signal detection, frequency tracking, etc.

The PLL is basically a device that tracks the phase and frequency of an incoming signal. It is a feedback system that adjusts the phase and frequency

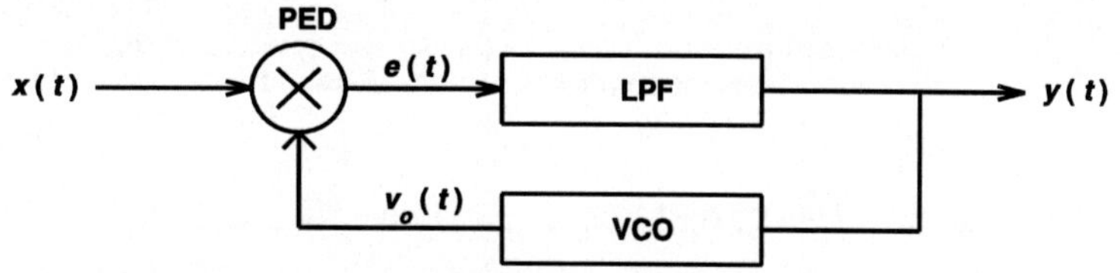


**Fig. 3.10** A block diagram of a generic analog phase-locked loop

slowly according to a feedback error signal. Because it adjusts slowly it is unable to follow perturbations due to noise—it is therefore quite resilient to the effects of noise. There are two major categories of PLLs, namely, analog PLLs (APLLs) and digital PLLs (DPLLs). Both of these are covered below.

## *3.4.1 Analog Phase-Locked Loops*

Figure 3.10 shows a block diagram of an analog phase-locked loop. A PLL is a feedback system that essentially tries to replicate the oscillation present in the input signal but shifted down by the carrier frequency, and without the random noise present in the input. It also tries to ensure that the output phase is as near as possible to the input phase. As will be seen subsequently, the PLL is very useful for demodulating frequency modulated (FM) signals.

The PLL consists of three major parts: the phase-error detector (PED), the loop filter which is a LPF which diminishes random noise, and the voltage-controlled oscillator (VCO).

Consider a frequency-modulated input signal with the following form:

$$x(t) = \cos[\phi(t)] = \cos[2\pi f_c t + \theta(t)] = \cos\left[2\pi f_c t + 2\pi\alpha \int_0^t m(t)dt\right] \tag{3.15}$$

where $m(t)$ is the message signal which must be transmitted in a communications system. The instantaneous frequency (IF) of the signal is given by:

$$f(t) = \frac{1}{2\pi}\frac{d[\phi(t)]}{dt} = f_c + \alpha m(t) \tag{3.16}$$

Hence, the frequency of the signal is varied in proportion to the message signal, i.e., $x(t)$ is a frequency modulated (FM) signal. Note that if the phase $\phi(t)$ were to vary linearly with the message, i.e., if $\theta(t) = \alpha m(t)$, then $x(t)$ would be a phase-modulated (PM) signal.

The VCO in the PLL is an oscillator whose output frequency varies linearly with its input voltage, as specified by the following equation:

$$v_o(t) = \sin[2\pi f_c t + \theta_o(t)] = \sin\left[2\pi f_c t + 2\pi\gamma \int_0^t y(t)dt\right] \tag{3.17}$$

where $\gamma$ is a constant. Hence, the VCO frequency is given by:

$$f_{\mathrm{vco}} = f_c + \gamma y(t) \tag{3.18}$$

Note that if $\theta_o(t) = \theta(t)$, then from Eqs. 3.16 and 3.18 it follows that:

$$y(t) = \frac{\alpha}{\gamma} m(t)$$

i.e., if the output phase follows the input phase the PLL will demodulate the FM signal. The following analysis will show that this is exactly what does happen—the output phase locks to the input phase and authentic FM demodulation occurs. This occurs (as will be seen in the analysis) provided that some non-restrictive conditions are met.

From Fig. 3.10, Eqs. 3.15, 3.17, and *Tables* (Useful Formulas—4), it is possible to write:

$$e(t) = x(t)v_o(t) = \frac{1}{2}\sin(4\pi f_c t + \theta_o + \theta) + \frac{1}{2}\sin(\theta_o - \theta)$$

The LPF rejects the high frequency component of $e(t)$ , giving the following filter output:

$$y(t) = \frac{1}{2}\sin(\theta_o - \theta) \tag{3.19}$$

Then from (3.18) and (3.19) one gets:

$$\frac{d\theta_o}{dt} = K\ \sin(\theta_o - \theta)$$

where $K = \pi\gamma$. For small differences between the input and output phases (i.e., for small phase errors) the following approximation can be made:

$$\frac{d\theta_o}{dt} \approx K\ (\theta_o - \theta).$$

Taking the Laplace transform of both sides yields:

$$s\Theta_o(s) = K\ (\Theta_o - \Theta). \tag{3.20}$$

From *Tables*–Laplace Transform Pairs, and Eqs. 3.16 and 3.18 one gets:

$$\frac{Y(s)}{s} = \frac{\Theta_o(s)}{2\pi\gamma} \quad \text{and} \quad \frac{M(s)}{s} = \frac{\Theta(s)}{2\pi\alpha}. \tag{3.21}$$

Equations 3.20 and 3.21 give:

$$Y(s) = \frac{-\alpha/\gamma K}{s - K} M(s) \tag{3.22}$$

For a large value of the loop gain, i.e., if $|K| \gg 1$, Eq. 3.22 can be re-written as:

$$Y(s) \approx \frac{-\alpha}{\gamma} M(s) \;\Rightarrow\; y(t) \approx \frac{-\alpha}{\gamma} m(t). \tag{3.23}$$

Equation 3.23 indicates that under the assumptions made in the previous paragraphs, the PLL produces a scaled version of the message signal at the output. Demodulation of the FM signal therefore occurs.

### *3.4.2 Digital Phase-Locked Loops*

Digital phase-locked loops (DPLLs) were introduced in the 1970s to alleviate some of the inaccuracies associated with analog systems. They are analogous in structure to analog PLLs (APLLs) (see Fig. 3.11). APLLs are still widely used, but DPLLs are attracting more attention due to the significant advantages of digital systems compared to their analog counterparts. These advantages include superior accuracy, faster locking speed, greater reliability, and reduction in size and cost. A more detailed discussion of the advantages brought by DPLLs is provided below.

1. APLLs suffer from the sensitivity of the VCO which decides the center frequency to temperature and power supply variations. They therefore need initial calibration and periodic adjustments. DPLLs do not suffer from such problems.
2. The most common error detectors used in APLLs utilize analog multipliers which are sensitive to d.c. drifts, a problem that does not exist in DPLLs.
3. DPLLs can operate at frequencies much higher than APLLs. Also, DPLLs can operate at very low frequencies, whereas the latter pose serious problems for APLLs. These problems are related to the reliable operation of the analog low-pass filter in extracting the lower frequency component. These problems also dictate that longer times are needed for effective filtering, and so the locking speed of the PLL is reduced.

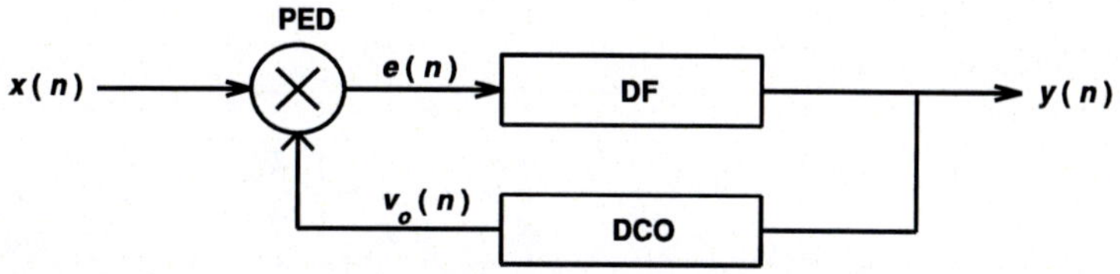


**Fig. 3.11** A block diagram of a generic digital phase-locked loop

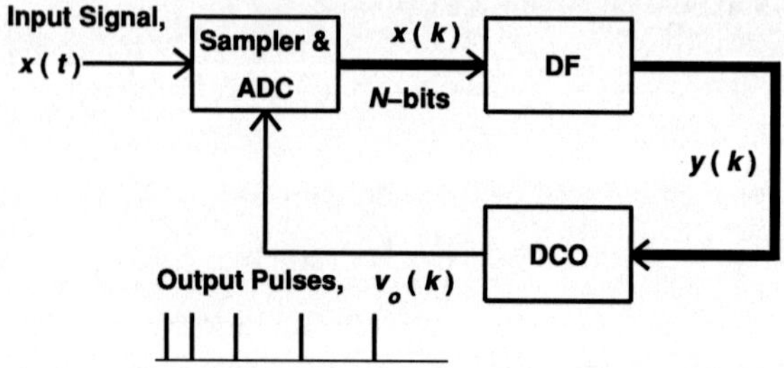


**Fig. 3.12** A block diagram of a generic sinusoidal digital phase-locked loop

There are many kinds of DPLLs. The interested reader is referred to [4, 5] for more details.

### 3.4.2.1 The Sinusoidal DPLL (SDPLL)

Among DPLLs, non-uniform sampling sinusoidal PLLs are particularly popular. Thy are simple to implement and suitable for relatively wide locking ranges. Figure 3.12 shows a block diagram of a generic SDPLL. It consists of a sampler/ADC unit which serves effectively as a PED, a digital low-pass filter (DF), and a digital voice-controlled oscillator (DCO) which produces a constant amplitude (but variable frequency) output pulses that control the sampling instants of the Sampler/ADC unit.

Different SDPLLs can be obtained by having different D-LPFs. First and second-order filters are both commonly used in practice for SDPLLs. The first-order SDPLL uses a digital filter of zero-order, i.e., it consists of just *a multiplicative constant* $G_1$. A second-order " typically uses a digital filter with the following transfer function (see also Fig. 3.13):

$$H(z) = G_1 + G_2 \frac{1}{1 - z^{-1}}.$$

where $G_1$ and $G_2$ are appropriately chosen constants. This transfer function can be implemented in parallel form as shown in Fig. 3.13. The output of this filter is given by:

$$y(k) = G_1 x(k) + G_2 \sum_{i=0}^{k} x(i)$$

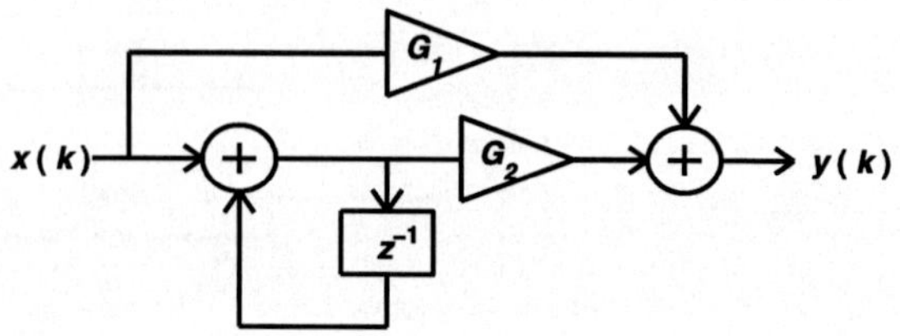


**Fig. 3.13** The digital filter of the 2nd-order SDPLL

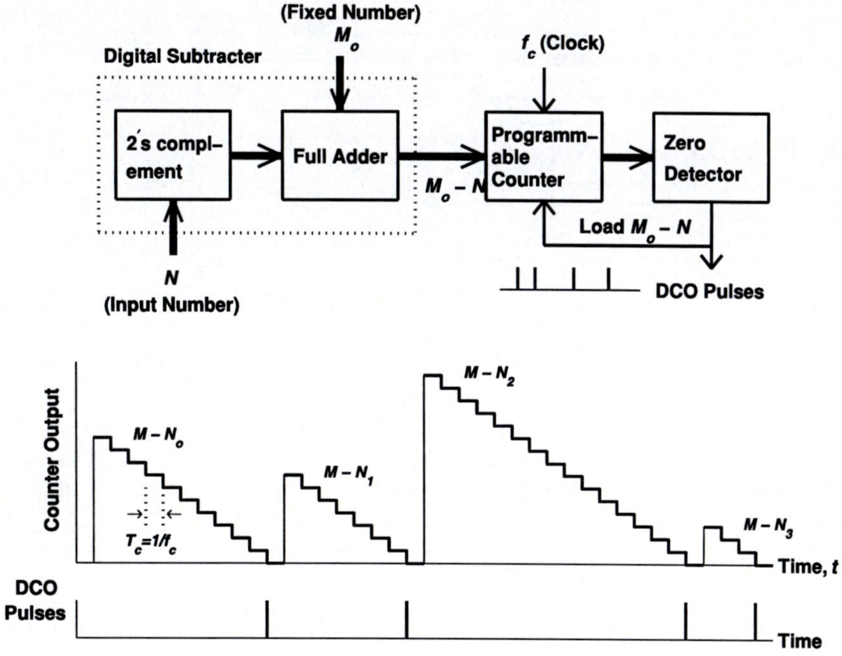


**Fig. 3.14** *Above* A block diagram of the DCO. *Below* The DCO output waveforms

The output of the DF modifies the input of the DCO to vary its phase (also its frequency) such that it *locks* onto the phase (and frequency) of the input analog signal $x(t)$. If no input is applied to the DCO, it has a *free running frequency* equal to $f_o$ Hz.

The Digital Controlled Oscillator

The Digital Controlled Oscillator consists of a programmable counter, a binary subtracter, and a zero detector as shown in Fig. 3.14. Subtraction is performed using a 2's complementer and a full adder.

When the input number is $N = 0$, the DCO period between pulses is $T_o = MT_c = M_o/f_c$, where $f_c$ is the clock frequency and $M_o$ is a constant number. Hence, the DCO frequency is $f_o = 1/T_o$, the free running frequency (or the center frequency). If $N \neq 0$, the DCO period is $T = (M_o - M)T_c = T_o - NT_c$, hence the DCO frequency is $f = 1/T$ (See Fig. 3.14). The SDPL adjusts itself such that, after a few sampling instants, the DCO frequency $f$ matches the input signal frequency.

### 3.4.2.2 Operation of the SDPLL

When the sampler takes a sample of $x(t)$ at the $k$th sampling instant, the ADC converts the analog value $x(k)$ into an $N$-bit digital value $x(k)$. Then the DF

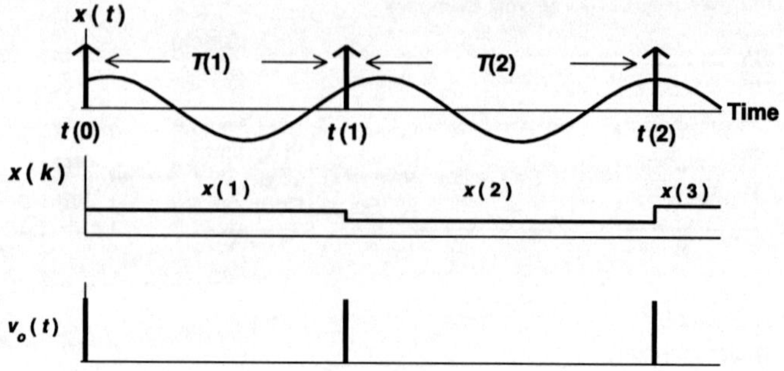


**Fig. 3.15** Waveforms associated with the SDPLL

modifies $x(k)$ to yield $y(k)$, which in turn modifies the *DCO input number*. As the DCO input is changed, so is the sampling period $T(k)$ of the ADC. Note that the sampling process is non-uniform, i.e., $f_s$ is not constant. Under certain conditions, *convergence or locking* will occur such that $T(k)$ approaches $T_m = 1/f_m$ (where $f_m$ is the carrier frequency of the input analog signal $x(t)$). Figure 3.15 shows the waveforms associated with the SDPLL.

## Analysis of the SDPLL

Assume that the input signal $x(t)$ is a sinusoidal signal with the form:

$$x(t) = A\ \sin(\omega t + \theta_o) + n(t),$$

where $A$ is the signal amplitude, $\omega = 2\pi f$ is the signal instantaneous frequency, $\theta_o$ is a constant phase, and $n(t)$ is additive noise. The *locking range* of the loop is the range of instantaneous frequencies that the loop can track. Since the locking range is dependent on the deviation of $\omega$ from the loop center frequency $\omega_o$, it is more convenient to write the above equation in the following form:

$$x(t) = A\ \sin[\omega_o t + \theta(t)] + n(t), \tag{3.24}$$

where $\theta(t)$ is the information-bearing phase given by:

$$\theta(t) = (\omega - \omega_o)t + \theta_o = \Delta\omega t + \theta_o.$$

After sampling, the input signal at the $k$th sampling instant $t(k)$ the input signal will have the following form:

$$x(k) = A\sin[\omega_o t(k) + \theta(k)] + n(k), \tag{3.25}$$

where the input phase at the $k$th sampling instant is defined as follows:

$$\phi(k) = \omega_o t(k) + \theta(k). \tag{3.26}$$

The sampling interval of the DCO at the $k$th sampling instant is given by:

$$T(k) = T_o - y(k-1) \tag{3.27}$$

(see Figs. 3.12 and 3.14).

Note that the $k$th output sample of the DF, $y(k)$, effectively determines the sampling period, $T(k+1)$. The $k$th sampling instant $t(k)$ is given by the cumulative sum of all sampling periods:

$$t(k) = t(0) + \sum_{0}^{k} T(i).$$

For simplicity it is assumed that the initial time instant is zero, i.e., $t(0) = 0$. Hence, the $k$th sampling instant is:

$$t(k) = kT_o - \sum_{0}^{k-1} y(i) \tag{3.28}$$

From Eqs. 3.26 and 3.28, the input signal phase at the $k$th sampling instant can be written as follows:

$$\phi(k) = \theta(k) - \omega_o \sum_{0}^{k-1} y(i). \tag{3.29}$$

Hence, at the $(k+1)$th sampling instant, the input signal phase is given by:

$$\phi(k+1) = \theta(k+1) - \omega_o \sum_{0}^{k} y(i). \tag{3.30}$$

The phase equations (3.29) and (3.30) will determine the system difference equation for the SDPLL of any order.

#### 3.4.2.3 The First-Order Noise-Free SDPLL

The first-order SDPLL is widely used, although it gives a non-zero steady-state phase error. Its digital filter transfer function is $H(z) = G_1$ (constant), hence its output is given by:

$$y(k) = G_1 x(k) = G_1 A \sin[\phi(k)] \tag{3.31}$$

From Eqs. 3.29 and 3.30 the following phase difference equation is obtained:

$$\phi(k+1) - \phi(k) = \theta(k+1) - \theta(k) - \omega_o y(k). \tag{3.32}$$

Using Eqs. 3.28 and 3.31 gives the following difference equation:

$$\phi(k+1) = \phi(k) - K_2 \sin[\phi(k)] + \Lambda_o \tag{3.33}$$

where $\Lambda_o = 2\pi(\omega - \omega_o)/\omega_o$ and $K_2 = \omega\, G_1 A$. If $K_1$ is defined to be $K_1 = \omega_o G_1 A$, and $W$ is defined as the frequency ratio $W = \omega_o/\omega$, then $K_2 = K_1\ (\omega/\omega_o) = K_1/W$. The parameters $W$ and $K_1$ control the system operation range as discussed below.

## Locking Conditions

If locking occurs, then $\phi(k+1) = \phi(k) = \phi_{ss}$ ($\phi_{ss}$ being the steady-state phase error). Using Eq. 3.33 it follows that:

$$\Lambda_o = K_2 \sin(\phi_{ss}) \Rightarrow \phi_{ss} = \sin^{-1}(\Lambda_o/K_2) \tag{3.34}$$

$$\therefore |\Lambda_o/K_2| < 1 \tag{3.35}$$

Equation 3.35 leads to the following condition (*prove as an exercise!*):

$$K_1 > 2\pi|1 - W| \tag{3.36}$$

Now the theory of *Fixed Point Analysis* states that the equation $g(\psi) = \psi$ has a solution $\psi^*$ only if the following condition is satisfied (see [6]):

$$|g'(\psi^*)| < 1 \tag{3.37}$$

Hence, Eq. 3.33 has a solution as in Eq. 3.31 only if:

$$|1 - K_2 \cos(\phi_{ss})| < 1$$

$$\therefore \left|1 - \sqrt{K_2^2 - \Lambda_o^2}\right| < 1$$

$$\therefore K_2^2 - \Lambda_o^2 < 4$$

which gives the following condition:

$$K_1 < \sqrt{(4 + 4\pi^2)W^2 - 8\pi^2 W + 4\pi^2} \tag{3.38}$$

Equations 3.35 and 3.38 specify the frequency locking region of the 1st-order SDPLL as illustrated in Fig. 3.16. Hence, if the frequency of the incoming signal is such that $\omega/\omega_o = 1/W \in \mathbf{R}$, where **R** is the *region of locking* in the ($W$,$K_1$) plane, then the SDPLL is capable of tracking this frequency. It is then also capable of eventually locking on the input phase. Note that as $K_1$ approaches 2, the locking range becomes wider. Practically, locking occurs when $|\phi(k+1) - \phi(k)| < \epsilon$, where $\varepsilon$ is a small positive number.

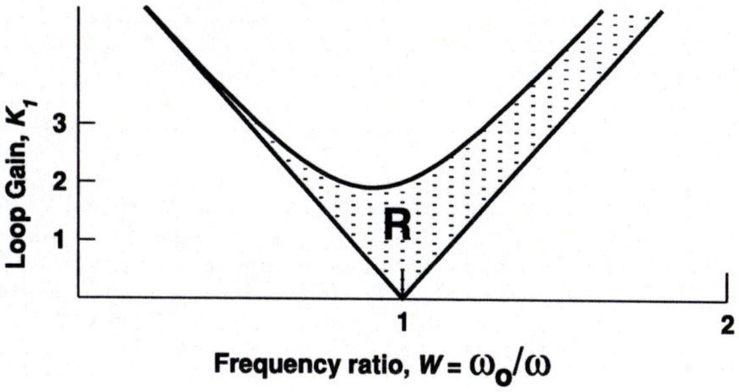


**Fig. 3.16** Locking region of the 1st-order SDPLL

## Phase Plane Diagram

Phase plane diagrams represent the phase at the $(k+1)$th sampling instant, $\phi(k+1)$, as a function of the phase at the $k$th sampling instant, $\phi(k)$. Such diagrams are useful for studying the convergence behavior of the SDPLL.

For the first-order SDPLL, the system equation is given by (see Eq. 3.33):

$$\phi(k+1) = \phi(k) - K_2 \sin[\phi(k)] + \Lambda_o.$$

If one puts $x = \phi(k)$ and $y = \phi(k+1)$, one gets the following simple equation:

$$y = g(x) = x - K_2 \sin(x) + \Lambda_o.$$

If one assigns a value for $\phi(0)$, then successive phase errors $\{\phi(k) | k = 1 \rightarrow \infty\}$ can be found and plotted (modulo $2\pi$) by successive projections on the curves $y = x$ and $y = g(x)$. This is shown in Fig. 3.17 for $\theta_o = 2.5$ rad, $t_o = 0$, $K_1 = 1.7$, center frequency $f_o = 1$ Hz, and input frequency $f_i = 0.83$ Hz (hence $W = 1.2$). The locking point is a point of *intersection* between the two curves. Note that the values of $W$ and $K_1$ are inside the locking region **R**. Figure 3.17 also shows the sampling process with the phase error process for the same parameters. Figure 3.18 depicts the frequency tracking process under the above circumstances.

## SDPLL in Noise

Figure 3.19 shows the frequency tracking probability density function for the same circuit parameters as above in the presence of *AWGN noise*. It is seen that the *pdf* ($f$) has a maximum at approximately $f = f_i$. Hence, the SDPLL can detect the input frequency in the presence of noise, even for low SNRs. Figure 3.20 shows the variance of this frequency estimate as a function of the SNR. As expected, it decreases when *SNR* increases.

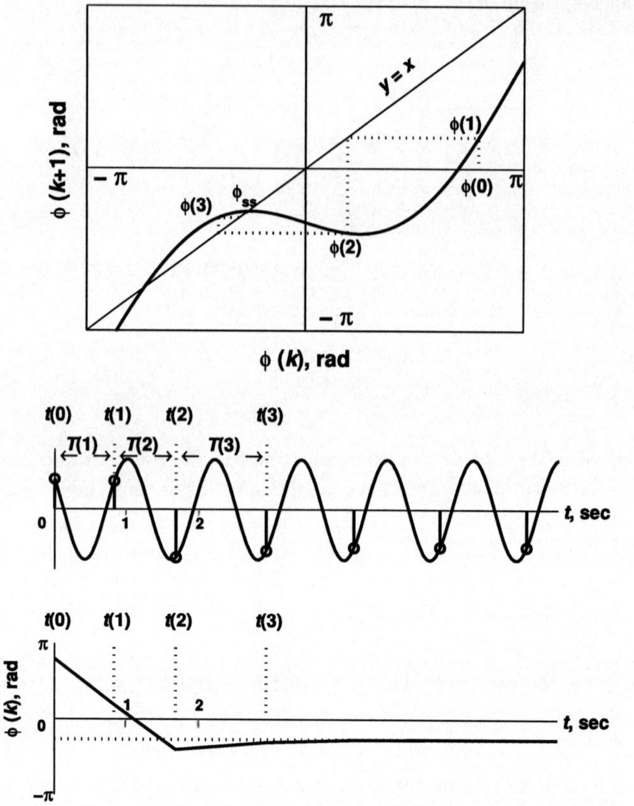


**Fig. 3.17** Locking process of the 1st-order SDPLL for $\theta_o = 2.5$ rad, $t_o = 0$, $K_1 = 1.7$, $f_o = 1$ Hz, and $f_i = 0.83$ Hz (hence $W = 1.2$). *Above* phase plane diagram. *Middle* sampling process. *Below* phase error process

**Fig. 3.18** Frequency tracking process using the 1st-order SDPLL with $\theta_o = 2.5$ rad, $t_o = 0$, $K_1 = 1.7, f_o = 1$ Hz, and $f_i = 0.83$ Hz (hence $W = 1.2$)

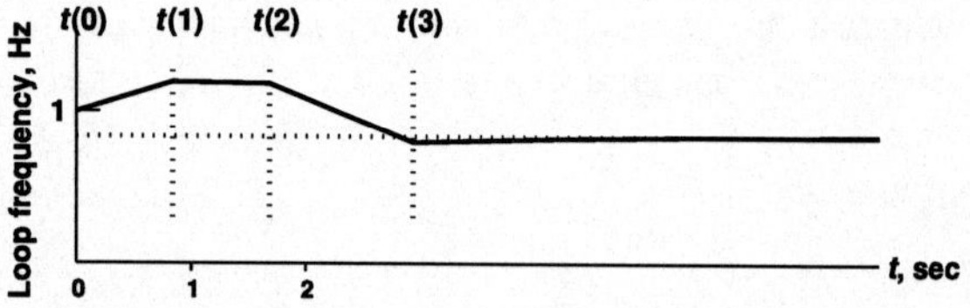


### 3.4.2.4 The Second-Order Noise-Free SDPLL

The first-order SDPLL locks on a *non-zero* steady-state phase error $\phi_{ss} = \sin^{-1}(\Lambda_o/K_2)$. This can be an *unwanted characteristic* when the PLL is used for synchronization. In such a case the second-order loop may be necessary as it locks on zero phase error. The *second-order SDPLL* utilizes a proportional and an accumulation path digital filter:

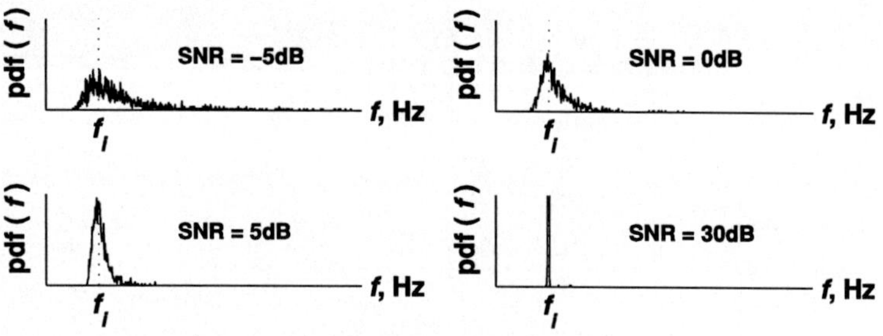


**Fig. 3.19** Frequency tracking pdf using the 1st-order SDPLL with $\theta_o = 2.5$ rad, $t_o = 0$, $K_1 = 1.7, f_o = 1$ Hz, and $f_i = 0.83$ Hz (hence $W = 1.2$) in AWGN for different SNRs

**Fig. 3.20** Variance of the 1st-order SDPLL estimation (in AWGN) of the input frequency $f_i$ as a function of the SNR for $\theta_o = 2.5$ rad, $t_o = 0$, $K_1 = 1.7$, $f_o = 1$ Hz, and $f_i = 0.83$ Hz (hence $W = 1.2$)

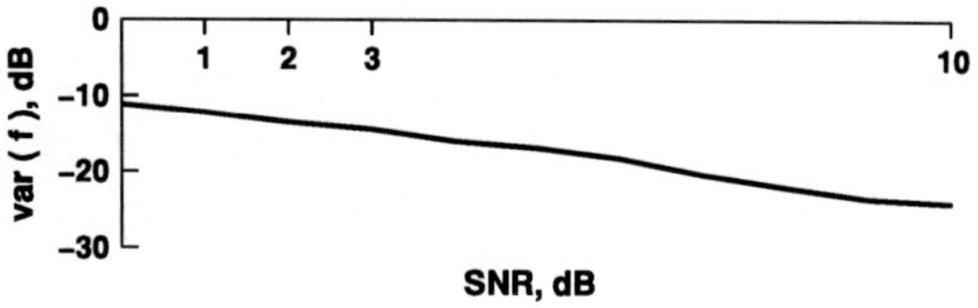


$$y(k) = G_1 x(k) + G_2 \sum_{i=0}^{k} x(i).$$

Following the same steps as for the first-order loop, it can be shown that:

$$\phi(k+2) - 2\phi(k+1) + \phi(k) = K_2 \sin[\phi(k)] - rK_2 \sin[\phi(k+1)] \tag{3.39}$$

where $r = 1 + \frac{G_2}{G_1}$. Recall that $K_2 = \omega G_1 A$, $K_1 = \omega_0 G_1 A$, $W = \omega_o/\omega$, $K_2 = K_1(\omega/\omega_o) = K_1/W$.

At locking it is true that $\phi(k+2) = \phi(k+1) = \phi(k) = \phi_{ss}$, hence the above equation becomes:

$$0 = K_2(1 - r) \sin[\phi_{ss}] \tag{3.40}$$

which implies that the locking phase $\phi_{ss}$ is zero.

## Lock Range

To apply *fixed-point analysis* as for the first order SDPLL, one must have an expression of the form $g(x) = x$. Since the 2nd-order SDPLL is characterized by a 2nd-order equation, one will need to have a vector expression: $g(x) = x$. To obtain this type of expression, one can write:

$$x_k = \begin{bmatrix} \phi(k) \\ \phi(k+1) \end{bmatrix} = \begin{bmatrix} u_k \\ v_k \end{bmatrix}$$

Then, using Eq. 3.40 one has:

$$\begin{aligned} x_k + 1 &= \begin{bmatrix} u_{k+1} \\ v_{k+1} \end{bmatrix} = \begin{bmatrix} \phi(k+1) \\ \phi(k+2) \end{bmatrix} \\ &= \begin{bmatrix} v_k \\ 2v_k - u_k + K_2 \sin(u_k) - rK_2 \sin(v_k) \end{bmatrix} = \begin{bmatrix} f(x_k) \\ h(x_k) \end{bmatrix} = g(x_k) \end{aligned}$$

The problem is now expressed in a form which can be analyzed with *fixed point analysis*:

$g(x) = x$ with $x = \begin{bmatrix} u \\ v \end{bmatrix}$ and $g(x) = \begin{bmatrix} f(x) \\ h(x) \end{bmatrix}$; where $f(x) = v$ and $h(x) = 2v - u + K_2 \sin(u) - rK_2 \sin(v)$.

As there is a function of two variables, one should use the *Jacobian* (instead of the derivative):

$$g'(x) = \begin{bmatrix} \partial f/\partial u & \partial f/\partial v \\ \partial h/\partial u & \partial h/\partial v \end{bmatrix} = \begin{bmatrix} 0 & 1 \\ -1 + K_2 \cos(u) & 2 - rK_2 \cos(v) \end{bmatrix}$$

The fixed point is $x^* = \begin{bmatrix} u^* \\ v^* \end{bmatrix} = \begin{bmatrix} 0 \\ 0 \end{bmatrix}$ (mod $2\pi$), at which:

$$g'(x^*) = \begin{bmatrix} 0 & 1 \\ -1 + K_2 & 2 - rK_2 \end{bmatrix}.$$

For this system to have a fixed point as above, one should have:

$$|\lambda_i| < 1 \quad \forall i$$

where $\lambda_i$ are the eigenvalues of $L = g'(x^*)$, i.e., roots of $|\lambda I - L| = 0$. Hence we get:

$$r > 1$$

and

$$0 < K_2 < \frac{4}{r+1}$$

which define the *locking range* of SDPLL2.

### 3.4.2.5 PM Demodulation Using the SDPLL

Using Eq. 3.29 for the 1st-order SDPLL yields:

$$\phi(k) = \theta(k) - \omega_0 \sum_{i=0}^{k-1} y(i) = \theta(k) - K_1 \sum_{i=0}^{k-1} \sin[\phi(i)] \tag{3.41}$$

If one puts $K_1 = 1$ and assumes that $\phi(i)$ is small near locking, then it follows that:

$$\theta(k) \approx \sum_{i=0}^{k} \phi(i) \tag{3.42}$$

Equation 3.42 suggests that the information-bearing phase $\theta$ $(t)$ can be *recovered* by summing the phase errors.

If $x(t)$ is a PM signal, then it has the following form:

$$x(t) = A \sin[\phi(t)] = A \sin[\omega_o t + \underbrace{\alpha \cdot m(t)}_{\theta(t)}]$$

where $m(t)$ is the message and $\alpha$ is a constant (the modulation index). It is clear that $\phi(t)$ is varying linearly with $m(t)$, and that $\theta(t) = \alpha{\cdot}m(t)$ is the information-bearing phase previously considered in the study of the SDPLL.

Now consider a single-tone message $m(t)$, i.e.,

$$m(t) = a \sin(\omega_m t).$$

In this case:

$$x(t) = \sin[\omega_o t + \underbrace{\beta \sin(\omega_m t)}_{\theta(t)}] \text{ (where } \beta = \alpha \cdot a).$$

The instantaneous frequency (IF) of any sinusoidal PM signal is given by:

$$\omega(t) = \frac{d\theta}{dt}.$$

Hence, for the signal $x(t)$ above:

$$\omega(t) = \omega_0 + \beta \cdot \omega_m \cos(\omega_m t).$$

The maximum and minimum values of this frequency are given by:

$$\omega_{\max} = \omega_o + \beta \cdot \omega_m \quad \text{and} \quad \omega_{\max} = \omega_o - \beta \cdot \omega_m.$$

From Eq. 3.36 using $K_1 = 1$ with a first-order loop:

$$\frac{2\pi}{2\pi + 1} < \frac{\omega}{\omega_0} < \frac{2\pi}{2\pi - 1} \Rightarrow \frac{2\pi}{2\pi + 1} < \frac{\omega_0 \pm \beta\omega_m}{\omega_0} < \frac{2\pi}{2\pi - 1}$$

$$\therefore 0 < \beta \frac{\omega_m}{\omega_0} < \frac{1}{2\pi + 1} \approx 0.13 \tag{3.43}$$

Hence, $\beta\, \omega_m$ should be specified by the above range, otherwise the SDPLL will not be able to demodulate the PM signal. Equation 3.38 is not effective here as its range exceeds $K_1 = 1$ (see Fig. 3.16).

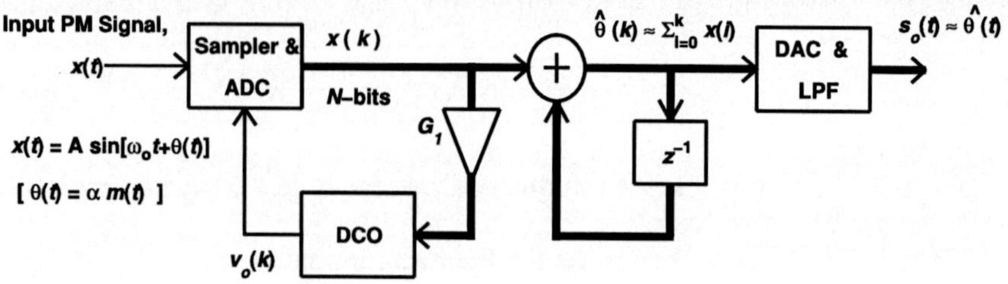


Fig. 3.21 Circuit diagram for PM demodulation using SDPLL1

Figure 3.21 shows the circuit diagram for PM demodulation using 1st-order SDPLL.

## 3.5 Linear Estimation and Adaptive Filtering

A transmitted signal $x(n)$ is normally corrupted by random noise during the transmission process so that the received signal $y(n)$ may be relatively difficult to detect. To enhance detectability, it is advisable to use the best possible means to reduce the effects of the noise. If one knows the sent signal exactly, one can use matched filtering. In many situations, however, only partial (as opposed to exact) information about the signal is available. The general process of forming an approximation to the noise free signal based on partial information is called *estimation*. This section deals with the issue of estimation, and in particular estimation based on the use of *linear* filters. Throughout the section the estimated signal will be denoted by $\hat{x}(n)$.

There are various different forms of partial information which may be available. One may, for example, know the spectral region occupied by the sent signal, and the spectral region occupied by the undesirable noise. Often the noise will occupy a wider bandwidth than the signal and so one can simply use a BPF to remove all noise outside the band of the sent signal—the filtered signal could then be considered to be the estimated signal. This is a very simplistic approach to estimation, and one that may well not be optimal. During the 1940s Norbert Wiener studied the problem of how to optimally estimate a sent signal, given a knowledge of the sent signal's spectrum and the received signal's spectrum. He found that the optimal estimate is given by the output of the filter whose transfer function is:

$$H_o(s) = \frac{S_{yx}(s)}{S_{xx}(s)}\, e^{\alpha s} \tag{3.44}$$

where $\alpha$ is the delay of the *Wiener filter* $H_o(s)$, $S_{xx}(s)$ is the PSD of the original signal, $S_{yx}(s)$ is the cross spectral density (CSD) between the original and observed signals.

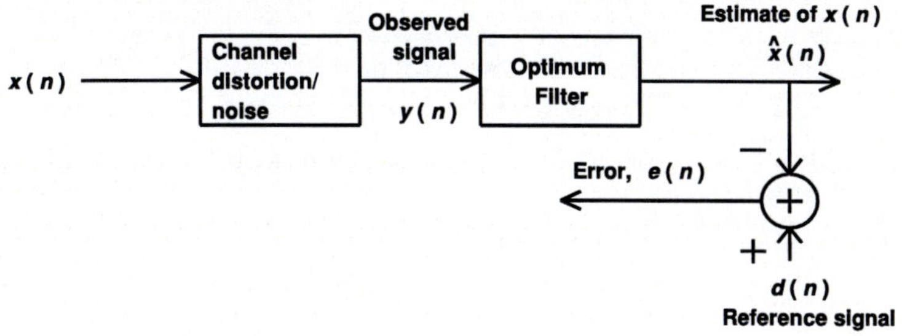


**Fig. 3.22** Generic linear estimation model

The so-called Wiener filtering formulation in (3.44) is a frequency domain solution, and is difficult to implement in practice because it requires a knowledge of both the sent signal's spectrum and the cross-spectrum between the sent and received signals. A time-domain implementation of Wiener filtering turns out to be much more amenable to practical implementation. This time-domain solution is discussed in the following paragraphs.

Because of the Wiener–Kinchin theorem, the spectral formulation given in (3.44) can be re-expressed with auto- and cross-correlation functions rather than spectra. This re-formulation is based on the minimization of the mean-squared error between the observed signal $y(n)$ and a reference signal $d(n)$, which is ideally equal to the original signal $x(n)$. In practice it often differs from $x(t)$ by some error signal. That is, the time domain approach still needs *prior knowledge* of the autocorrelation function of the observed signal autocorrelation $R_y$ and the observed/desired signal's cross-correlation, $R_{y_d}$. The time-domain FIR Wiener filtering solution is given by:

$$\mathbf{h}_{\text{opt}} = \left[\boldsymbol{R}_{yy}^{-1}\boldsymbol{R}_{yd}\right]^T \tag{3.45}$$

In itself, the solution presented in (3.45) has limited usefulness because of the unavailability of the autocorrelation functions needed a priori to implement (3.45). By using adaptive implementations however, it is possible to adaptively estimate the initially unknown quantities and so arrive at a solution which is very close to that provided in (3.45). In what follows, both non-adaptive and adaptive implementations of Wiener filter will be considered.

Figure 3.22 shows a general model for linear estimation.

### *3.5.1 Non-adaptive FIR LMS Filter*

There are two kinds of filters for implementing the Wiener solution. These are FIR nd IIR versions of the so-called least mean-square (LMS) filters. This book will concentrate on the FIR LMS filter, largely because of the advantage they have with respect to stability. If the filter length is $M + 1$, then from Fig. 3.22 it follows that:

$$\hat{x}(n) = \sum_{k=0}^{M} h(k)y(n-k) \tag{3.46}$$

Often it is convenient to write the above equation in matrix form as follows:

$$\hat{x}(n) = \mathbf{h}\mathbf{y}^T(n) = \mathbf{y}(n)\mathbf{h}^T \tag{3.47}$$

where

$$\mathbf{h} = [h(0)\ h(1)\ h(2),\ldots,h(M)],$$

which is the impulse response vector, and

$$\mathbf{y}(n) = [y(n)\ y(n-1)\ y(n-2),\ldots,y(n-M)],$$

which is the observed signal vector at the time instant $n$. The error is given by:

$$e(n) = d(n) - \hat{x}(n) \tag{3.48}$$

and the mean-squared error (MSE) is given by:

$$e_{\text{mse}} = \mathcal{E}\{[e(n)]^2\} = \mathcal{E}\{[d(n) - \hat{x}(n)]^2\} = \mathcal{E}\{[d(n) - \sum_{k=0}^{M} h_k y(n-k)]^2\} \tag{3.49}$$

To minimize $e_{\text{mse}}$ it is necessary that:

$$\frac{\partial e_{\text{mse}}}{\partial h_j} = 0 \quad \forall j \tag{3.50}$$

where for simplicity, the designation $h_j = h(j)$ is made. From Eqs. 3.47, 3.49, and 3.50 it is possible to write:

$$\begin{aligned}\frac{\partial e_{\text{mse}}}{\partial h_j} &= \mathcal{E}\left\{2e(n).\frac{\partial e(n)}{\partial h_j}\right\} \\ &= -2\mathcal{E}\{e(n)y(n-j)\} \\ &= -2\mathcal{E}\{y(n-j)[d(n) - \mathbf{y}(n)\mathbf{h}^T]\} = 0; \quad j = 0,1,\ldots,M\end{aligned} \tag{3.51}$$

Now Eq. 3.51 can be written in matrix form as follows:

$$\begin{aligned}-2\mathcal{E}\{\mathbf{y}^T(n)[d(n) - \mathbf{y}(n)\mathbf{h}^T]\} &= 0 \\ \therefore \mathcal{E}\{\mathbf{y}^T(n)d(n)\} - \mathcal{E}\{\mathbf{y}^T(n)\mathbf{y}(n)\mathbf{h}^T]\} &= 0 \\ \therefore \boldsymbol{R}_{yd} &= \boldsymbol{R}_{yy}\mathbf{h}^T\end{aligned}$$

where $\boldsymbol{R}_{yy}$ is the observed signal autocorrelation, while $\boldsymbol{R}_{yd}$ is the observed signal cross-correlated with the desired signal. Hence, the optimal filter coefficients (in the MSE sense) are given by:

$$\mathbf{h}_{\text{opt}} = \left[ \boldsymbol{R}_{yy}^{-1} \boldsymbol{R}_{yd} \right]^T \tag{3.52}$$

which is known as the *Wiener–Hopf Equation.*

### 3.5.2 Adaptive Filters

An adaptive filter is a *programmable filter* whose coefficients (i.e., impulse response values, $\{h(k)\}$) are *changed or adapted.* Normally one seeks to adapt the coefficients so as to give an optimal estimate $\hat{x}(n)$ of the original signal $\{x(n)\}$ at the time instant $n$, and an adaptive feedback algorithm is used to update the filter coefficients. The measure of optimality is typically based on the information available for the given application. Illustrative examples will be presented later in this Sect. 3.5.7.

Because of their ability to adapt the coefficients, adaptive filters cannot only be used for "homing in" on the optimal Wiener solution for time-invariant scenarios, they are also useful for applications in which conditions are continuously varying. They can, for example, be used to compensate for time-varying channel conditions. Figure 3.23 shows a generic adaptive filter structure.

### 3.5.3 Choice of the Desired Signal

In communications systems, a "training sequence" is sent before transmission of data. The receiver knows this signal and utilizes a copy of it as the desired signal $d(n)$. As such the adaptive filter can adapt coefficients to the point of near optimality during the short period in which the training sequence is transmitted.

In noise cancelers, $d(n)$ can be selected to be the observed data $y(n)$, or a delayed version thereof. This signal differs from the true signal $x(n)$ in that it contains additive random noise. In such a scenario, the adaptive filter is usually

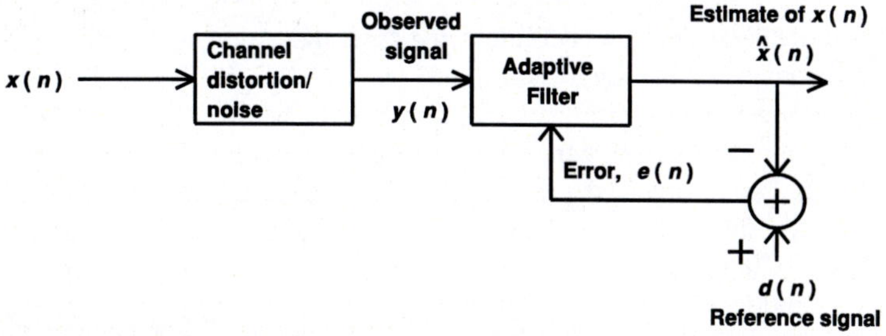


**Fig. 3.23** Generic adaptive filter configuration

specified to adapt slowly, so that it cannot respond very effectively to the rapid changes inherently present in the noise component of the signal, only to the relatively slowly changing components of the true desired signal. For this reason the choice of $y(n)$ as a reference signal often serves as a very reasonable replacement for the ideal signal $x(n)$.

### 3.5.4 The Adaptive LMS Algorithm

The adaptive LMS algorithm is probably the most popular adaptive filtering technique is use today. In this algorithm the filter coefficients are adapted according to the 6-line procedure given in the following paragraph. This algorithm was developed by *Widrow & Hoff*, and can be shown to realize the optimal *Wiener* filtering solution which minimizes the MSE: $e_{\text{mse}} = \mathcal{E}\{[e(n)]^2\} = \mathcal{E}\{[d(n) - \hat{x}(n)]^2\}$ at every time instant $n$ (see [7]).

*Definition of the Adaptive LMS Algorithm*:

$\mathbf{h}(k) = [h_o(k)\ \ h_1(k)\ \ h_2(k), \ldots, h_M(k)]$, the filter coefficients at the $k$th instant.

$\mathbf{y}(k) = [y(k)\ \ y(k-1)\ \ y(k-2), \ldots, y(k-M)]$, observed signal vector at $k$th instant.

The algorithm can be described in vector form (*MATLAB-like code*) as follows:

---

$\mathbf{h}(0) = \mathbf{0}$; % Initialize the filter coefficients.

for n = 1: N % N = length(y);

  $\hat{x}(n) = \mathbf{h}(n-1)\mathbf{y}^T(n)$; % Filter output (this is matrix multiplication).

  $e(n) = d(n) - \hat{x}(n)$;

  $\mathbf{h}(n) = \mathbf{h}(n-1) + \mu * e(n)y(n)$; % $\mu$ is the step-size.

end

---

### 3.5.5 Choice of Adaptation (Convergence) Coefficient and Filter Length

The choice of the adaptation coefficient $\mu$ affects the estimation accuracy and the convergence speed of the algorithm. Small values of $\mu$ give better final accuracy but slower convergence. Large values do the contrary. Very small or very large values for $\mu$ can cause significant errors, and hence, a compromise between these two extremes is desirable. Because quick convergence is achieved by making $\mu$ large, and good final accuracy is attained by making $\mu$ small, many authors have proposed that $\mu$ itself be adapted as the filter runs. At the start of the algorithm $\mu$ is made large, and after (approximate) convergence is reached, $\mu$ is made small. For the sake of simplicity, however, this book will assume that $\mu$ is invariant once the algorithm commences.

As well as having to select $\mu$, one needs to select an appropriate filter length. A larger filter length $M + 1$ is likely to give better estimation, but it also introduces more delay into the output. Again, a compromise is desirable.

### 3.5.6 Hardware Implementation of Adaptive FIR Filters

The adaptive LMS algorithm can be implemented in either digital hardware or software. A digital hardware implementation is depicted in Fig. 3.24. Note that adaptive filters really only became practical with the advent of digital signal processing—the intelligence required to implement these algorithms is generally problematical for analog hardware.

### 3.5.7 An Example of LMS Filtering

*Example* Consider a **2-tap** FIR adaptive LMS filter. If $y(4) = 0.25$, $y(5) = 0.5$, $d(4) = 1.03$, $d(5) = -0.27$, and $\mathbf{h}(4) = [1.2 \ \ 3.7]$, use the LMS algorithm with $\mu = 0.02$ to update the impulse response of the filter.

Solution: Applying the LMS filter algorithm to the above data specifications gives:

$$\begin{aligned}
\mathbf{h}(5) &= \mathbf{h}(5) + \mu[0.5 \ \ 0.25]e(5) \\
e(5) &= d(5) - \hat{x}(5) = d(5) - \mathbf{h}(4)\mathbf{y}^T(5) \\
&= -0.27 - [1.2 \ \ 3.7]\begin{bmatrix} 0.5 \\ 0.25 \end{bmatrix} = -1.79 \\
\mathbf{h}(5) &= \mathbf{h}(4) + 0.02[0.5 \ \ 0.25](-1.79) \\
&= [1.2 \ \ 3.7] - [0.0179 \ \ 0.0089] = [1.18 \ 3.69].
\end{aligned}$$

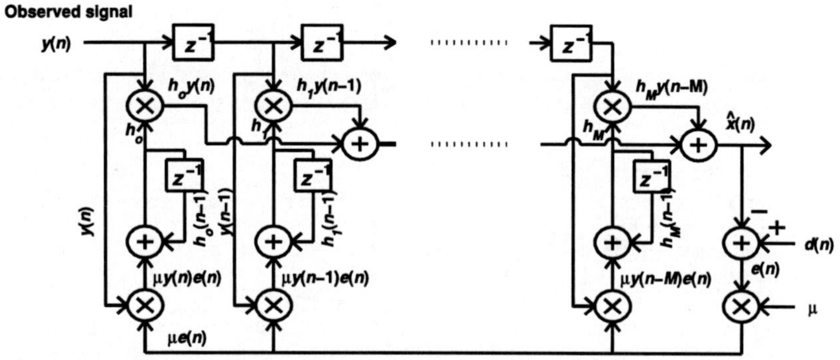


**Fig. 3.24** Hardware implementation of the adaptive LMS filter

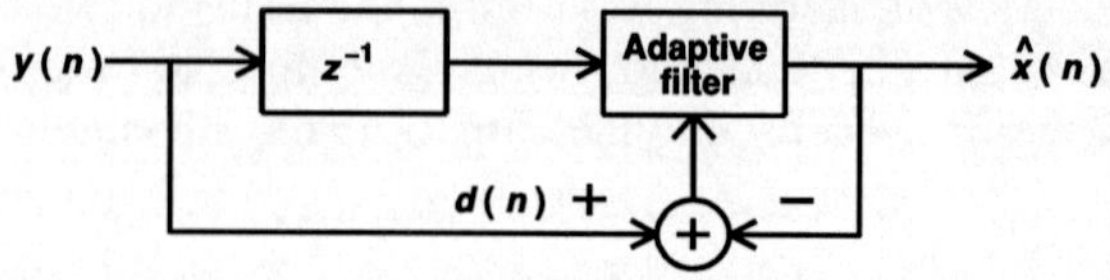


**Fig. 3.25** Noise reduction using an adaptive filter

## 3.5.8 *Application of Adaptive Filtering to Noise Reduction in Narrow-Band Signals*

For noise removal in narrowband signals of unknown frequency one can use adaptive LMS filters. In such applications one can set the desired signal $d(n)$ to be equal to the observed signal ($d(n) = y(n)$), as shown in Fig. 3.25. Note that with this arrangement: i) the filter operates only on previous samples of the signal (not the current one) and ii) the filter coefficients are adapted until the current output from the filter matches the current observed sample as closely as possible. Now ideally the desired signal should not be derived from the (noisy) observed signal but from a noise free version of the observed signal. However, if the filter coefficients are forced to adapt slowly, the filter will be unable to adapt to the rapid fluctuations of the noise component of the observation—they will adapt mostly to the noise-free version of the observed signal. The adaptive filter therefore functions reasonably effectively. The arrangement is illustrated in Fig. 3.26 which shows adaptive noise reduction for a sinusoid with *unknown frequency*.

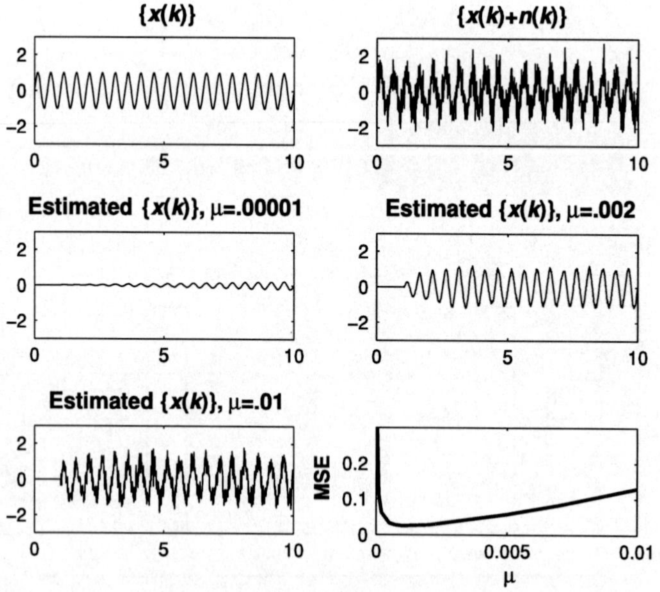


**Fig. 3.26** Noise reduction for a sinusoid of unknown frequency using a 100-taps adaptive FIR filter

## 3.5.9 *Application of Adaptive Filtering to Channel Equalization*

### 3.5.9.1 Reducing Intersymbol Interference

A communication channel can introduce noise and other forms of distortion which adversely affect the transmitted data. One kind of distortion which is particularly important in communication systems is so-called *intersymbol interference.*

Consider the scenario where it is desired to transmit a square pulse across a communications channel. This pulse has a *sinc function* spectrum, which has an *infinite* bandwidth. The square pulse is therefore compact in time, but infinite in frequency extent. If the square pulse enters a communications channel which acts as an ideal low-pass filter, the output signal will be *bandlimited in frequency, but it will be stretched in the time-domain.* In fact it will have infinite extent in the time-domain, due to the fact that the output is the convolution of the input with the filter impulse response, and the latter is infinitely long. This is in accord with the fact that any signal with finite frequency spectrum is infinitely long in the time domain. As will be seen in the following paragraphs, this fact has significant implications for digital pulse transmission in communication systems.

A communication channel has normally a limited bandwidth, $B_c$, *and hence acts like a LPF* (with transfer function $H_c$ ($f$)). Consider for simplicity a baseband antipodal signal transmission, where rectangular data bits are transmitted as $\pm$ 1, with symbol duration $T$ (see Sect. 3.2.2). The channel will cause each data symbol to be stretched in time beyond its allocated interval $T$, and this stretching will interfere with the receiver's ability to reliably detect the true data value at any given time. This kind of interference is referred to as intersymbol interference (ISI). ISI tends to become worse as the data rate $R = 1/T$ bps increases. The number of other 'parasitic' symbols incorporated into the current symbol period depends on the data rate $R$ and the channel bandwidth $B_c$. The higher the data rate the worse the ISI, and the lower the bandwidth the more the ISI becomes a problem.

In practice, there is a significant effect on ISI from only a finite number of symbols on either side of the current symbol. The practical effect of the ISI can therefore be modeled via a distortion or filtering with a finite impulse response. i.e. one can model the ISI as a filtering produced by a tapped delay line with a finite number of taps $\{h_o, h_1, \ldots, h_M\}$. If the channel is distortionless, then $h_o = 1$ and $h_k = 0 \ \forall k \neq 0$. For practical channels which have some distortion, at least some of the $h_k \ \forall k \neq 0$ are non-zero.

In the case of baseband antipodal binary transmission, the distortionless channel output (i.e., the observed signal seen at the matched filter detector) would be $y(k) = \pm 1 +$ noise, assuming a sampling rate at the receiver similar to that at the transmitter. A decision here can be made as follows: if $y(k) \geq 0$, then +1 was transmitted, otherwise $-1$ was transmitted. This decision is not reliable if significant if ISI exists; and the matched filter is insufficient for detection.

In practice, ISI is negligible in telephone channels for data rates less than *2400 bps*. For higher data rates, ISI is a problem that degrades the system performance.

One way to reduce the effect of ISI is to smooth the corners of the rectangular pulses (that represent the symbols) to reduce the effective bandwidth of each symbol. This is called *pulse shaping*, and this shaping is done by filtering the original rectangular pulses with special-purpose filters such as the *raised-cosine filter*. This filter has a transfer function which is given by:

$$H(f) = \begin{cases} T & 0 \le |f| \le \frac{1-a}{2T} \\ \frac{T}{2}[1 + \cos\frac{\pi T}{a}(|f| - \frac{1-a}{2T})] & \frac{1-a}{2T} \le |f| \le \frac{1+a}{2T} \\ 0 & |f| > \frac{1+a}{2T} \end{cases}$$

where the constant $0 \le a \le 1$ is called the *roll-off factor* and *T* is the symbol period. This filter is useful in reducing ISI as long as the data rate $R = 1/T < 2W$, $W$ being the channel bandwidth. The rate $R = 2W$ is called the *Nyquist rate*.

The impulse response *h*(*t*) of the raised-cosine filter is given by:

$$h(t) = \mathrm{sinc}\left(\frac{t}{T}\right)\frac{\cos\left(\frac{\pi\beta t}{T}\right)}{1 - \frac{4\beta^2 t^2}{T^2}}$$

Not only does the raised cosine filter reduce the time extent of the ISI, but it also shapes the interference intelligently. In particular, the raised cosine filter transforms the rectangular pulse into an oscillating pulse which happens to be zero at all symbol detection points before and after the current data symbol as shown in Fig. 3.27. That is, whenever a data symbol needs to be detected, the interference from all prior and subsequent symbols is zero at that point. This is so because the impulse response *h*(*t*) is zero at all $nT$ (where $n$ is an integer), except at $n = 0$. Therefore, if the transmitted waveform is correctly sampled at the receiver at $t = nT$, the original symbol values can be recovered exactly in the absence of noise. Note that a matched filer is necessary at the receiver to reduce the effect of noise.

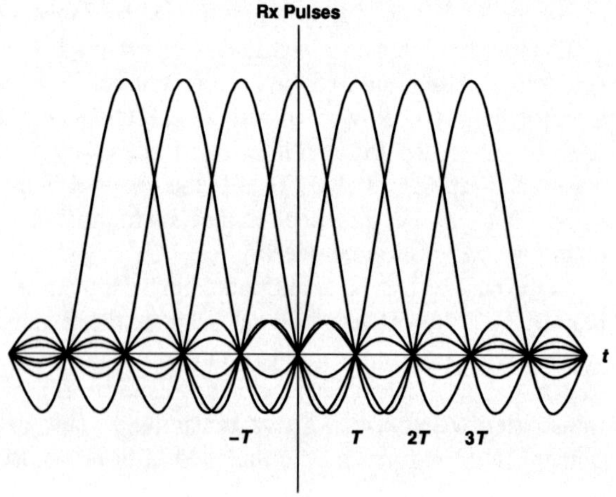


**Fig. 3.27** Successive data impulses (train of 1s) filtered using a raised-cosine filter

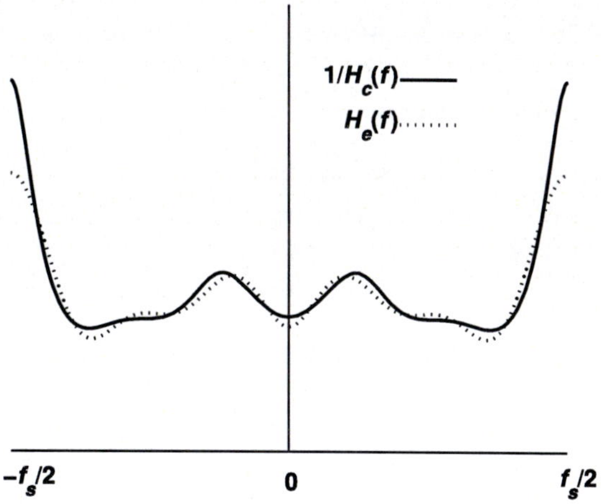


**Fig. 3.28** Channel estimation using an 11-taps adaptive LMS filter

### 3.5.9.2 The Adaptive Channel Equalizer

Pulse shaping using a raised-cosine filter can alleviate the ISI problem, but it *does not adequately deal with noise*. As discussed in Sect. 3.5.8, an effective way to reduce noise (as well as ISI) is to use an LMS adaptive filter to equalize the effects of the channel distortion and noise.

An efficient approach to design an equalizer is to utilize an adaptive FIR Filter. Such filters are called adaptive *transversal filters*) in communications, and they adapt their coefficients during the transmission of a known initial training signal to counterbalance the effect of the channel, i.e., they adapt the filter so as to produce an equalizing transfer function $H_e(f) = 1/H_c(f)$.

The adaptive LMS filter algorithm can be used to design an efficient channel equalizer, with the length $M$ of the filter depending on the number of symbols the ISI can span. Of course, larger $M$ tends to give better equalization, but more delay, so a compromise is necessary. Figure 3.28 shows channel estimation/equalisation using an adaptive LMS filter with $M = 11$, SNR $= 30$ dB, and a data length of 500 symbols. The transmitted symbols are a pseudo-random (training) sequence of the form $\pm 1$, known to the receiver, and hence are used as the desired signal, $d(n)$.

After the initial adjustment (*training mode*), the receiver switches to the *decision-directed mode*, where the receiver starts to process actual data rather than simply training the equalizer.

## 3.6 Sigma-Delta Modulation & Noise Shaping

In this section the topic of single-bit signal processing is addressed, along with the related topics of quantization and oversampling.

## 3.6.1 *Quantization*

Sampling is the first step to preparing an analog signal for processing with digital techniques. After a signal is sampled it is often referred to as a *pulse amplitude modulation* (PAM) signal. Sampling alone is not enough to fully prepare an analog signal for digital processing, since the amplitude can take an infinite number of possible values; quantization therefore needs to be subsequently applied. This process involves representing the analog amplitudes using a finite number of levels.

After quantization, data is typically *encoded into a multibit binary sequence* before it is transmitted. This process is known as *pulse code modulation* (PCM), and the resulting signal is known as PCM signal. In PCM, each sample is represented by one of $2^M$ code words, each of length $M$ bits. If $f_s$ is the sampling rate, the data rate is $Mf_s$ bps.

Normally multibit analog-to-digital converters (ADCs) perform all of the above-mentioned operations: sampling, quantization and binary encoding. The accuracy (resolution) of multibit ADC depends on the number of bits, $M$. This accuracy can be increased by increasing the number of bits, $M$, but this obviously requires more sophisticated hardware.

Quantization can be *uniform or nonuniform*, as discussed below.

### 3.6.1.1 Uniform Quantization

In uniform quantization, the quantization levels are uniformly distributed over the full allowable range. For example, if the input voltage is limited to $\pm V$ volts, then the whole range $2V$ is divided into $N = 2^M$ steps that represent any analog voltage between $-V$ and $+V$. The approximation accuracy is dependent on the quantization step $\Delta = 2V/N$. The quantization error $e$ is given by:

$$-\frac{\Delta}{2} < e < \frac{\Delta}{2}.$$

If it is assumed that the analog signal takes all values between $-V$ and $+V$ with equal probability, then the quantization error $e$ is uniformly distributed over $(-\Delta/2, \Delta/2)$. Hence, the pdf of $e$ is:

$$p(e) = \frac{1}{\Delta}.$$

The mean of this error is zero since:

$$m_e = \int_{-\Delta/2}^{\Delta/2} e \cdot p(e) de = \int_{-\Delta/2}^{\Delta/2} e(1/\Delta) de = 0 \tag{3.53}$$

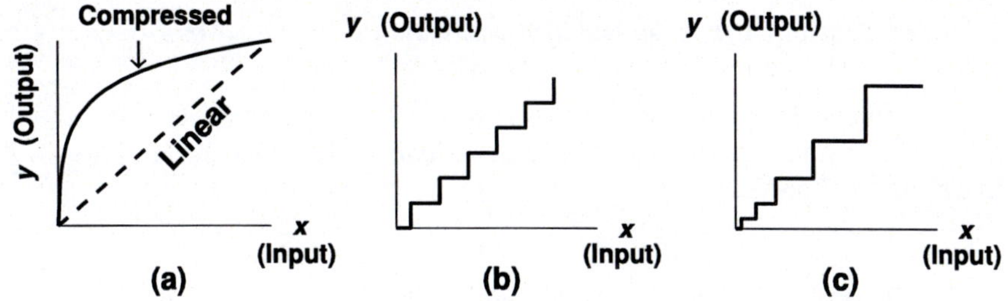


**Fig. 3.29** Nonuniform quantization: compression followed by uniform quantization

Since the mean is zero, the quantization noise power is equal to the variance, and is given by:

$$\sigma^2{}_q = \int_{-\Delta/2}^{\Delta/2} e^2 p(e)de = \int_{-\Delta/2}^{\Delta/2} e^2(1/\Delta)de = \frac{\Delta^2}{12} \tag{3.54}$$

which is independent of the number of bits $M$.

#### 3.6.1.2 Nonuniform Quantization

For some applications (such as speech communications), uniform quantization is not efficient. In voice communication channels, very low speech amplitudes are common, while large amplitudes are rare. From Eq. 3.54 above, the quantization noise power $\sigma_q^2$ is constant for constant step-size $\Delta$, hence, the quantization signal-to-noise ratio $\mathrm{SNR}_q$ is worse for low amplitudes than for large amplitudes. In speech communications, therefore, smaller quantization steps should be used for weaker amplitudes, and larger steps for larger amplitudes. This is the basis of *nonuniform quantization*. It can be achieved by first distorting the original signal using a logarithmic compression transfer function as shown in Fig. 3.29a, then using a uniform quantizer with a linear transfer function as shown in Fig. 3.29b. The final transfer function is shown in Fig. 3.29c. The compression curve has steeper slope for weak amplitudes, hence, it magnifies them more than large ones.

The compression law currently used in North America for speech is given by:

$$y/y_{\max} = \frac{\ln(1 + \mu|x|/x_{\max})}{\ln(1 + \mu)} \mathrm{sgn}(x),$$

where $\mu$ is a parameter whose standard value = 225. The sgn function is used to handle negative inputs. In the receiver, inverse compression, also called *expansion*, is performed to counteract the distortion of data by compression.

## *3.6.2 Oversampling and Its Applications*

In many applications it is advantageous to increase the sampling rate $f_s$ of a signal above the Nyquist rate. In this subsection two applications for this type of 'oversampling' are explored.

### 3.6.2.1 Quantization SNR Improvement

It was shown in Sect. 3.6.1.1 that the (uniform) quantization noise power is given by $p_n = \sigma_q^2 = \Delta^2/12$. Quantization noise, like other kinds of noise, is broadband. After sampling, all of the noise power will be spread over one frequency period, with this period being equal to $f_s$. If it is assumed that there is a flat spectrum for the quantization noise, then the quantization noise PSD will be as follows:

$$G_q(f) = \frac{(\Delta^2/12)}{f_s}.$$

Now if the sampling frequency in increased to $f_{s2} = 2f_s$, then the spectrum will be of period $f_{s2} = 2f_s$. Hence, the noise power (which is unchanged) will be spread over $2f_s$ rather than $f_s$, and the new PSD of the noise will be given by:

$$G_{q2}(f) = \frac{(\Delta^2/12)}{f_{s2}} = \frac{(\Delta^2/12)}{2f_s} = \frac{1}{2} G_q(f).$$

If one passes this noise through a low-pass filter with pass band in the interval $(-f_s/2, f_s/2)$, the output noise power will be:

$$p_{n2} = \int_{-f_s/2}^{f_s/2} G_{q2}(f)df = \frac{1}{2} \int_{-f_s/2}^{f_s/2} G_q(f)df = \frac{1}{2} p_n.$$

Hence, the quantization noise power is *halved* by oversampling by two and then low-pass filtering. An oversampling ratio (*OSR*) of *R* followed by LPF will decrease the quantization noise power by *R*, and will hence improve the $\mathrm{SNR}_q$ by *R*.

### 3.6.2.2 Relaxing Conditions on the Anti-Aliasing Filter

Suppose one wishes to sample an analog signal at $f_s = 2B$Hz. It is then ideally necessary to use an anti-aliasing filter to ensure that all frequencies above *B*Hz are removed, while all frequencies below *B*Hz are removed. The most obvious

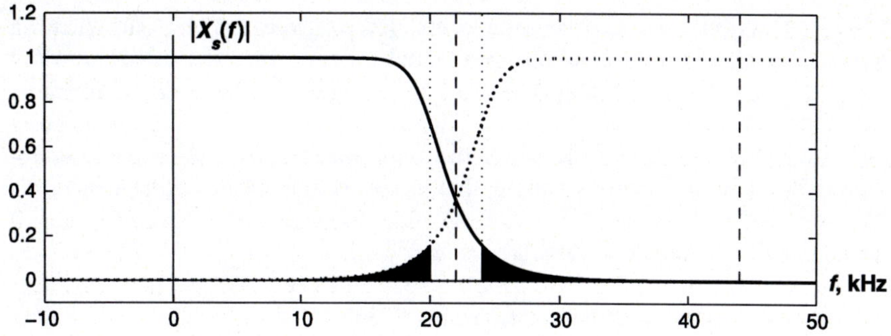


**Fig. 3.30** Anti-aliasing cutoff and attenuation in CD. *Shaded areas* should be removed by an analog filter, while the area in between can be removed later by a digital filter

way to achieve this filtering is use an ideal analog 'brick wall' filter, but such filters are impossible to implement in practice. Fortuitously, increasing the sampling rate relaxes the conditions on the LPF, enabling a lower order filter to be used.

As an illustration of the advantages of oversampling, consider the audio compact disc (CD). The audible spectrum extends up to 20 kHz. Hence, one needs a sampling frequency of at least 40 kHz. CD's use $f_s = 44.1$ kHz and 16-bits quantization. The stop band for this system extends from 20 to 22.05 kHz, and within this band the filter must produce an 80 dB drop (so as to prevent aliasing). This requires a 50th-order analog Butterworth filter (with a cutoff frequency of 20 kHz). Such a large order Butterworth filter is highly impractical. In addition to space problems, large filter orders are very difficult to realize because they are more vulnerable to errors due to the tolerances of the filter components.

The solution to the above problem is to increase the sampling rate to $4f_s = 176.4$ kHz. Now to prevent aliasing, there needs to be an 80 dB roll-off between 88.2 and 20 kHz. This requires only a *5th-order* Butterworth filter with a cutoff frequency of 20 kHz, which is a significant reduction in analog components. After the analog anti-aliasing filtering is applied a digital filtering is applied to achieve a very sharp cutoff. Subsequent to that filtering, downsampling to $f_s = 44.1$ kHz occurs. This downsampling is necessary for reduction of storage requirements without any loss of information. Figure 3.30 explains the above process.

### *3.6.3 Delta Modulation*

In multibit A/D conversion the signal samples are quantized and encoded. The samples are usually coded into $M$ bit code words, where $M$ is typically 16. As an alternative to conventional multi-bit conversion, one can quantize just the difference between the *current* sample and the *previous* sample; with such an approach, a *smaller* number of bits is required for encoding. The disadvantage of using this type of differential quantization is that a memory is needed to maintain knowledge of the previous sample. A slightly different approach involves comparing the current sample to its predicted value (determined from previous samples) rather than the previous sample, then encoding the difference. The encoding here is called differential PCM (DPCM). The disadvantage of DPCM is that it needs a feedback loop and associated memory.

One way to achieve DPCM is the delta modulation (DM) system shown in Fig. 3.31. This DM is a first-order analog feedback system that consists of a quantizer clocked at the sampling rate $f_s$ and an integrator (I). The integrator does a linear prediction $\hat{x}(n)$ of the current sample $x(n)$, based on the previous sample $x(n-1)$ and the most recent error. The quantized output is a *comparator* that gives either 1 (logic-1) if $x(n) \geq \hat{x}(n)$ and $-1$ (logic-0) if $x(n) < \hat{x}(n)$. Only a *1-bit word* is needed for encoding, since the output is either logic-0 or logic-1. The integrator has a scaling factor $\Delta$, which is referred to as the *step-size*.

*The operation of the DM can be further explored with an example.* Consider the input analog signal shown as a dotted curve in Fig. 3.32. It is $x(t) = 0.8\sin[2\pi(0.1)t + 0.1]$ volt. The sampling frequency is $f_s = 10$ Hz (i.e., $T_s = 0.1$

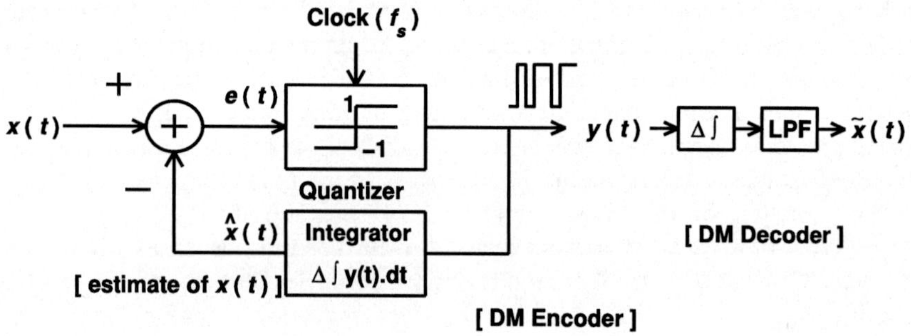


**Fig. 3.31** Delta modulator and demodulator

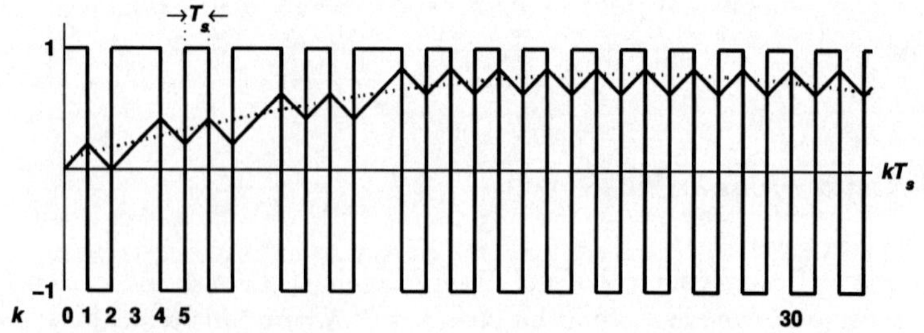


**Fig. 3.32** Delta modulation waveforms

sec), and the quantization step is $\Delta = 0.21$ volt. It is assumed that at $t = 0$, the integrator output is zero (i.e., $\hat{x}(0) = 0$), and the quantizer output is 1. At $t = 0$, the integrator starts integrating, and since its input is 1, the output grows linearly with time. Its output is $\Delta \int_0^t 1.dt = t$. At the next sample instant $t = T_s$, the error is $e(t) = x(t) - \hat{x}(t) < 0$ since the estimate $\hat{x}(t) = \Delta = 0.21$ is more than the signal $x(t) = \sin(0.2\pi \cdot T_s + 0.1) = 0.162$. Accordingly, the quantizer output flips to $-1$ and remains at that value until the next sampling instant $t = 2T_s$. In the meanwhile, because the integrator input is negative, its output decreases linearly with time until $t = 2T_s$. At $t = 2T_s, x(t) > \hat{x}(t)$, hence $e(t) > 0$, and the quantizer output flips to +1. This causes the integrator to linearly increase again. At $t = 3T_s$, the error remains positive and so there is no change in the quantizer output, etc.

The 1-bit output of the DM effectively turns out to be a pulse width modulated waveform. That is, the greater the amplitude of the input signal, the more often the binary output is "on". To demodulate the output signal one needs to take the signal estimate $\hat{x}(n)$, and then use a low-pass filter to remove the heavy quantization noise.

The accuracy of A/D conversion increases when either (i) the step-size is decreased, or (ii) when successive samples are highly correlated as is normally achieved if the samples are closely-spaced in time. Now the samples will be closely spaced if a high sampling rate is used, and for this reason DMs and other low-bit quantization systems usually use high sampling rates. These rates are typically much higher than the Nyquist rate corresponding to the input signal, hence the name *oversampling A/D converters*.

Because timing can be controlled much more easily and accurately than voltage levels, increasing the sampling rate tends to be less expensive than increasing the word length. For this reason 1-bit ADCs are less expensive than multibit ADCs, and are widely used for audio applications.

If the step-size is decreased to the point where the slope of the DM integrator (i.e., $\Delta/T_s$) is less than the slope of the input signal, a *slope-overload distortion* will occur. In this case, the DM is unable to correctly encode the signal, as illustrated in Fig. 3.33 for the scenario where $x(t) = 2\sin[2\pi(0.1)t + 0.1]$, $f_s = 10\,\text{Hz}$, and step-size $\Delta = 0.1$. On the other hand, if a large $\Delta$ is used such that the DM slope is much larger than the signal slope in some interval, another kind of

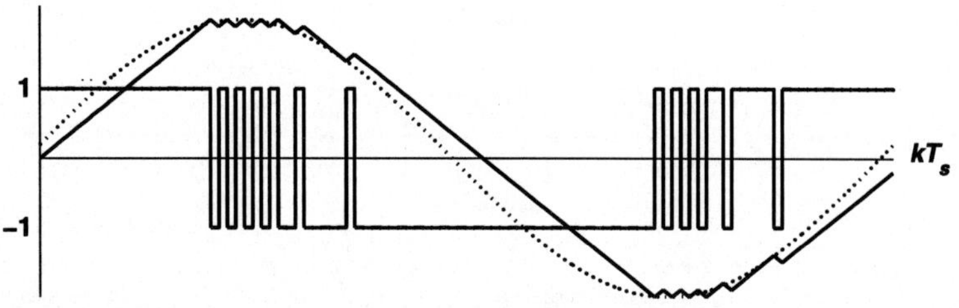


**Fig. 3.33** Slope-overload distortion in DM

distortion known as *granular noise*) occurs. The solution for these two problems is to use a DM with adaptive step-size.

### 3.6.3.1 Digital DM System

The discrete version of the DM system operates in a similar fashion to the analog one, and is depicted in Fig. 3.34.

### 3.6.3.2 Sigma-Delta Modulation

An sigma-delta modulator (SDM) is a 1-bit processing system which is based on the DM, but which has greater resilience to slope overload and granular noise. Recall that slope overload occurs when successive samples are not sufficiently correlated. Now low frequency signals tend to have low correlation between samples, but high frequency ones do not. To reduce problems with the processing of high frequency signals, one can simply place an integrator in front of the DM. The integrator tends to attenuate high frequency components, so that they are more amenable to coding with a single bit. A block diagram for the Sigma DM (SDM) is shown in Fig. 3.35.

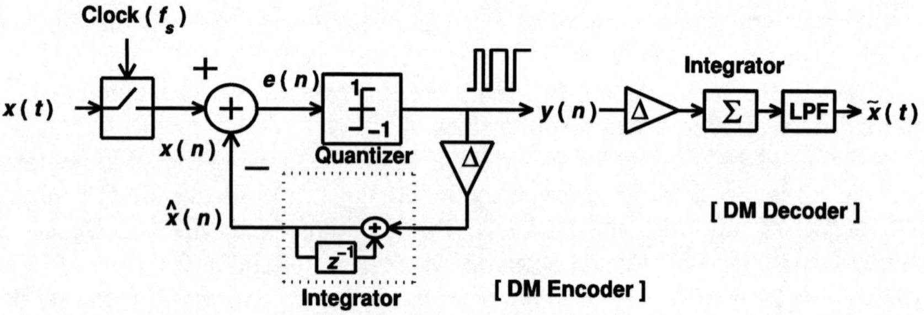


**Fig. 3.34** First-order digital delta modulator and demodulator

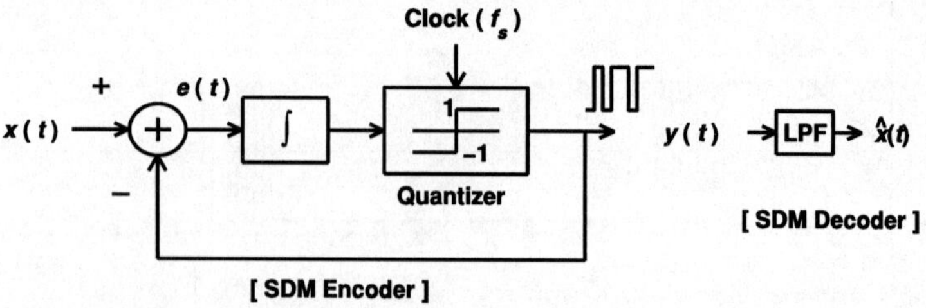


**Fig. 3.35** Sigma-delta modulation system

It is not hard to show that the representation in Fig. 3.35 has an equivalent input/output relationship to that in Fig. 3.35; the former is more hardware efficient. and so it more frequently used in practice for implementing SDMs. An analysis of the SDM is performed below. It will be seen that the SDM has the remarkable property of being able to shape the quantization noise away from the spectral band occupied by the signal. This means that the SNR in the band of interest can be made very high.

The SDM can be represented in the s-domain as shown in Fig. 3.36. The quantization noise can be represented as additive noise with transfer function $N(s)$. Since $\int_{-\infty}^{t} r(t)dt \xrightarrow{\mathcal{L}} \frac{1}{s}R(s)$ [*Tables*–Laplace Transform Pairs and Theorems], the integrator is represented as $1/s$.

From Fig. 3.36 one can write:

$$Y(s) = \frac{E(s)}{s} + N(s) = \frac{X(s) - Y(s)}{s} + N(s)$$

Hence,

$$Y(s) = \frac{1}{1+s}X(s) + \frac{s}{1+s}N(s) = H_1(s)X(s) + H_2(s)N(s),$$

where the signal transfer function $H_1(s) = \frac{1}{1+s}$ is a LP function, while the noise transfer function $H_2(s) = \frac{s}{1+s}$ is a HP function (see Fig. 3.37). Since the input signal is restricted to having low-frequency content, the band of interest is the low frequency band. From this it is clear that the SDM passes the signal energy but attenuates the quantization noise in the band of interest. That is, the noise transfer function is a HP function, and the quantization noise accumulates outside the band of interest, as shown in Fig. 3.38. This highly advantageous property of the SDM is known as *noise shaping*.

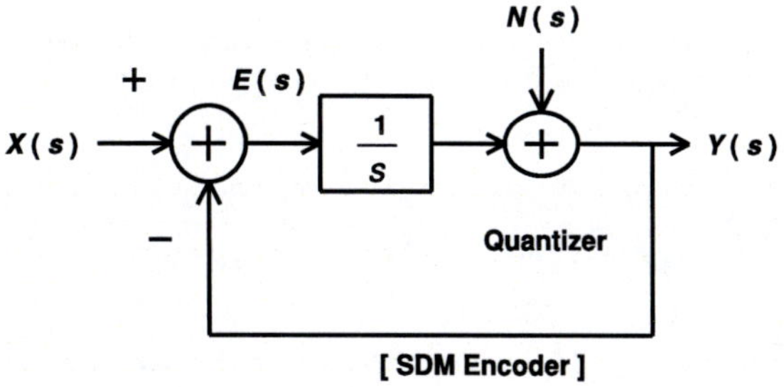


**Fig. 3.36** Sigma-delta modulation system in the s-domain

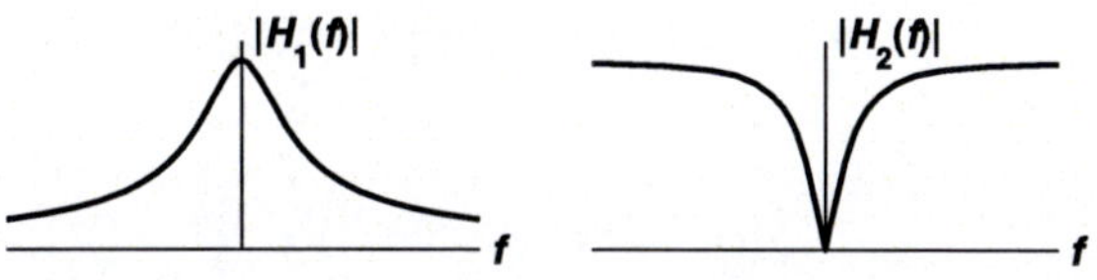


**Fig. 3.37** Signal and noise transfer functions in SDM

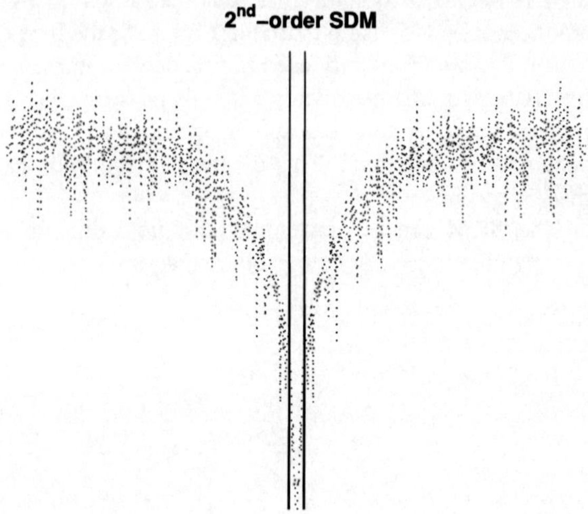


**Fig. 3.38** Signal and noise magnitude spectra in SDM for a sinusoidal input

## 3.7 Non-Stationary Signal Analysis

The Fourier transform (FT) is an important signal processing tool. In computation, the FT can convert the somewhat complicated convolution operation into the relatively simple multiplication operation. For signals, the FT acts like a *prism* that reveals the frequency spectrum of the signal and the weight of its frequency components. Note that the frequency spectrum given by the FT is the average spectrum over all time, from $-\infty$ to $+\infty$. This is quite useful for signals whose frequency content is stationary (not changing with time). i.e., it is useful for single-tone sinusoids, a sum of single-tones, a square wave, a rectangular pulse, etc. However, for frequency modulated signals, speech, and biomedical signals, *the FT often fails to reveal the time-varying characteristics* of the signal. The averaging process inherent in the FT smears the time varying spectral features. Although the FT is still important in studying these signals, it may not be sufficient in itself.

### *3.7.1 The Need for Time-Frequency Analysis*

*The finite-length linear frequency-modulated* signals:

$$x(t) = \sin[2\pi(f_o t + et^2)]\Pi_T(t - T/2)$$

and

$$y(t) = \sin[2\pi\{f_o(T - t) + e(T - t)^2\}]\Pi_T(t - T/2),$$

where $f_o = 1$ Hz, $T = 10$ s, and the modulation index is $e = 0.2$. The *instantaneous frequency* (IF) of $x$ is $f_x = (1/2\pi)d\phi_x/dt = f_o + 2et$ (which is a linear function of time), while the instantaneous frequency of $y$ is $f_y = |(1/2\pi)d\phi_y/dt| = f_o + 2e(T - t)$. The magnitude of the FT's of these two signals are *identical*, as shown in Fig. 3.39.

In this case the magnitude spectrum is *not sufficient* to reveal the underlying characteristics of the signal. No information is in the FT, since one can reconstruct the signal using the inverse FT. The time information is hidden in the phase response, but it is *not easy* to deduce much information about the time-varying nature of the signal from the complicated phase-magnitude relations. Using a time-frequency representation (TFR), however, it is possible to obtain the contour plots shown in Figs. 3.40 and 3.41, which reveal the true IF laws of the two Linear FM signals. (The particular time-frequency distribution is the Choi–Williams distribution, with parameter $\sigma = 11$ [8].)

*In addition* to the above dilemma of the FT, there is a more important problem that the FT cannot deal with. If the signal is composed of *more than one FM component*, it is even more difficult to observe the underlying structure from the FT. Many practical sounds do indeed consist of multiple time-varying frequency components (eg. EEG, ECG, and animal sound signals).

In particular, consider *first* the two *sinusoidal pulses* with the same frequency occurring at different times as shown in Fig. 3.42. It is evident that the magnitude Fourier spectrum cannot reveal the time-frequency structure of the signal, while

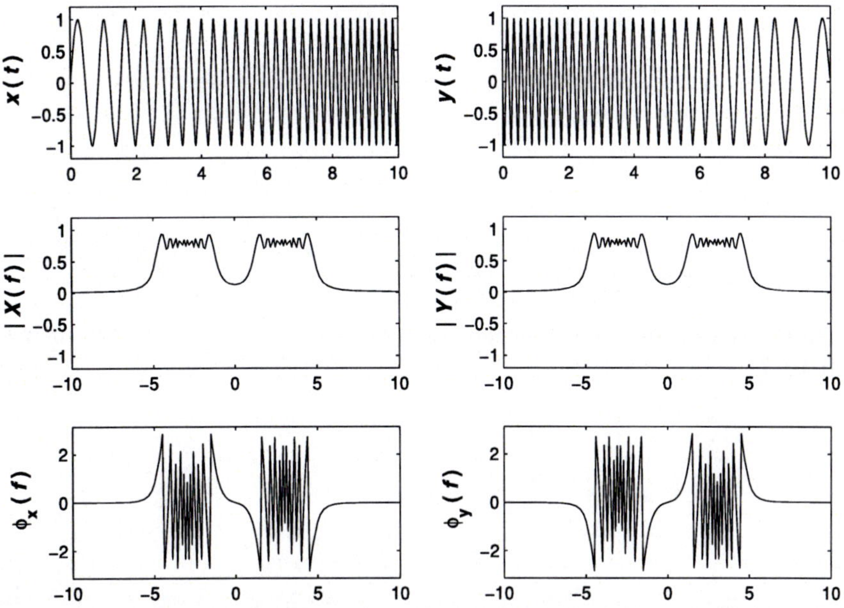


**Fig. 3.39** Two different Linear FM signals with identical Fourier magnitude spectra

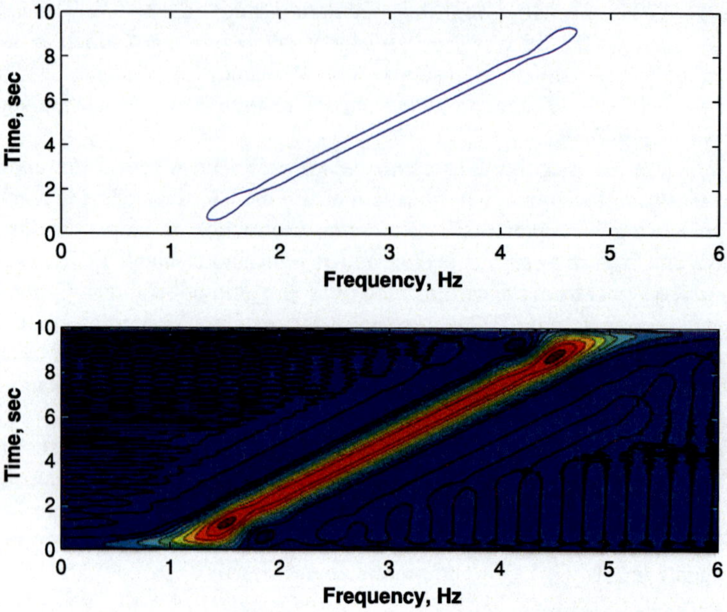


**Fig. 3.40** TFR of the FM signal $x(t) = \sin[2\pi(f_o t + et^2)]\Pi_T(t - T/2)$ with $f_o = 1$ Hz, $T = 10$ s, and $e = 0.2$. *Top* contour plot of the TFD. *Bottom* the TFR, with *red color* representing highest amplitude

the time-frequency distribution shows the number of components as well as their frequencies and durations.

*Second*, consider the sum of the two Linear FM signals $r(t) = x(t) + y(t)$, where $x(t) = \sin[2\pi(f_1 t + e_1 t^2)]\Pi_T(t - T/2)$, $y(t) = \sin[2\pi(f_2 t + e_2 t^2)]\Pi_T(t - T/2)$, $f_1 = 2\,\text{Hz}, f_2 = 3\,\text{Hz}, e_1 = 0.1, e_2 = 0.2$, and $T = 10$ s. The FT cannot reveal the multicomponent nature of this signal, while the *TFR* can, as shown in Fig. 3.43. Here the *Choi–Williams distribution* is used, with $\sigma = 41$.

*Third*, consider a bird sound represented in Fig. 3.44. The TFR can reveal the IF laws of the signal components, while one cannot get this information from the FT.

### 3.7.2 Some Important TFRs

A TFR is a two-dimensional transform of the signal into the time-frequency ($t$–$f$) plane. Signal components are typically observed in the $t$–$f$ plane by ridges around the IF laws of the components.

Below some of the commonly used TFRs are introduced.

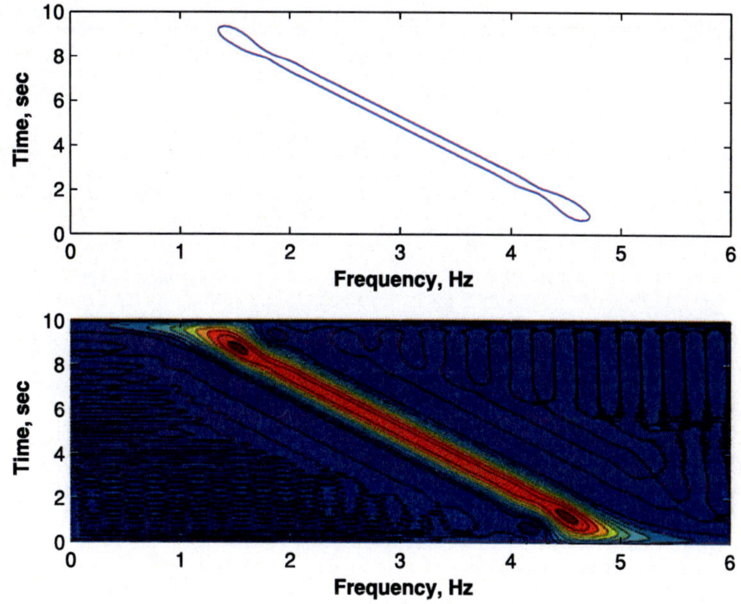


**Fig. 3.41** Time-frequency distribution of the linear FM signal $y(t) = \sin[2\pi\{f_o(T-t)+ e(T-t)^2\}]\Pi_T(t-T/2)$ with $f_o = 1$ Hz, $T = 10$ s, $e = 0.2$. *Top* contour plot of the TFR. *Bottom* the TFR, with *red color* representing highest amplitude

#### 3.7.2.1 The Short-Time Fourier Transform

The short-time FT, or windowed FT, is the simplest TFR. Instead of transforming the whole signal $s(t)$ all at once using the FT, it is transformed on a *block-by-block basis* using a *moving time-window*, centered at the time instant $t$, The STFT is defined as:

$$\rho(t,f) = \int_{-\infty}^{\infty} s(\lambda)h(\lambda - t)e^{-j2\pi f\lambda}d\lambda = \underset{\lambda\to f}{\mathrm{FT}}\{s(\lambda)h(\lambda - t)\}$$

where $h(t)$ is a suitable time-window such as a hamming window. Usually one refers to the above TFR as the STFT($t$, $f$).

The spectrogram is obtained from the STFT by simply taking the squared magnitude of the STFT.

#### 3.7.2.2 Cohen's Class of TFRs

Many of the TFRs used in common practice can be shown to be members of a generalized class known as *Cohen's class*, or the *quadratic class*. The general formula for this class is:

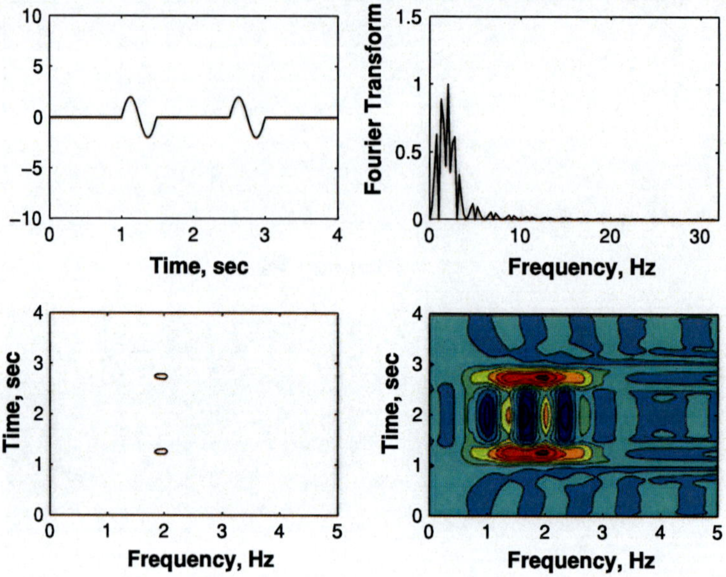


**Fig. 3.42** Two sinusoidal pulses (*Top-Left*) and their Fourier (*Top-Right*) and time-frequency (*Bottom-Right*) spectra. *Bottom-Left* represents the contour plot of the TFR in *Bottom-Right*

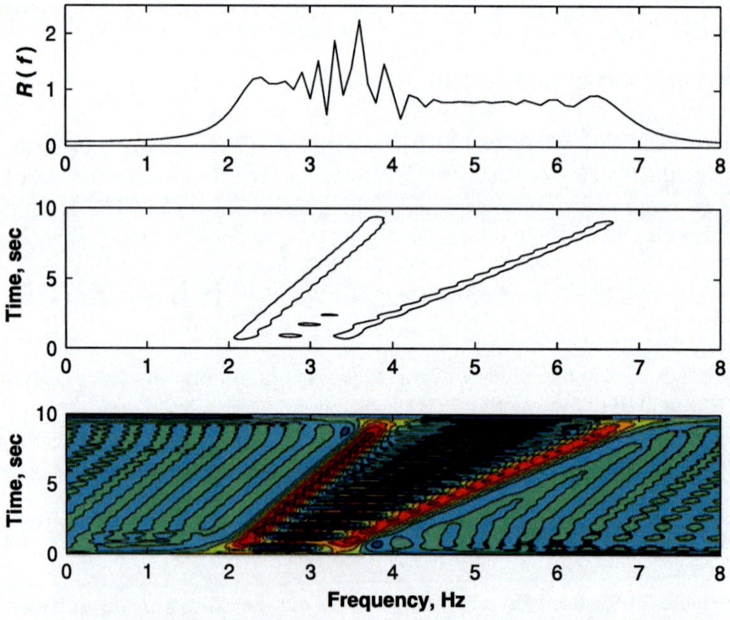


**Fig. 3.43** FT (*Top*) and TFR (*Bottom*) of a sum of two LFM signals. *Middle* is the contour plot of the TFR in the Bottom

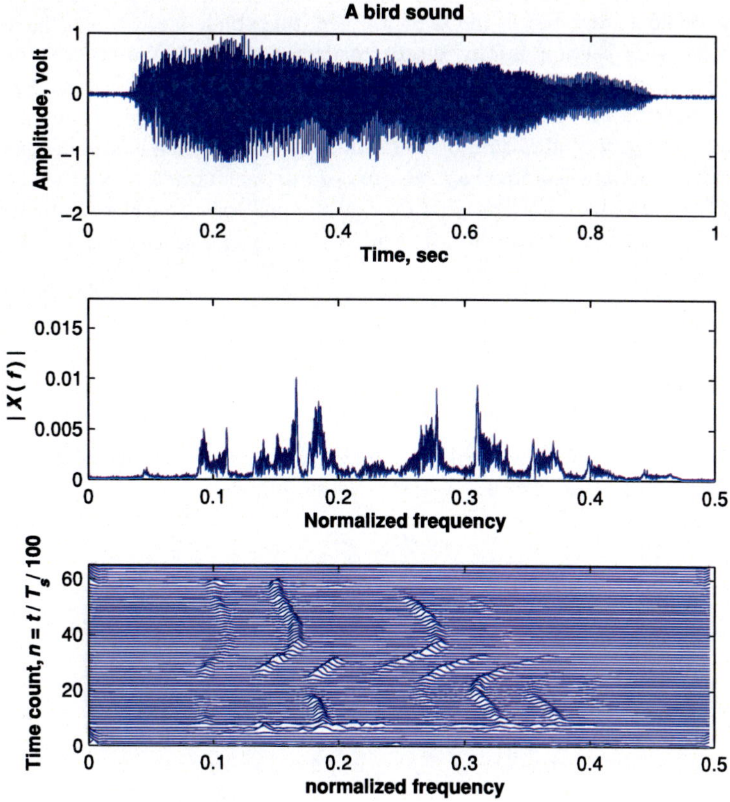


**Fig. 3.44** A bird sound signal (*Top*) with its FT (*Middle*) and TFD (*Bottom*)

$$\rho(t,f) = \underset{\tau\to f}{\mathrm{FT}}\{p(t,\tau)\}$$

where: $p(t,\tau) = G(t,\tau) \underset{(t)}{*} K_z(t,\tau), K_z(t,\tau) = z(t+\tau/2)z^*(t-\tau/2)$ is the instantaneous autocorrelation, $z(t)$ is the analytic signal associated with the original signal $s(t)$, and $G(t, \tau)$ is known as the kernel of the TFR. The kernel completely characterizes the particular TFR and is in general a symmetric 2D low-pass shaped function in the time-lag $(t, \tau)$ domain. Further details on Cohen's Class and TFRs in general can be found in the book [8].

### *3.7.3 The Discrete Cosine Transform*

*In multimedia communications* it is often necessary to transmit a large amount of data in a short period of time. This is one of the challenges for 4G mobile communications, and one of the ways to meet this challenge is to compress data efficiently. This section briefly discusses data compression using the discrete cosine transform (DCT).

Many popular transforms in signal processing decompose a given time-domain signal *x(t)* into a weighted sum of elementary functions or signals. This group of elementary functions is known as the *basis* of the transform. Often transforms are chosen to have basis functions which are orthogonal (or uncorrelated), because of the many advantages that such basis functions bring. These advantages include:

i. reduced computation compared to that required with non-orthogonal basis functions,
ii. good immunity to noise,
iii. invertibility, and
iv. compact representations for many naturally occurring signals.

The general class of *invertible orthogonal transforms* is defined using an *orthogonal basis* $\{\phi(k)|k = 0, 1, \ldots, N-1\}$ as follows:

$$X(k) = \sum_{n=0}^{N-1} x(n)\phi_k^*(n)$$

$$x(n) = \frac{1}{N}\sum_{k=0}^{N-1} X(k)\phi_k(n)$$

where:

$$\frac{1}{N}\sum_{n=0}^{N-1} \phi_k(n)\phi_m^*(n) = \begin{Bmatrix} 1, & m = k \\ 0, & m \neq k \end{Bmatrix}.$$

For the DFT $\phi_k(n) = \exp(j2\pi kn/N)$. Note that in the case of the DFT, *X(k)* is a *complex* function even if *x(n)* is *real*. In some applications, it is preferable to have a real transform for a real signal. One such real transform is the *Discrete Cosine Transform* (DCT). The definition for the DCT and its inverse is:

$$X(k) = \sqrt{\frac{2}{N}} w(k) \sum_{n=0}^{N-1} x(n) \cos\left[\frac{\pi k(2n+1)}{2N}\right] \quad k = 0, 1, \ldots, N-1$$

$$x(n) = \sqrt{\frac{2}{N}} \sum_{n=0}^{N-1} w(k)X(k) \cos\left[\frac{\pi k(2n+1)}{2N}\right] \quad n = 0, 1, \ldots, N-1$$

where $w(k) = \begin{Bmatrix} 1/\sqrt{2}, & k = 0 \\ 1, & k = 1, \ldots, N-1 \end{Bmatrix}$.

This transform is given in MATLAB© as *dct(x)* and *idct(X)*. It is closely related to the DFT/IDFT and can be calculated using a similarly efficient algorithm. It is worth noting that *Parseval's Theorem* is preserved by the DCT as reflected in the following equation:

$$\sum_{n=0}^{N-1} x^2(n) = \sum_{k=0}^{N-1} X^2(k).$$

### 3.7.3.1 An Application of the DCT: Data Compression

The DCT is used in speech and image compression due to its tendency to compress much of the signal energy in its low-indexed (informally, low-frequency) coefficients. This enables higher indexed coefficients to be discarded or transmitted with a reduced number of bits. In many applications, more than 50% compression of the data can be achieved without virtually any loss of information.

To transmit the FFT information, one needs *N/2 real numbers* from the real part *and N/2 from the imaginary* part (using the symmetry of the FFT), or a *total of N numbers*. However, for the DCT, one needs only *N/2 real numbers*. In audio and video compression, since the human eye and ear are *less sensitive to high* frequencies, it is often possible to transmit just N/5 of the coefficients or even less, as shown in Fig. 3.45

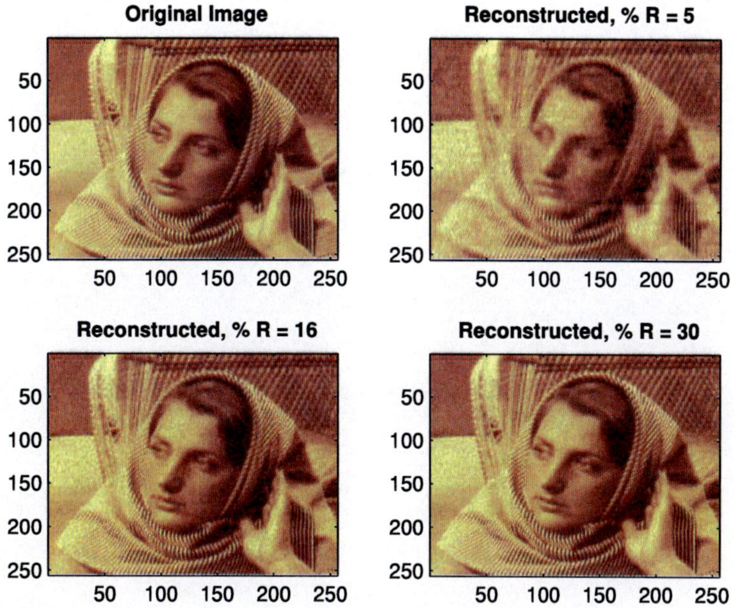


**Fig. 3.45** Image compression using the DCT, with $R$ indicating the % ratio of the number of transmitted coefficients to the total number of DCT coefficients, assuming noise-free channel. The image is obtained from MATLAB using "load woman"

## References


1. Peebles, P.: Probability, Random Variables, and Random Signal Principles. McGraw-Hill, New York (2000)
2. Proakis, J.G., Salehi, M.: Contemporary Communication Systems. Brooks/Cole, Pacific Grove, CA (2000)
3. Ipatov, V.: Spread Spectrum and CDMA: Principles and Applications. Wiley, Chichester (2005)
4. Osborne, H.C.: Stability analysis of an Nth power digital phase-locked loop—parts I & II. IEEE Trans. Commun. **COM-28**(8), 1343–1364 (1980)
5. Gill, G.S., Gupta, S.C.: First-order discrete phase-locked loop with applications to demodulation of angle-modulated carrier. IEEE Trans. Commun. Technol. **Com-20**(7), 454–462 (1972)
6. Jeffrey, A.: Advanced Engineering Mathematics. Harcourt Academic Press, San Diego, (2002)
7. Haykin, S.: Adaptive Filter Theory. Prentice Hall, Englewood Cliffs (2001)
8. Cohen, L.: Time-Frequency Analysis: Theory and Applications. Prentice-Hall, Englewood Cliffs, NJ (1995)

# Part III
# Advanced Topics

# Chapter 4
# The Impact of Finite Wordlength Implementation

## 4.1 Introduction

In the practical implementation of DSP systems, whether in hardware or software, one needs to take into account the effects of finite wordlength. If the wordlength is very long (say 32 bits) then these effects may be minimal, but for a shorter wordlength the limitations can be significant. In assessing the importance of finite wordlength effects three different factors need to be considered, namely:

i. the intended application and the cost of failure of the DSP system,
ii. the type of host architecture, and
iii. the arithmetic format used in the implementation, e.g., fixed- point or floating- point.

There are various sources of errors arising from finite wordlength effects. The first chief source of errors is the quantization error which arises when A/D conversion occurs. As seen previously, the power of the quantization noise is determined by the number of bits of the A/D converter. The second chief source of error is "roundoff or truncation" errors that arise when arithmetic operations are performed inside the DSP processor(s). For instance, an error typically occurs when one takes the product of a signal sample *rounded* to $b$ bits and a filter coefficient, which is also *rounded* to $b$ bits. The overall product is $2b$ bits length, but the product is typically *rounded* to $b$ bits to accommodate the internal register/data bus size. This overall error is dependent upon the type of number format (i.e., fixed-point or floating-point representation). Note that there are also various different forms of fixed-point representation, with each yielding its own particular quantization error contribution.

## 4.2 Overview of Number Formats

Typically, DSP processors, general-purpose computers, and special-purpose digital VLSI systems use binary number representations. Generally, the binary number

Z. M. Hussain et al., *Digital Signal Processing*,
DOI: 10.1007/978-3-642-15591-8_4, 

format consist of two parts: the integer part and the fractional part, with a binary point separating them:

$$\underbrace{10111}_{\text{integer}} \quad \diamond \quad \underbrace{101}_{\text{fraction}} \tag{4.1}$$

where $\diamond$ represents the binary point. Generally, the decimal equivalent of a binary number $a_{B-1}\, a_{B-2} \ldots a_0 \diamond a_{-1}\, a_{-1} \ldots a_{-b}$ consisting of $B$ integer bits and $b$ fractional bits is obtained as follows

$$\sum_{n=-b}^{B-1} a_i 2^i, \tag{4.2}$$

where $a_i \in \{0,1\}$. There are two main types of binary representations: fixed-point and floating-point, with these being discussed below.

### 4.2.1 Fixed-Point Format

The binary point in this representation is fixed at a specific location as suggested by its name. This representation is often used in implementing DSP algorithms because of its simplicity. In implementing DSP algorithms, fixed-point numbers are frequently scaled to be represented as *fractions*, as shown in Fig. 4.1. The scaling followed by fractional representation helps to ensure that overflow and loss of crucial information does not occur when arithmetic operations are performed. The leftmost bit $s$ represents the 'sign' bit, where the number is positive if $s = 0$ and it is negative if $s = 1$. The $b$ bits to the right of the binary point represent the other signal information.

The exact signal value represented by the non-sin bit in fixed point representation depends on the type of format used. There are several fixed-point conventions commonly used, namely, the 2's complement, the sign-magnitude and the offset binary conventions [1]. The offset binary convention, which is mainly used in bipolar D/A converters, can be converted to the 2's complement convention by simply complementing the sign bit.

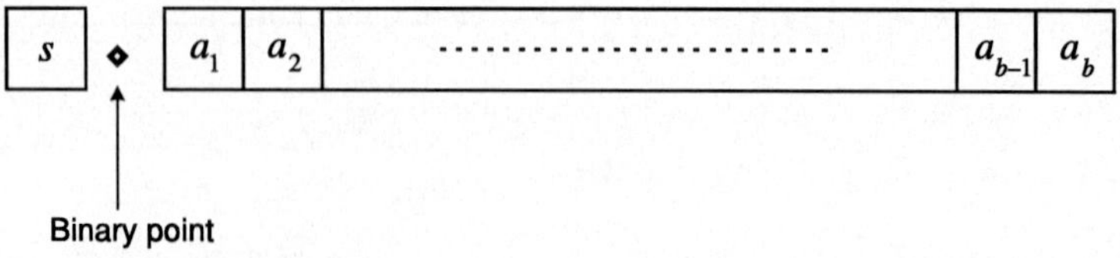


**Fig. 4.1** Fractional fixed-point format

The dynamic range of a fixed- point number conforming to the format in Fig. 4.1 ranges between $-1$ and $+1$. The least-significant bit (LSB), which also corresponds to the quantization step $\Delta$ is equal to $2^{-b}$.

### 4.2.2 *Floating-Point Format*

One can expand the effective dynamic range of numbers by using floating-point format. That is, for the same number of bits, floating-point representation provides wider dynamic range than is possible with fixed-point format. This improved range is due to the fact that floating-point representation provides a variable resolution. The general format for floating-point number representation is

$$\beta = M2^E, \tag{4.3}$$

where $M$ is a fractional part ($M \in [-1,1)$) that represents the so-called mantissa, and $E$ is the exponent. Both the mantissa and the exponent are signed numbers. Therefore, the dynamic range is determined in large part by the exponent $E$.

IEEE standard 754 floating-point formats for 32-bits (single-precision) and 64-bits (double-precision) are widely adopted in most modern floating-point DSP processors. Figure 4.2 shows the IEEE single precision format, where, in addition to the sign bit, there is a 23 bit mantissa and a $B_E = 8$ bit exponent. When the sign bit (bit 32) $S = 0$, the number is positive while it is negative when $S = 1$. The mantissa $M$ is a 2's complement positive binary fraction (bits 0–22) such that it occupies the range $0 \leq M < 1$. The exponent field contains the value, $E - 127$ and is 8 bits long, i.e., the exponent field supports exponent values between $-127$ and 126. Then, using the IEEE single-precision floating-point scheme, a number can be represented as follows

$$\beta = (-1)^S \quad (1 \cdot M) \quad 2^{E-127}, \tag{4.4}$$

where $0 < E < 255$. With this format, the range of a number is from $1.18 \times 10^{-38}$ to $3.4 \times 10^{+38}$.

## 4.3 The Quantization Process

Quantization can be defined as the process of mapping infinite-precision (or high-precision) numbers into lower-precision ones, with the purpose of creating more

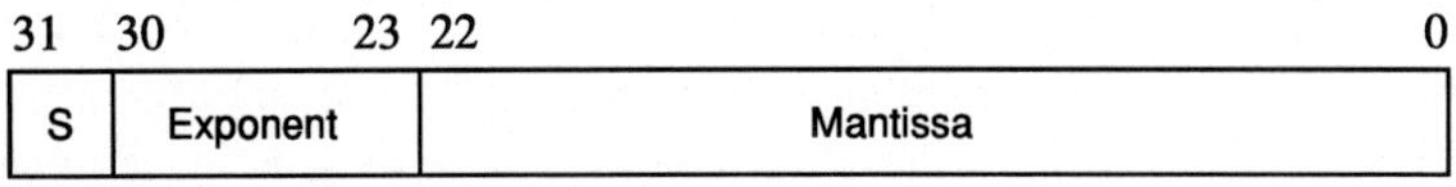


**Fig. 4.2** IEEE single-precision floating-point format

compact representations. This compactness is achieved at the expense of an induced error (quantization noise), which is expressed as

$$e_q(n) = Q[x(n)] - x(n), \tag{4.5}$$

where $Q[x(n)]$ represents the quantized value of $x(n)$.

The quantization error depends on the representation format (in particular, fixed-point or floating-point) and on the method of quantization being used, specifically, whether truncation or rounding is used [1].

### 4.3.1 *Quantization of Fixed-Point Numbers*

A common practice in DSP is to represent data in a digital machine either as a fixed-point fraction or as a floating-point binary number with the mantissa as fraction. In each of these formats, three different forms can be used to represent a negative number.

As indicated in Fig. 4.1, the fixed-point format assumes that a number is represented with $b + 1$ bits, with the most significant bit (MSB) conveying the sign of the number and the remainder of the bits constituting a binary fraction. That is, the binary point is just to the right to the sign bit. With this format the LSB is equivalent to $2^{-b}$ and represents the quantization step-size.

Several different variants of fixed-point representation are commonly used. These variants include 1's complement, 2's complement, sign-magnitude, and offset binary (see, for example, Mitra [1]). When one converts from one format to another, it is frequently necessary to quantize. This quantization is usually performed with a simple rounding or truncation, as discussed more fully below.

#### 4.3.1.1 The Rounding Method

In rounding, the number is quantized to the nearest quantization level. Assuming $b$ bits are used to represent the number's magnitude, then the quantization step is $2^{-b}$. After rounding, therefore, the maximum rounding error is $2^{-b}/2$. Consequently, the range of the rounding error $e_{qr}$ is

$$-\frac{1}{2}(2^{-b} - 2^{-\lambda}) < e_{qr} \le \frac{1}{2}(2^{-b} - 2^{-\lambda}). \tag{4.6}$$

Figure 4.3a shows the transfer characteristics of the rounding quantization method, where $\Delta = 2^{-b}$ is the quantization step. Notice that the rounding error is independent of the format being used to represent the negative fractions. This is so because the rounding operation is decided by the magnitude of the number.

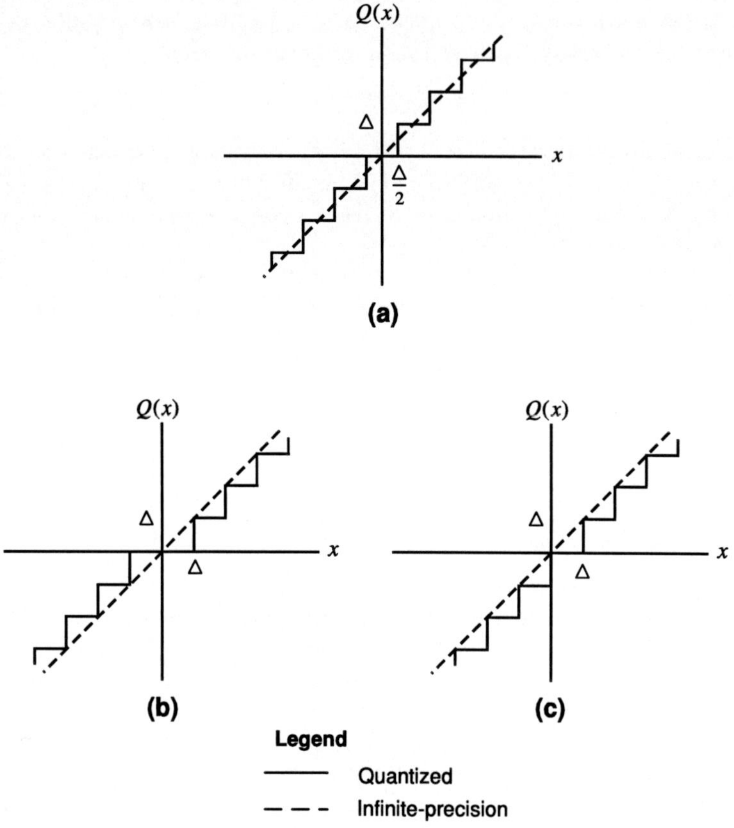


**Fig. 4.3** Transfer characteristics of a quantizer for **(a)** rounding; **b** 1's complement and sign-magnitude truncation; **c** 2's complement truncation.True signal is *dashed line*, quantized signal is *full line*

#### 4.3.1.2 Truncation Method

This method is achieved by truncating a fixed-point number of wordlength $\lambda + 1$ bits to $b + 1$ bits. The quantization error ($e_{qt} = Q[x] - x$) would depend on the polarity of the number $x$. For positive $x$, the error $e_{qt}$ will be always less than or equal to 0 (i.e., $e_{qt} \leq 0$). Therefore, the range of the error is given as

$$-(2^{-b} - 2^{-\lambda}) \leq e_{qt} \leq 0. \tag{4.7}$$

On the other hand, for negative number $x$, the truncation error depends on the negative number convention used to represent $x$. Three negative number conventions are commonly adopted: 1's complement, 2's complement, or sign-magnitude notation.

It can be shown for negative fraction $x$ in 1's complement notation that the truncation error is always positive and spread over the range

$$0 \leq e_{qt} \leq (2^{-b} - 2^{-\lambda}). \tag{4.8}$$

The same truncation error is found in the sign-magnitude notation as the magnitude of the truncated number $Q[x]$ is smaller than that of the original negative number $x$. The truncation error of the 1's complement and sign-magnitude notations is shown in Fig. 4.3b.

However, negative numbers represented in 2's complement notation has different truncation error as it is always negative. It can be shown that the error range in this case is

$$(2^{-b} - 2^{-\lambda}) \leq e_{qt} \leq 0, \tag{4.9}$$

as shown in Fig. 4.3c.

### 4.3.2 *Quantization of Floating-Point Numbers*

In fixed-point variables, the increment between adjacent numbers is always the same. In floating-point format, on the other hand, the increment between adjacent numbers varies considerably over the allowable number range. Therefore, it is much more informative to consider what is known as the relative quantization error $e_f$. Considering the floating-point format in Fig. 4.2, it is apparent that the quantization error is given by $Q(x) = 2^E Q(M)$. The relative quantization error, $e_f$, is defined as

$$e_f = \frac{Q(M) - M}{M}. \tag{4.10}$$

It can be shown that the *relative rounding error* $e_{fr}$ of a floating-point number (regardless of whether one uses 1's complement, 2's complement, or sign-magnitude format) has the range

$$-\Delta < e_{fr} \leq \Delta, \tag{4.11}$$

for all positive and negative numbers. On the other hand, the relative truncation error using the 1's complement and sign-magnitude conventions is given by

$$-2\Delta < e_{fr} \leq 0, \tag{4.12}$$

for all positive and negative numbers. Finally, the relative truncation error for 2's complement convention is

$$e_{fr} = \begin{cases} -2\Delta < e_{fr} \leq 0, & \text{for } x > 0 \\ 0 \leq e_{fr} < 2\Delta, & \text{for } x < 0 \end{cases} \tag{4.13}$$

For more details, see Porat [3].

### 4.3.3 Impact of Quantization on DSP System Implementation

Quantization is known to have several undesirable effects on the practical implementation of LTI systems. These effects are conveniently explained by considering an example. Consider the first-order IIR digital filter as shown in Fig. 4.4a, whose constant coefficient difference equation is

$$y(n) = \alpha y(n-1) + x(n), \tag{4.14}$$

where the input signal $x(n)$ is a digitized version of a bandlimited analog signal $x_a(t)$. The output signal $y(n)$ is ultimately fed into an ideal reconstruction filter to yield the bandlimited analog signal $y_a(t)$, while $\alpha$ is a gain parameter. Ideally, all discrete-time parameters and signal variables would have infinite-precision. In such a case, the corresponding transfer function would be given by

$$H(z) = \frac{1}{1 - \alpha z^{-1}}. \tag{4.15}$$

In real-world DSP implementations, the system shown in Fig. 4.4a is commonly implemented using a finite-wordlength fixed-point digital machine. This discretization process, in fact, transitions the system from being linear to nonlinear. Figure 4.4b shows a realistic model of that non-linear system, in which various

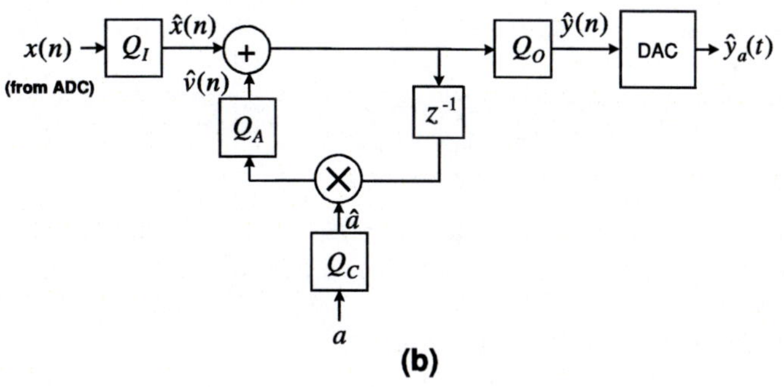


**Fig. 4.4** DSP system implementation. **a** Ideal, **b** practical

quantizers have been introduced to represent the appropriate round-off/truncatoin effects. The origins of these different types of quantization sources are:

1. A/D conversion $Q_I$.
2. D/A conversion $Q_O$.
3. Coefficient quantization $Q_C$.
4. Arithmetic operation quantization $Q_A$.

Because of the nonlinearity introduced into DSP systems via quantization, spontaneous oscillations can sometimes occur, and these oscillations are often referred to as limit cycles. Conveniently, these limit cycles only tend to occur in recursive (IIR) digital filters with poles that are operating close to the margin of stability.

Unfortunately, precise analysis of the model given in Fig. 4.4b is almost impossible, since the quantization is a nonlinear process and depends on the input. The latter, of course, is unknown a priori. In addition, the possibility of overflow exacerbates the difficulty of analysis. One can perform an approximate analysis through the adoption of some assumptions based on using a stochastic model that eventually linearizes the problem. This, however, is only an approximation and needs to be used with care.

**MATLAB:** The Fixed-point numeric object `fi` can be used to convert the double-precision representation normally used in MATLAB to fixed-point format. The `fi` object has the following three general types of properties:

- Data Properties.
- Mathematical (`fimath`) Properties.
- Numeric type (`numerictype`) Properties

Each one of these classes of properties automatically generates several relevant options that can be accessed to set the required fixed-point operations.

*Example 1* Design an FIR digital lowpass filter using fixed-point representation with 12 bit wordlength and a 32-bit accumulator. The required filter specifications are as follows:

Sampling frequency fs = 4000 Hz
The passband-edge frequency, fpass = 300 Hz
The stopband-edge frequency, fstop = 1000 Hz
maximum peak-to-peak ripple, Rpass = 0.015
minimum stopband attenuation, Rstop = 0.2

*Solution* : The first step is to design the filter in the default MATLAB double-precision representation. The Parks-McClellan optimal equiripple FIR order estimator `firpmord` can be used to find the approximate order M (see MATLAB signal processing toolbox documentation):

```
[M,fo,A,W] = firpmord([fpass fstop],[1 0],[Rpass Rstop],fs);
```

```
h = firpm(order,fo,A,W);
```

Note that if the filter does not satisfy the required specifications, then the order should be increased.

Next, convert the filter coefficient to fixed-point representation with the required fixed-point parameters:

```
   reset(fipref);

hfi = fi(h,1,12);

   F = fimath('ProductMode', 'KeepLSB', ...

'ProductWordLength', 32,...

'SumMode', 'KeepLSB', ...

'SumWordLength', 32,...

'OverFlowMode', 'wrap',...

'Roundmode', 'nearest');


   hfi.fimath = F;
```

Display the fixed-point coefficients:

```
hfi
```

## 4.4 Coefficient Quantization Error in Digital Filters

When one seeks to implement a desired digital filter transfer function $H(z)$ (FIR or IIR) one generally implements a modified transfer function $\hat{H}(z)$, due to the effects of quantization of the filter parameters. For this reason it is important to try and design digital filter structures that exhibit minimal sensitivity to coefficient quantization errors. The following subsections discuss this issue in more detail.

### *4.4.1 Coefficient Quantization Error in IIR Filters*

Quantization of the coefficients of a rational function results in the movement of the poles and zeros of that transfer function from their original, infinite-precision locations. Consequently, the actual system frequency response will be different from the desired one. An extreme scenario is that some of the poles overshoot the unit circle boundary in the z-transform, resulting in an unstable IIR structure.

A very important question to answer is: How can the harmful effects due to quantization of IIR filter coefficients be minimized?

To begin to answer this question, it is useful to recall the expression for the Direct Form transfer function of an IIR filter, assuming infinite-precision coefficients:

$$H(z) = \frac{\sum_{i=0}^{M-1} b_i z^{-i}}{1 - \sum_{i=1}^{N-1} a_i z^{-i}}. \tag{4.16}$$

Now, the modified transfer function which arises as a result of quantized coefficients is given by:

$$\hat{H}(z) = \frac{\sum_{i=0}^{M-1} \hat{b}_i z^{-i}}{1 - \sum_{i=1}^{N-1} \hat{a}_i z^{-i}}. \tag{4.17}$$

where $\hat{a}_i = a_i + \Delta a_i$ and $\hat{b}_i = b_i + \Delta b_i$ are the quantized coefficients. Note, the locations of the zeros and poles are affected by the errors $\Delta b_i$ and $\Delta a_i$, respectively. Detailed analysis in Ref. [2] shows that direct-form implementations such as those given in (4.17) exacerbate problems due to coefficient quantization. These problems are particularly significant when the poles/zeros are tightly clustered. Moreover, as the order of the IIR transfer function increases, the sensitivity to coefficient quantization errors increases accordingly.

It has been found that one can significantly improve the problems due to coefficient quantization by realizing the overall filter as a conglomerate of first- and second-order filter sections. Note that one needs *both* first and second order sections in general so as to avoid complex filter coefficients.

The realization of the filter can be achieved by either (i) doing a partial fraction expansion of the overall transfer function, and then operating all of the first and second order sections in parallel, or (ii) factorizing the overall transfer function into first and second order sections and then operating all of these sections in sequential cascade.

When an overall filter is implemented as a conglomerate of first and second order sections, each pole or pair of poles (zeros) is realized independently of the remaining poles (zeros), and thus the movement occuring in a certain pole or pair of poles does not affect the others. Hence, cascade-form implementation is much less sensitive to coefficient quantization errors than direct-form implementation.

Now, in considering the coefficient quantization sensitivity, one may also wonder whether the type of structure used to implement the second-order section itself would make any difference. The answer is 'yes'. To understand this more fully, consider the following example.

*Example 2* A second—order IIR filter has the transfer function:

$$H(z) = \frac{1}{1 - a_1 z^{-1} + a_2 z^{-2}}. \tag{4.18}$$

One can easily show that this second-order system is stable as long as $|a_1| < 2$ and $|a_2| < 1$. To graphically appraise the sensitivity of the filter implementation, one can do the following:

(a) Assume Direct Form—I realization: Plot all the possible locations of the complex stable poles when the coefficients $a_1$ and $a_2$ are represented in sign-magnitude format with 5-bit word-length.
(b) Realize the second-order IIR filter using coupled form and repeat the z-plane plot as in (a).

This second-order system is stable as long as $|a_1| < 2$ and $|a_2| < 1$. There are a number of possibilities for implementing this filter. The first way to implement it is with a Direct Form—I realization, with this realization being illustrated in Fig. 4.5. An alternative way to implement the filter is with the so-called coupled form realization. This form uses the real and imaginary parts of the filters pole locations explicitly in the transfer function expression. The real part of the pole pair is $r\sin(\theta)$, the imaginary part of the pole pair is $r\sin(\theta)$ and the coupled-form expression is:

$$H(z) = \frac{r\sin(\theta)}{1 - 2r\cos(\theta)z^{-1} + r^2z^{-2}}. \tag{4.19}$$

The coupled-form implementation structure is shown in Fig. 4.6. To implement this structure, the coefficients $r\cos(\theta)$ and $r\sin(\theta)$ must be quantized. A plot of all possible stable pole positions is shown in Fig. 4.7, assuming that 5-bit quantization is used. As can be seen in the figure, the pole positions are distributed uniformly within the unit circle. On the other hand, Fig. 4.8 shows the set of all possible pole locations when one uses 5-bit quantization in a direct form realization. The pole locations are seen to be non-uniformly distributed, i.e., there is a bias in the distribution of pole locations when one uses direct form filter implementations. For this reason, coupled- form representations give rise to more reliable implementations than direct form realizations. However, there is a price to be paid for this advantage, namely, more hardware complexity, as the multipliers are doubled in number when compared to the direct-form implementation.

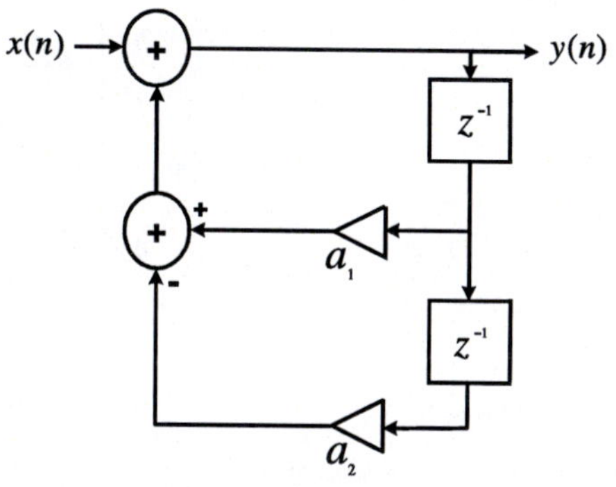


**Fig. 4.5** Direct form—i realization of an IIR filter

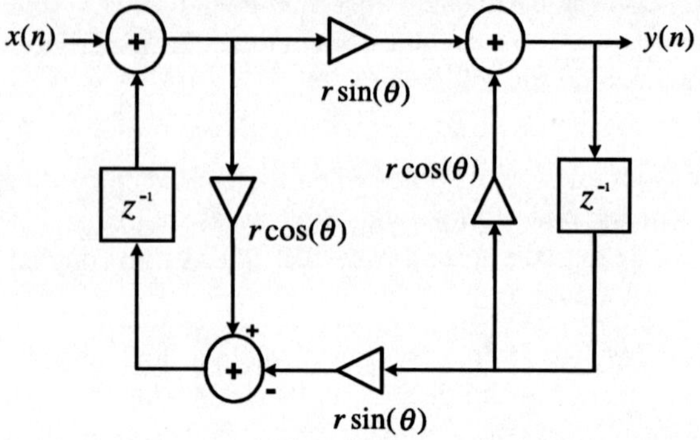


**Fig. 4.6** Coupled-form realization of a second-order IIR filter

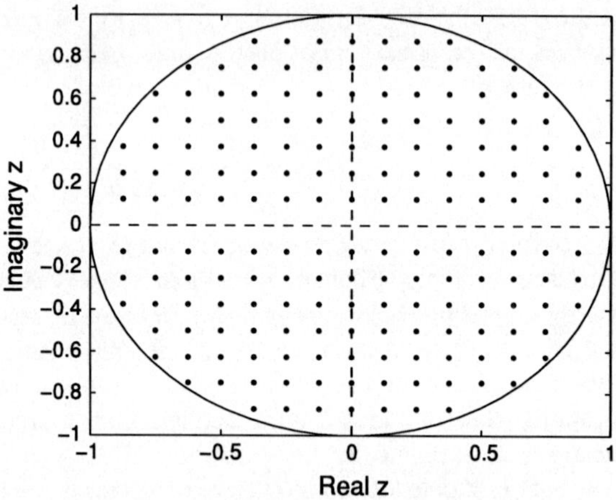


**Fig. 4.7** Possible locations of all stable pole pairs, assuming 5-bit quantization: coupled-form realization

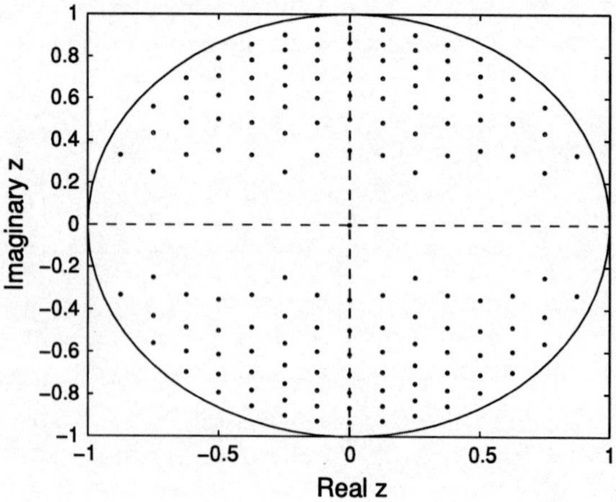


**Fig. 4.8** Possible locations of all stable pole pairs, assuming 5-bit quantization: direct form realization

### 4.4.2 Coefficient Quantization Error in FIR filter

As an FIR filter contains only zeros, one only needs to be concerned about the zeros, rather than *both* the poles and zeros. Consider an *M*th order linear-phase FIR filter with infinite-precision impulse response coefficients $\{h(i)\}_{i=0}^{i=M-1}$. Suppose the quantized version of this impulse response is the set $\{\hat{h}(i)\}$. Then the corresponding quantized transfer function can be expressed as

$$\hat{H}(z) = \sum_{m=0}^{M-1}(h(m) + q(m))z^{-m} = H(z) + Q(z), \quad (4.20)$$

where $q(m)$ represents the quantization error. A reasonable assumption is to let $q(m)$ be an uncorrelated zero mean random Gaussian process with variance $\sigma_q^2$. This means that the quantized coefficients can be modeled as a linear combination of the original coefficients and their quantization error (see Fig. 4.9). This suggests that the actual filter is comprised of an ideal filter plus an 'error filter'. The frequency response of this error FIR filter is

$$Q(e^{j\omega}) = \sum_{m=0}^{M-1} q(m)e^{-j\omega m}. \quad (4.21)$$

This filter has a number of interesting properties. First, the mean of the filter coefficients in the frequency domain is zero:

$$\mathcal{E}\{Q(\omega)\} = \mathcal{E}\left\{\sum_{m=0}^{M-1} q(m)e^{-j\omega m}\right\} = 0, \quad (4.22)$$

where $\mathcal{E}\{.\}$ is the expectation operator. Second, since the original, infinite-precision FIR filter is symmetrical (due to its linear-phase property), $Q(\omega)$ will also have this property. Therefore, because $M$ is odd, (4.22) can be re-written as

$$Q(\omega) = e^{-j(\frac{M-1}{2})\omega}\left\{2\sum_{m=0}^{\frac{M-3}{2}} q(m)\cos\left[\left(\frac{M-1}{2} - m\right)\omega\right] + q\left(\frac{M-1}{2}\right)\right\}. \quad (4.23)$$

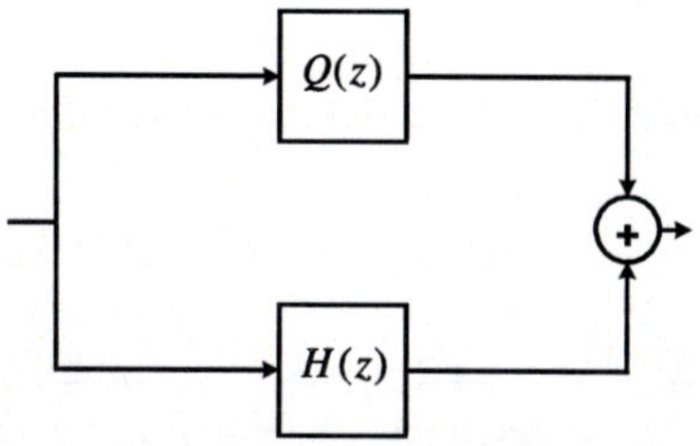


**Fig. 4.9** Quantized-coefficient model of an FIR filter

The variance of the error filter frequency response $\sigma_Q^2$ can then be calculated as

$$\begin{aligned}\sigma_Q^2 &= \mathcal{E}\{Q(\omega)Q^H(\omega)\} = \sigma_q^2\left[1+4\sum_{m=0}^{\frac{M-3}{2}}\cos^2\left[\left(\frac{M-1}{2}-m\right)\omega\right]\right] \\ &= \sigma_q^2\left[M-1+\frac{\sin(M\omega)}{\sin(\omega)}\right]\end{aligned} \tag{4.24}$$

where ${\sigma^2}_q$ is the variance of the error as a result of coefficients quantization. It can be shown that (4.24) reduces to

$$\sigma_Q^2 \leq \sigma_q^2(2M-1), \tag{4.25}$$

(See Ref. [2])

Equation 4.25 gives us a bounding formula for the error in the frequency response. Assume now that each coefficient is quantized to $b$ bits via rounding, and that there is a uniformly distributed error in each of the frequency domain coefficients given by $\sigma_q^2 = \Delta^2/12$, where $\Delta = \text{Full-scale}/2^b$ is the quantization step. With these assumptions, the error bounding relation becomes

$$\sigma_Q^2 \leq (2M-1)\frac{\Delta^2}{12} \tag{4.26}$$

The bound given in (4.26) is useful for estimating accuracies in filter design. It can be seen that for a certain level of tolerable error in the frequency response, the larger the FIR filter length $M$ the finer the quantization step $\Delta$ should be. Similarly, as in the case of high-order IIR filters, improved high-order FIR filter operation can be achieved by realizing the filter as a cascade of short (second-order) sections rather than with one large FIR filter. However, as usual, the price paid to obtain the cascade structure is more multiplication operations compared to that of the direct form structure. Additionally, the errors induced by quantization are less dangerous than in the case of IIR filters, because they do not have the potential to send the filter into unstable modes.

**MATLAB:** The Matlab object `dfilt` provides a convenient means to realize and simulate both FIR and IIR filters in a variety of structures. The possible structures include direct-form, second-order sections, lattices, and state-space formulations. Readers interested in further detail are referred to Mitra [1].

## 4.5 Quantization Errors in Arithmetic Operations

The key elemental processing operations used in DSP are multiplication and accumulation. Note that accumulation is effectively addition to an existing sum. As these mathematical operation are commonly carried out using fixed-point digital machines, additional errors can be introduced with each operation. As each weighted sample accumulates, the result becomes larger and hence needs more integer bits for perfect accommodation. When overflow is possible, there is a need to intervene. To prevent accumulator overflow, there needs to be truncation (shift-right operation) to a smaller wordlength.

For analysis purposes, the following assumptions are adopted. First, the arithmetic error sequence $\{q_A(n)\}$ is considered to be a wide-sense stationary white process which is uniformly distributed over the range of the quantization error. Second, $\{q_A(n)\}$ is uncorrelated with both any relevant input sequences and all other quantization error sources. To facilitate understanding, the possible sources of arithmetic errors are introduced in the following subsections.

## *4.5.1 Multiplier and Accumulator Errors in Fixed-Point Arithmetic*

### 4.5.1.1 Multiplier Error

When a sample is multiplied by some constant - coefficient value, there will be a scaling of the noise present on the signal sample. Additionally, there is a possible error due to the fact that the product is often truncated after a multiplication. This truncation occurs because the multiplication of two arbitrary numbers creates a product which is the sum of the wordlength of the multiplier and the multiplicand.

### 4.5.1.2 Accumulator Error

The accumulation of error due to the addition of noisy samples depends on the type of noise involved. Generally, the addition operation of two binary numbers would result in one bit increase in the integer part, no increase in the number of bits of the fractional part, and the impact on the noise will depend on the nature of the noise. If the noise is not a white random process, the signal-to-quantization ratio (SQR) could be very substantially degraded. Additionally, when digital registers cannot accommodate all the bits of the final accumulated result, the least-significant bit(s) of the result must be discarded. Therefore, if some of these truncated bits are noise- free bits, the SQR can deteriorate accordingly.

## *4.5.2 Scaling in Fixed-Point Arithmetic*

A major pitfall in practical implementation of digital filters in fixed-point arithmetic is the possibility of overflow. An overflow in 2's complement arithmetic, for instance, causes polarity reversal, which can lead to very harmful consequences. Therefore, careful attention should be paid to mitigate against overflows. One also needs to devise strategies to deal with overflows properly if and when they do take place. It is worth noting that floating-point implementation of digital filters overcomes the problem of overflow. However, fixed-point implementation is still the more popular implementation option due to its attractiveness in terms of cost and speed.

By scaling in fixed-point arithmetic, one seeks to ensure that the signals do not overflow the dynamic range permitted by the number system. This scaling can effectively be implemented by adopting the *fixed-point fractional* representation.

#### 4.5.2.1 Scaling of Direct Form IIR Filter

To illustrate the scaling technique, consider a first-order IIR filter as shown in Fig. 4.10. To be consistent with the fixed-point fractional format, let $x(n)$ be in the range $[-1,1)$ and let its impulse response and transfer function be $h(n)$ and $H(z)$, respectively. Then, the output signal $w(n)$ at node $d$ is given by the convolution

$$w(n) = \sum_{k=0}^{\infty} f(n-k)x(k), \tag{4.27}$$

where $f(n)$ is the impulse response from the input to the node $d$ (which is equals $h(n)$ in this simple case). The signal $w(n)$ will not in general be in the required range. One may put a general bound for $w(n)$ as follows

$$|w(n)| \leq \sum_{k=0}^{\infty} |x(n-k)||f(k)| \leq x_{\max} \sum_{k=0}^{\infty} |f(k)|, \tag{4.28}$$

where $x_{\max}$ represents the upper bound of the input signal.

Now in the case at hand, $x_{\max} < 1$, and so the bound reduces to

$$|w(n)| \leq \sum_{k=0}^{\infty} |f(k)|. \tag{4.29}$$

Now, for $|w(n)| < 1$, the summation $\sum_{k=0}^{\infty} |f(k)|$ must be less than unity, and therefore the node $d$ must be scaled down by a factor of $c$ such that

$$c = \frac{1}{\sum_{k=0}^{\infty} |f(k)|}. \tag{4.30}$$

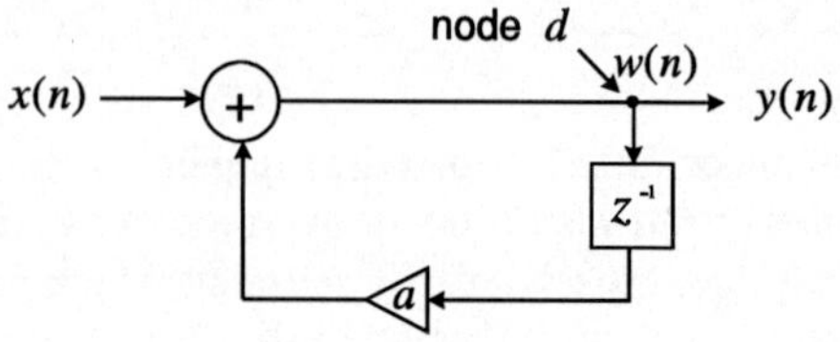


**Fig. 4.10** First-order IIR filter

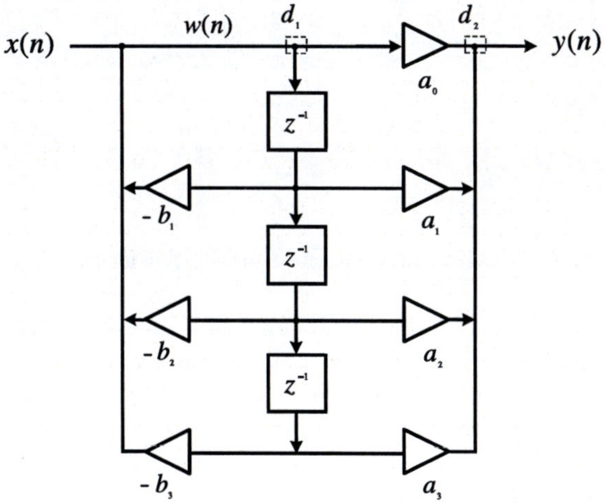


**Fig. 4.11** Nodes to be scaled

For instance, in the first-order IIR filter shown in Fig. 4.10, $f(n) = \alpha^n\, U(n)$. Then the summation in (4.30) is reduced to

$$c = \frac{1}{\frac{1}{1-|\alpha|}}. \tag{4.31}$$

For $\alpha = 0.975$, the scale will be $c = 1/40$. The price to be paid for this avoidance of overflow is a reduction in the SNR which accompanies the scaling.

In practice, the digital filter structures are much more sophisticated than the first-order IIR filter considered above. Therefore, one typically needs to scale several nodes to ensure no computational overflow. So, to put (4.30) in a general form, let $f_m(n)$ and $F_m(z)$ denote the impulse response from the input to the $m$th internal node and its corresponding transfer function, respectively. Then, the scaling factor for the $m$th node is given by

$$c_m = \frac{1}{\sum_{k=0}^{\infty} |f_m(k)|}. \tag{4.32}$$

If all nodes satisfy this scaling condition, then the entire structure is said to be *scaled*. However, it can be shown that normally, only those nodes that are inputs to multipliers must be scaled [4]. Consider the system depicted in Fig. 4.11, for example. In this system every "multiplier" is merely a delay version of the signal $w(n)$. Hence, it is adequate to simply scale node $d_1$ along with the output node $d_2$. It is worth mentioning that there are several approaches for scaling. The scaling technique described in (4.32) is both necessary and sufficient to guarantee a complete overflow free structure, however, it is relatively strict. A less stringent scaling technique can be realized by ensuring that the following condition is verified

$$c_m = \frac{1}{\sum_{k=0}^{\infty} |f_m(k)|^2}. \tag{4.33}$$

While the above structure is less strict than that in (4.32), overflows are still possible [3].

#### 4.5.2.2 Scaling of Cascade-Form IIR Filters

For the cascade implementation of a high-order IIR filter, scaling is required to avoid overflow in individual second and first-order sections in addition to the final output node. A systematic rule, known as the 'pole-zero pairing rule', has been developed to minimize the output noise power. The idea behind this rule is that grouping a pole with an adjacent zero tends to reduce the peak gain of the relevant section (Fig. 4.12).

In brief, the pole-zero pairing rule states the following Porat [3]. First, in the z-plane, the complex pole pair that is closest to the unit circle should be paired with its nearest complex zero pair. This procedure of matching pole and zeros pairs is repeated until all the poles and zeros are paired. Second, the second-order sections developed from the pairing process are ordered according to their peak magnitude response either in descending or ascending order.

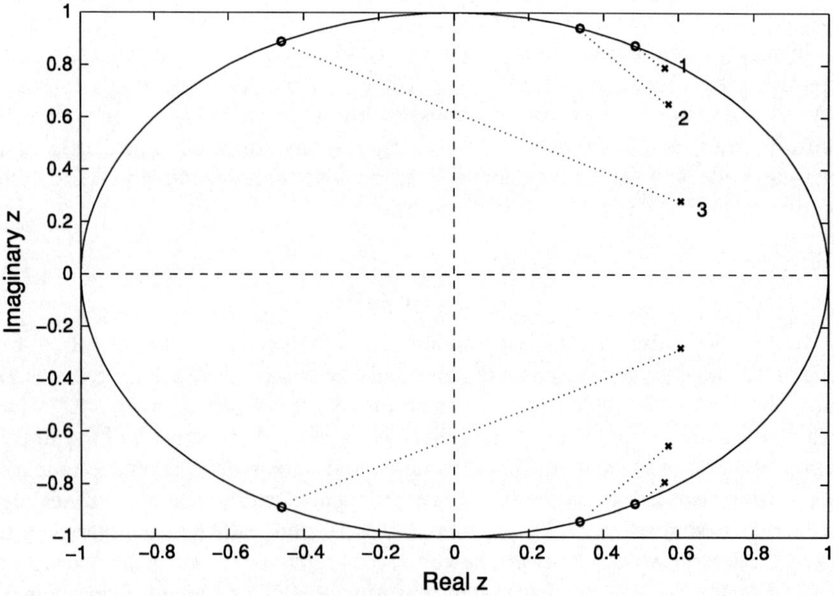


**Fig. 4.12** Pole-zero pairing and ordering rule

**MATLAB:** The conversion of the zero-pole-gain filter parameters to second-order sections form with optimum pole-zero pairing can be achieved by using the MATLAB function `zp2sos`. This function generates the following $L \times 6$ matrix

$$\text{sos} = \begin{bmatrix} b_{01} & b_{11} & b_{21} & 1 & a_{11} & a_{11} \\ b_{02} & b_{12} & b_{22} & 1 & a_{12} & a_{22} \\ \cdot & \cdot & \cdot & \cdot & \cdot & \cdot \\ \cdot & \cdot & \cdot & \cdot & \cdot & \cdot \\ \cdot & \cdot & \cdot & \cdot & \cdot & \cdot \\ b_{0L} & b_{1L} & b_{2L} & 1 & a_{1L} & a_{1L} \end{bmatrix}$$

where the rows contain the numerator coefficients ($b$'s) and the denominators coefficients $a$'s of the second-order sections that constitute $H(z)$.

*Example 3* A sixth-order elliptic low-pass filter has the following specifications:

Pass-band peak-to-peak ripple = 0.4 dB
Minimum stop-band attenuation = 40 dB
The pass-band-edge frequency = 0.3
Implement this filter using second-order sections.

*Solution*: First, the required elliptic filter can be designed to determine the poles, zeros, and the gain. This can be achieved readily using the MATLAB function `[Z, P, G] = ellip (6, 0.4, 40, 0.3)`. Then, convert zero-pole-gain filter parameters to second-order sections form using the function `sos = zp2sos (Z, P, G)`. Now, the matrix `sos` contains the coefficients of three second-order sections. The overall filter transfer function is then given by

$$H(z) = \prod_{i=1}^{3} H_i(z), \tag{4.34}$$

where,

$$\begin{aligned} H_1(z) &= \frac{1 + 0.9163z^{-1} + z^{-2}}{1 - 1.215z^{-1} + 0.4474z^{-2}} \\ H_2(z) &= \frac{1 - 0.6758z^{-1} + z^{-2}}{1 - 1.1499z^{-1} + 0.7568z^{-2}} \\ H_3(z) &= \frac{1 - 0.9713z^{-1} + z^{-2}}{1 - 1.1301z^{-1} + 0.9444z^{-2}}. \end{aligned} \tag{4.35}$$

#### 4.5.2.3 Scaling of Direct-Form FIR Filters

The scaling rules are similar to those already discussed for IIR filters. However, the the case of FIR filter design is simpler because the transfer function is confined to zeros only. i.e. there are no poles.

Consider the FIR direct-form structure shown in Fig. 4.13(a), which assumes that every coefficient multiplier output is quantized before the addition operation is performed. According to the linear noise model shown in Fig. 4.13(b), the total noise will be equal to the superposition of all the noise components. Accordingly, the output noise variance is given by

$$\sigma_t^2 = M\sigma_q^2, \tag{4.36}$$

where $\sigma_q^2$ is the quantizer noise variance. Recall that under the assumption of round-off based arithmetic, $\sigma_q^2 = 2^{-2b}/12$. For a symmetric linear-phase FIR filter, the output noise variance would be approximately half of that value, as the number of required multipliers is halved.

As the input to each multiplier branch in the FIR structure is just a delayed version of the original input signal $x(n)$, the coefficients could be scaled if an overflow is expected. Otherwise, only the output node needs to be scaled. Cascade form implementation of FIR filters is reduced to ordering of zero-pair sections (since FIR filters have no poles), rather than pairing and ordering, as was the case for IIR filters.

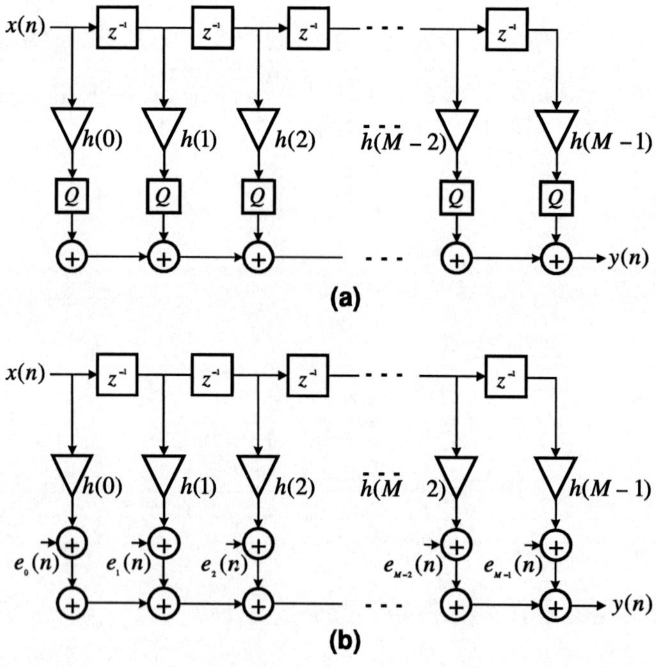


**Fig. 4.13** Direct-form implementation of FIR filter, **a** the effect of arithmetic quantization, **b** the linear noise model

Modern digital signal processors are equipped with optimized hardware structures to deal with multiply-accumulate operations. These structures include such things as double-length accumulators and pipelined multipliers.

## 4.6 Limit Cycle Phenomena

In the absence of any applied signal, the output of a stable IIR filter, implemented with infinite-precision arithmetic, decays asymptotically to zero. On the contrary, the same IIR filter implemented with finite-wordlength arithmetic can under some circumstances, exhibit sustained oscillations with periodic patterns. The potential instabilities exist due to the fact that quantization is a nonlinear process.

The phenomena of instability in the absence of applied input is sometimes called 'zero-input limit cycles', and tends to take place when there is feedback in the filter. For that reason, limit cycles do not tend to appear in FIR filters; this fact constitutes a significant advantage of FIR filters over IIR filters.

Limit cycles typically manifest as periodic patterns or oscillations and are highly undesirable. They are particularly problematical in audio applications as they can be audible and highly distracting.

There are two basic types of limit cycles: quantization and overflow limit cycles. Both of these types are described further below.

*Limit cycles due to quantization* arise as a result of successive truncation or rounding of products of an IIR structure. This type is also known as 'granular' limit cycles. The following example provides an illustration of the phenomenon.

*Example 4* Consider the first-order IIR filter shown in Fig. 4.13, which can be represented by the following infinite-precision difference equation

$$y(n) = a\,y_q(n-1)] + x(n), \quad |a| < 1. \tag{4.37}$$

Use signed 5-bit fractional arithmetic to implement this filter. Compare the output of the filter after and before quantization. Assume the quantization type is rounding.

*Solution*: The above difference equation becomes nonlinear and can be re-written as

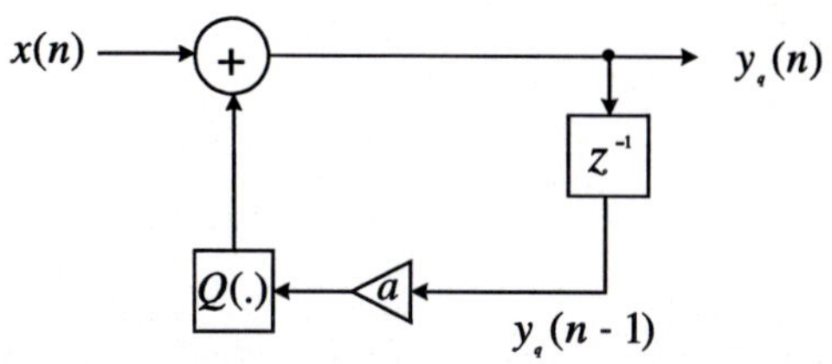


**Fig. 4.14** A first-order IIR filter with product rounding quantizer $Q(.)$

$$y_q(n) = Q[a\,y_q(n-1)] + x(n), \tag{4.38}$$

where $Q[.]$ represents the rounding operation with a quantization step of $\Delta = 2^{-(5-1)=2^{-4}}$. Figure 4.14 shows the equivalent structure. In this case, both the actual output $y_q(n)$ and the coefficient $a$ are represented by signed fractional 5-bit numbers. Their product is a signed 9-bit number being rounded to a signed 5-bit number to accommodate the internal registers length. For $a = 0\diamond1101 = 0.8125$, Fig. 4.15a shows the infinite-precision output which is decaying asymptotically to zero. Figure 4.15b, on the other hand, depicts a steady-state oscillatory output $y_q(n)$ of period 1 and magnitude 0.125.

For $a < 0$, Fig. 4.16a and b show the zero-input response of infinite-precision arithmetic implementation of a first-order IIR filter and its quantized version, respectively. The limit cycle period in Fig. 4.16b is 2 with magnitudes of $\pm0.125$.

It is worth mentioning that the appearance of limit cycles in $y_q(n)$ can be represented by a linear system with a pole on the unit circle, specifically at $z = \text{sgn}(a)$. The amplitude intervals to which the limit cycles are restricted are known as 'dead bands'. To define the dead band of the first-order IIR filter under consideration, recall (4.38) which can be re-written as

$$Q[a\,y_q(n-1)] = \text{sgn}(a) \quad y_q(n), \tag{4.39}$$

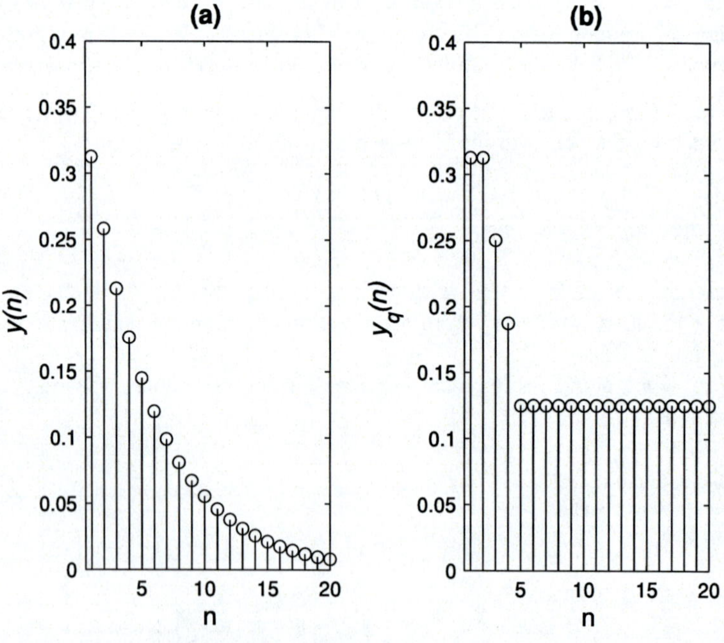


**Fig. 4.15** Zero-input response of a first-order IIR filter with $a = 0.815$, **a** infinite-precision, **b** fixed-point 1 + 4 bits

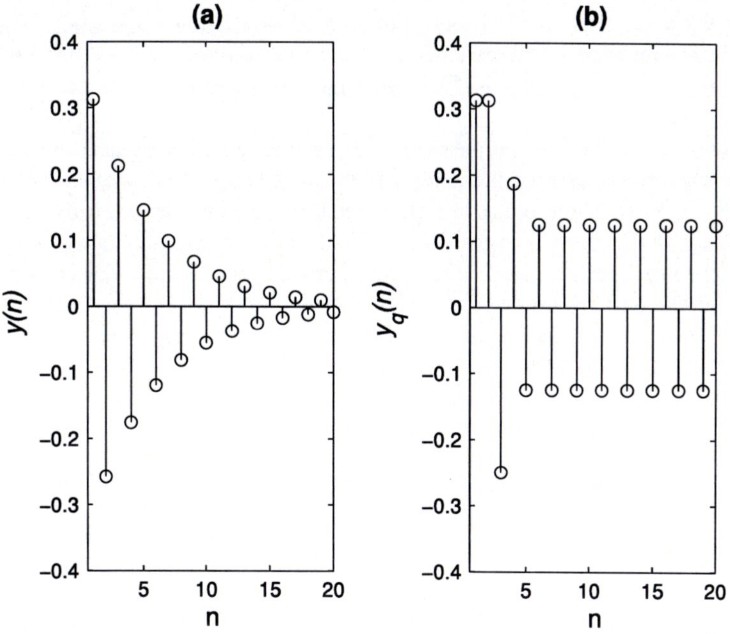


**Fig. 4.16** Zero-input response of a first-order IIR filter with $a = -0.815$, **a** infinite-precision, **b** fixed-point 1 + 4 bits

as $x(n) = 0$. Since the rounding error is bounded by $\pm\Delta/2$, then

$$|Q[a\,y_q(n-1)] - a\,y_q(n-1)| \leq \frac{\Delta}{2}. \tag{4.40}$$

Substituting (4.39) into (4.40), yields

$$|y_q(n-1)| \leq \frac{\Delta}{2(1-|a|)}. \tag{4.41}$$

or, for a steady-state output $y_q$

$$|y_q| \leq \frac{\Delta}{2(1-|a|)}. \tag{4.42}$$

This relation imposes an upper limit on the magnitude of granular limit cycles. Recall that in Example (4), $y_q(n-1) \leq 0.1689$. This result satisfies the condition given in (4.42). Notice that granular limit cycles occur due to the small error introduced by rounding quantization. According to (4.42), the magnitude of this oscillation is proportional to the quantization step size $\Delta$, and this magnitude can therefore be reduced by increasing the number of precision bits.

*Limit Cycles Due to Overflow* This type of oscillation takes place when the quantizer input exceeds the dynamic range. The magnitude of these limit cycles is large and can not be mitigated by adding more bits of precision as with granular limit cycles.

Recall that in the 2's-complement arithmetic system, if one adds two numbers whose sum is greater than the allowable dynamic range, the carry will propagate into the sign bit. Therefore, overflow tends to create very large errors due to the switching of the result's sign. To avoid this type of oscillation, an alternative addition characteristic, called 'saturation-overflow', is adopted. With this approach steps are taken to ensure that the size of the error does not increase unexpectedly. Saturation-overflow is achieved simply by clipping the result of accumulation if an overflow is detected. This strategy limits the overflow error and prevents a change of sign. This method is commonly used in signal processors and A/D converters that use 2's-complement numbers.

## References

1. Mitra, S.: Digital Signal Processing: A Computer Based Approach, 3rd edn. McGraw-Hill, New York (2006)
2. Oppenheim, A.V., Schafer, R.W., Buck, J.R.: Discrete-Time Signal Processing, 2nd edn. Prentice Hall Inc., NJ (1999)
3. Porat, B.: A Course in Digital Signal Processing. Wiley, New York (1997)
4. Vaidyanathan, P.P.: Multirtae Systems and Filter Banks, Prentice Hall Inc., NJ (1993)

# Chapter 5
# Multirate Digital Signal Processing

## 5.1 Introduction

The digital signal processing systems presented to date in this book have been *single-rate* systems, as the sampling rate has been fixed. There are, however, many applications where it is necessary to change the sampling rate of the signal at different stages in the signal processing chain. Such discrete-time systems are referred to as *Multirate Systems*. There are varying reasons for wanting to change the sample rate. In some applications it is changed to reduce computation, while at other times it is changed to improve accuracy and mitigate against quantization errors. At other times again, the sample rate may be changed to conserve bandwidth. In all sampling rate alterations, however, it will be assumed that the Nyquist criterion is always met and that there is therefore no aliasing.

Multirate systems play a particularly pivotal role in the areas of filter banks, digital to analog conversion, de-noising, and compression.

This chapter reviews the fundamentals of multirate signal processing and introduces some important multirate DSP applications.

## 5.2 Basic Elements of Multirate Processing

Multirate DSP systems require sampling-rate conversion at various stages of the processing chain, and this conversion is usually achieved with either decimation (i.e., sample rate reduction) or interpolation (sample rate increase). To perform decimation and interpolation one typically uses *three* basic building blocks, namely linear time invariant (LTI) *low-pass filters*, *down-samplers* and the *up-samplers*.

The following sub-sections discuss the implementation of decimation and interpolation, in both the time domain and frequency domains. Throughout these sub-sections it will be assumed that the sampling rate of the original signal is $f_s$.

Z. M. Hussain et al., *Digital Signal Processing*,

DOI: 10.1007/978-3-642-15591-8_5, 

### 5.2.1 The Down-Sampler and the Up-Sampler

A reduction in the sampling rate by a factor of $M$ (where $M$ is a positive integer) is achieved by retaining every $M$th sample and discarding the other samples. The device that performs this operation is called a down-Sampler, and the output of this downsampler is a sequence whose sampling rate is $1/M$ times that of the input sequence. Figure 5.1 shows a block diagram of the down-sampler. Its operation in the time-domain can be described mathematically as

$$y(n) = x(nM). \tag{5.1}$$

According to the above equation, all input samples with indices equal to integer multiple of $M$ are kept, while all others are discarded.

**MATLAB:** down-sampling can be achieved using either the "downsample" command, or equivalently, the vector command: $\mathbf{y} = \mathbf{x}(1 : M : \text{length}(\mathbf{x}))$. Figure 5.2 shows a sinusoid with frequency = 0.0356 Hz downsampled by a factor of $M = 3$. It is obvious that the sampling interval $\check{T}_s$ of the downsampled signal (Fig. 5.2b) is $M = 3$ times larger than the the original signal period, $T_s$.

An increase in the sampling rate by a factor of $L$ (where $L$ is a positive integer) is achieved by inserting $L - 1$ zero samples between each of the existing input signal samples. Mathematically, up-sampling can be expressed as

$$\hat{y}(n) = \begin{cases} x(n/L), & n = 0, \pm L, \pm 2L, \ldots \\ 0, & \text{otherwise} \end{cases} \tag{5.2}$$

From (5.2) it is apparent that the number of samples in the upsampled signal is $L$ times the number of input signal samples. Figure 5.3 shows the block diagram of an up-sampler.

**MATLAB:** Up-sampling (expanding) can be implemented in MATLAB using the command "up-sample", or alternatively, by using the following vector command scripts:

$$\hat{y} = \text{zeros}(1, L * \text{length}(x));$$
$$\hat{y}(1 : L : \text{length}(\hat{y})) = x;$$

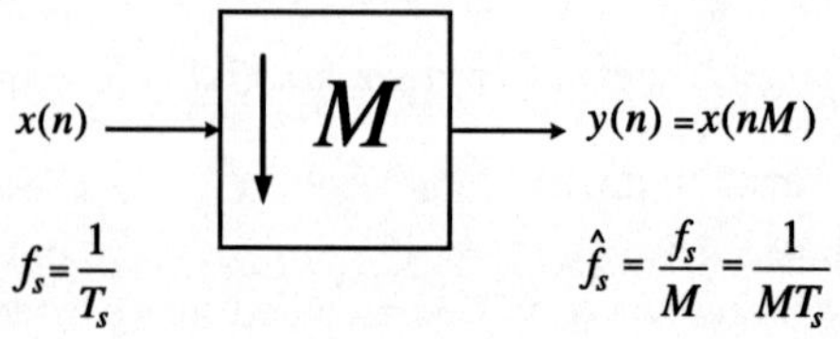


**Fig. 5.1** Block diagram of a down-sampler

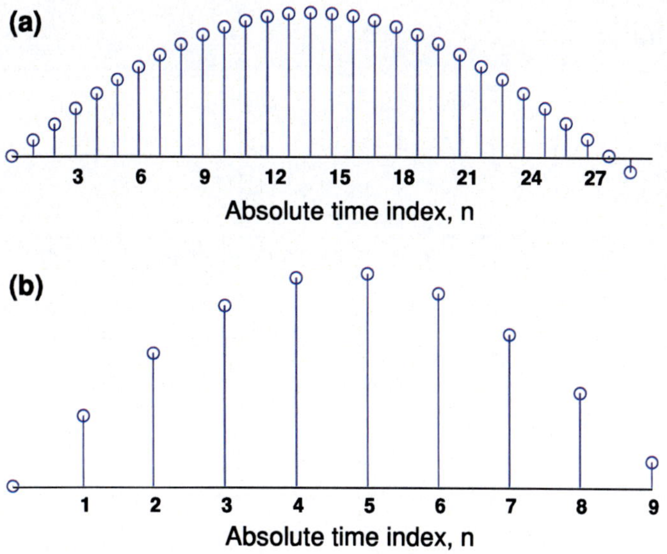


**Fig. 5.2** Downsampling process ($M = 3$)

**Fig. 5.3** Block diagram of an up-sampler

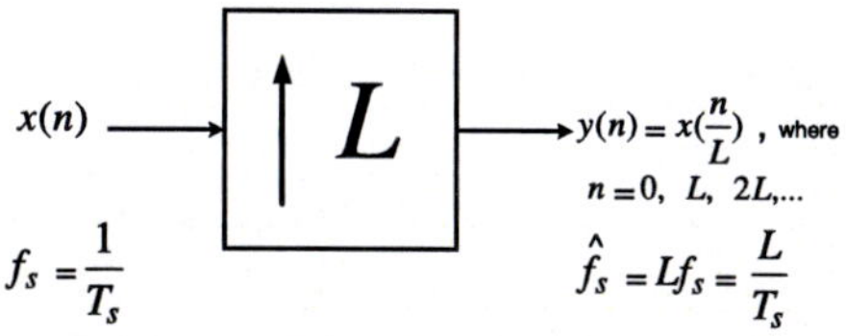


As an illustration of the upsampling process, Fig. 5.4 depicts the input sequence $x(n) = \sin(0.05\pi n)$ being up-sampled by a factor of $L = 3$.

Important properties of down-sampling and up-sampling are summarized as follows.

Upsampling and downsampling are:

- *linear* operations.
- *time variant* operations.
- *commutative* provided that $M$ and $L$ are relatively prime.

Note that the integers $M$ and $L$ are said to be "relatively prime" if they share no common factor in their prime factorization.

Down-sampling and up-sampling can be viewed as modifications of the existing sampled signal by eliminating sample values and inserting zero valued samples, respectively.

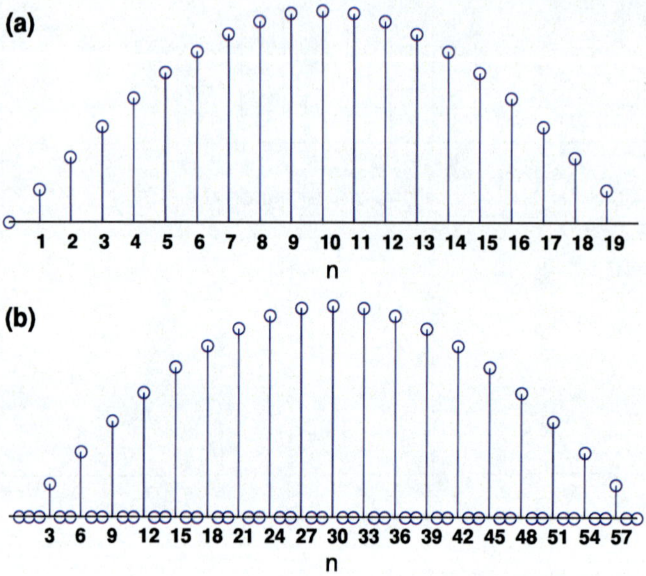


**Fig. 5.4** Up-sampling process ($L = 3$)

### *5.2.2 Frequency-Domain Representation*

As mentioned earlier, although down-sampling and up-sampling are linear operations, they are time variant. They therefore do not conform to standard LTI transfer function relationships. It is nonetheless informative to examine the input and output spectra of the down-sampling and up-sampling processes. Consider the z-transform of the downsampler defined in (5.1), yields

$$Y(z) = \sum_{n=0}^{N-1} x(Mn)z^{-n}. \tag{5.3}$$

One can not get a useful input-output relationship from (5.3) because of the way the input signal is expressed. To solve this problem, one may introduce the sampling sequence $c_M(n)$ as

$$c_M(n) = \sum_{i=-\infty}^{\infty} \delta(n - iM) = \begin{cases} 1, & \text{for } n = 0, \pm M, \pm 2M, \ldots \\ 0, & \text{otherwise} \end{cases} \tag{5.4}$$

which can equivalently be expressed as (See Miscellaneous DSP Problems-D, Q1)

$$c_M(n) = \frac{1}{M}\sum_{k=0}^{M-1} e^{j\frac{2\pi}{M}kn}. \tag{5.5}$$

Then, the down-sampled signal can be expressed as

$$\check{y}(n) = x(Mn)\, c_M(Mn), \tag{5.6}$$

and its $z$-transform is

$$\check{Y}(z) = \sum_{n=-\infty}^{\infty} x(Mn)\, c_M(Mn) z^{-n} \tag{5.7}$$

As $c_M(n)$ is zero for any $n$ that is not an integer multiple of $M$, (5.7) can be simplified to

$$\check{Y}(z) = \sum_{n=-\infty}^{\infty} x(n)\, c_M(n) z^{-\frac{n}{M}}. \tag{5.8}$$

Substituting (5.5) into (5.8) yields

$$\check{Y}(z) = \frac{1}{M} \sum_{n=-\infty}^{\infty} \sum_{k=0}^{M-1} x(n)\, e^{j\frac{2\pi}{M}kn} z^{-\frac{n}{M}} \tag{5.9}$$

which can be simplified to

$$\check{Y}(z) = \frac{1}{M} \sum_{k=0}^{M-1} \left\{ \sum_{n=-\infty}^{\infty} x(n)\, (z^{1/M} e^{j\frac{2\pi}{M}k})^{-n} \right\} \tag{5.10}$$

That is,

$$\check{Y}(z) = \frac{1}{M} \sum_{k=0}^{M-1} X(z^{1/M} e^{j\frac{-2\pi}{M}k}) \tag{5.11}$$

To clarify the implications of (5.11), it is convenient to switch to the Fourier transform representation by letting $z = e^{j\omega}$

$$\check{Y}(e^{j\omega}) = \frac{1}{M} \sum_{k=0}^{M-1} X(e^{j\frac{\omega - 2\pi k}{M}}) \tag{5.12}$$

According to (5.12), the spectrum of the down-sampled signal $\check{Y}(e^{j\omega})$ is composed of a scaled (by $1/M$) summation of $M$ shifted and frequency stretched versions of $X(e^{j\omega})$. This implies that if the input spectrum $X(e^{j\omega})$ extends from $-\pi$ to $\pi$ as shown in Fig. 5.5a, then its down-sampled spectra $X(e^{j\frac{\omega-2\pi k}{M}})$ with $M = 3$ suffers from aliasing as shown in Fig. 5.5b. Then, the output spectrum $\check{Y}(e^{j\omega})$ is different from the original spectrum shape due to the aliasing effect.

Because of the inherent potential for aliasing when downsampling is used, it is necessary to force a prior bandlimiting $X(e^{j\omega})$ to $\omega \in [-\pi/M, \pi/M]$ by using low-pass (anti-aliasing) filtering. This situation is illustrated in Fig. 5.6 where the output spectrum of $M = 3$ fold down-sampling is alias-free because the input

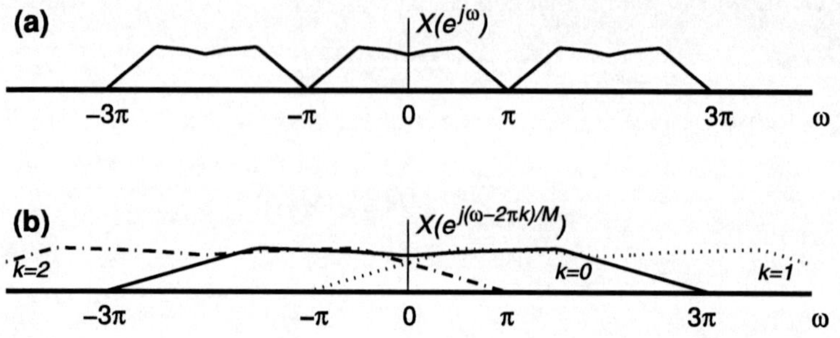


**Fig. 5.5** The effect of down-sampling on signal spectrum: **a** original, **b** down-sampled

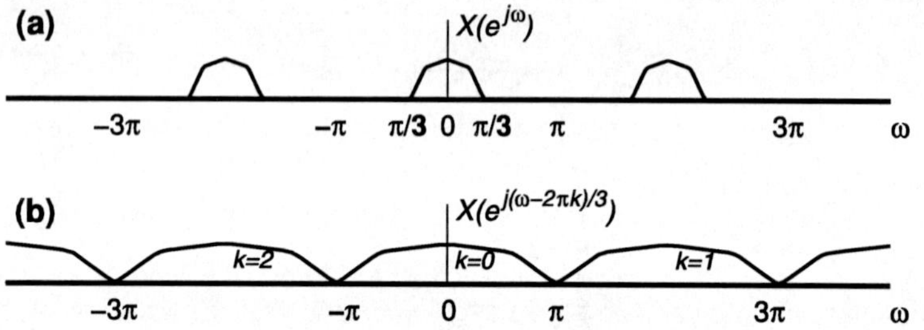


**Fig. 5.6** Properly bandlimited spectrum prior to down-sampling

spectrum that is limited to $\pm\pi/3$. This situation is similar to case of using an anti-alias filter to band limit the continuous-time signal before sampling.

Generally, to avoid aliasing when downsampling low-pass signals, the input signal bandwidth before down-sampling $\omega_B$ is kept within the limit $\omega_B \in [-\omega_B/M, \omega_B/M]$. The same principle applies for the case of bandpass signals with asymmetrical shape spectrums. (See Miscellaneous DSP Problems-D, Q2).

In contrast to down-sampling, up-sampling process is much easier to analyze in the frequency domain. Referring to (5.2), the $z$-transform of a signal which is up-sampled signal by a factor of $L$ is

$$\hat{Y}(z) = \sum_{n=-\infty}^{\infty} \hat{y}(n) z^{-n} \tag{5.13}$$

As the non-zero samples occur only when $n = \pm mL$ (where $m$ is an integer) then (5.13) can be re-written as

$$\hat{Y}(z) = \sum_{m=-\infty}^{\infty} x(m) z^{-mL} = X(z^L). \tag{5.14}$$

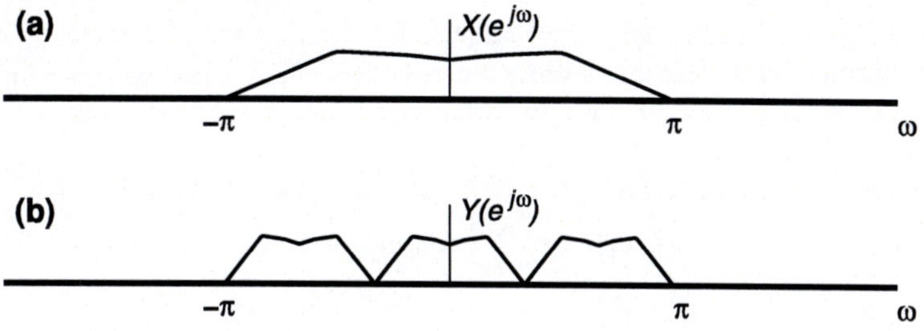


**Fig. 5.7** Up-sampling by a factor of $L = 3$

By letting $z = e^{j\omega}$, (5.14) becomes $\hat{Y}(e^{jw}) = X(e^{jwL})$, which implies that the up-sampled spectrum is compressed and repeated $L$ times in the frequency range $[-\pi, \pi]$. This is illustrated in Fig. 5.7 for $L = 3$.

The spectrum repetition phenomenon which occurs in upsampling is called "imaging", as it contains $L - 1$ undesired replicas in the baseband. These images need to be removed with an anti-image low-pass filter or "interpolation filter". This low-pass filtering has the effect of "interpolating" the zero-valued samples that have been inserted during the up-sampling process.

Fractional changes *L*/*M* of the sampling rate can be achieved by combining a down-sampler with factor *M*, followed by an up-sampler with factor *L*. as will be discussed later.

## 5.3 Sampling Rate Conversion Using Multirate Structures

### *5.3.1 Decimation*

A decimator includes both a down-sampler and a lowpass filter to ensure that no aliasing occurs when the sampling rate is changed. A block diagram of the decimator is shown in Fig. 5.8. The output of the decimator is at a lower sampling rate

**Fig. 5.8** Decimator

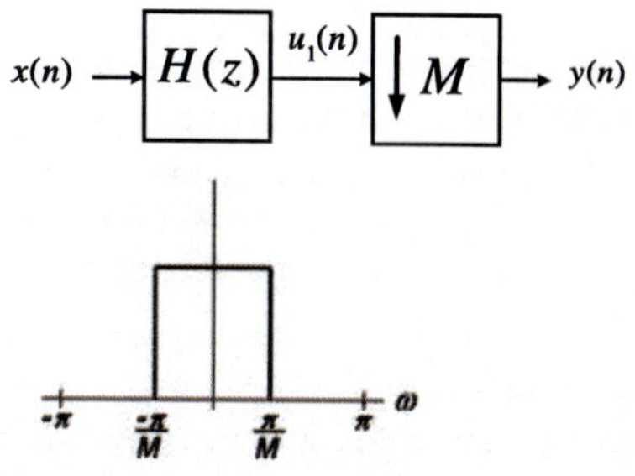

than the input, and thus has a lower bandwidth ($\check{f}_s/2$) than the input ($f_s/2$). To avoid aliasing and thus ensure correct reproduction of the signal spectrum in $\check{f}_s/2$, the input signal bandwidth must be limited to $|\omega_B| < \pi/M$. The output of the decimation filter is given by

$$u_1(n) = \sum_{i=-\infty}^{\infty} x(i)h(n-i), \tag{5.15}$$

where $h(n)$ is the impulse response of the LTI low-pass filter $H(z)$. According to (5.1), $\check{y}(n) = u_1(Mn)$. Then, the decimator output in the time-domain can be expressed as

$$\check{y}(n) = \sum_{i=-\infty}^{\infty} x(i)\, h(Mn-i) = \sum_{i=-\infty}^{\infty} h(i)\, x(Mn-i). \tag{5.16}$$

Compared to traditional convolution, the operation in (5.16) corresponds to a reduction computation by a factor of $M$ (since only 1-out-of every $M$ outputs is needed from the filter).

From (5.11), the decimator output in the $z$-domain can be simplified to obtain

$$\check{Y}(z) = \frac{1}{M}\sum_{k=0}^{M-1} H(z^{1/M} e^{j\frac{\pi k}{M}})\, X(z^{1/M} e^{j\frac{2\pi k}{M}}). \tag{5.17}$$

*Example 1* An audio signal is sampled at $f_s = 22{,}050$ Hz. Its spectrum is shown in Fig. 5.9a. This signal is to be decimated by a factor of $M = 3$. Therefore, prior to down-sampling, it is necessary to limit its spectrum to $\pm\check{f}_s/2 = \pm f_s/(2 \times 3) = \pm 3675$ Hz. This is done by passing the signal through an appropriate low-pass filter as shown in Fig. 5.9a (the dotted line). The threefold decimated spectrum is depicted in Fig. 5.9b which represents a zoomed in version of the original spectrum in the frequency band $\pm 3675$ Hz. If one listens to the decimated signal, the loss of high frequency components can clearly be detected, but the decimated signal is faithful to the lowpass portion of the original one.

### 5.3.2 *Interpolation*

The reverse process to decimation is "interpolation". The first stage in the interpolation process is up-sampling by an integer factor $L$, which is realized by inserting $L - 1$ zero samples between adjacent input signal samples. This up-sampled signal has a sampling rate which is $L$ times the original sampling rate ($\hat{f}_s = Lf_s$), and a spectrum that correspondingly has $L$ times as many repeated images of the original signal spectrum (see Fig. 5.4). To correctly obtain the interpolated signal, one needs to low-pass filter the upsampled signal. This filtering eliminates all the extra spectral images introduced by the upsampling.

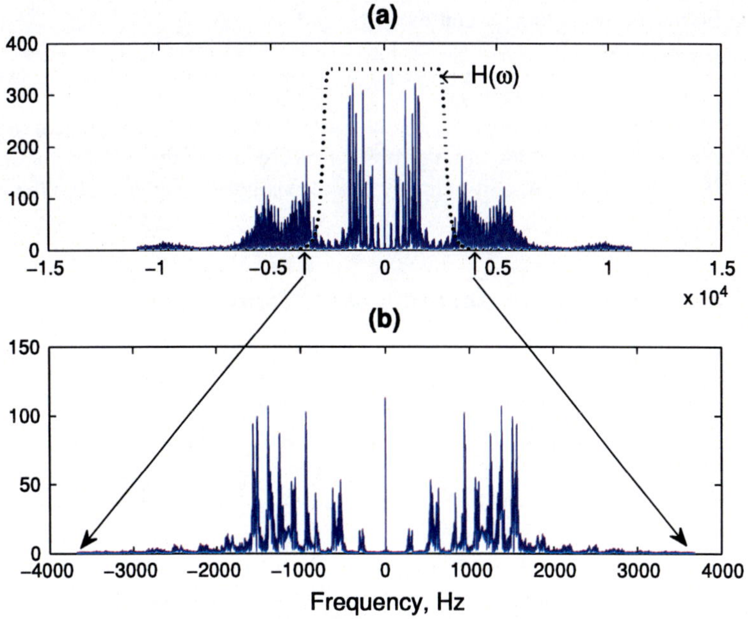


**Fig. 5.9** Example 1: decimation of an audio signal

This so-called interpolation filter should be an ideal low-pass filter with a cutoff frequency of $\pi/L$. The block diagram of an interpolator is shown in Fig. 5.10.

In the time-domain the $L$-times up-sampling operation is characterised in (5.2). This operation can be re-written as

$$u_2(Lp) = x(p), \quad \text{where } p = 0, \pm 1, \pm 2, \ldots \tag{5.18}$$

The final interpolation filter output is

$$\hat{y}(n) = \sum_{p=-\infty}^{\infty} x(p)\, h(n - Lp). \tag{5.19}$$

**Fig. 5.10** Interpolator

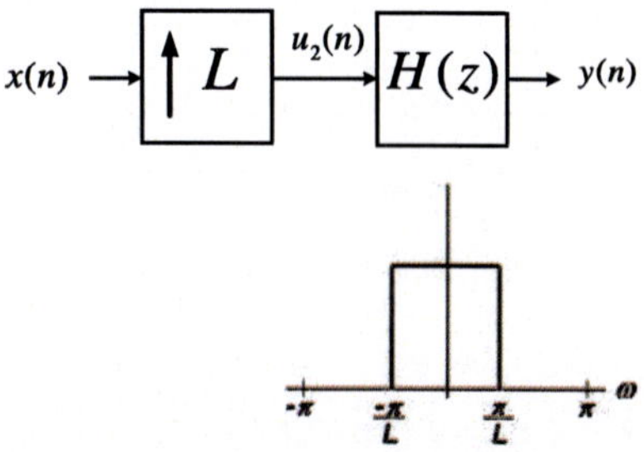

The output in the $z$-domain can be expressed as

$$\hat{Y}(z) = H(z)\,X(z^L). \tag{5.20}$$

As with the case of decimation, interpolation filters can be implemented relatively efficiently. Many of the input samples are zero and so there is a reduction by a factor $L$ in the number of add-multiply operations needed for realizing the filter.

### 5.3.3 Rational Number Sampling Rate Conversion

The sampling rate conversions that have been considered so far have involved integer changes in sampling rate (via either decimation or interpolation). There are many applications, however, that require the alteration of the sampling rate by some rational number, i.e., by $L/M$, where both $L$ and $M$ are arbitrary positive integers. There are even some applications (such as the pitch control of audio signals) that require irrational factor sampling rate conversion. This subsection will address only rate conversion by rational numbers. As shown in Fig. 5.11, a sampling rate change of $L/M$ should be realized by cascading an $L$-fold interpolator with an $M$-fold decimator. As decimation destroys information while interpolation does not, the decimator should be preceded by the interpolator. In this case the time-domain expression of the output is

$$Y(z) = \sum_{p=-\infty}^{\infty} h(Mn - pL)\,x(p), \tag{5.21}$$

where $p$ is as defined in (5.18).

For computationally efficient implementation of the structure shown in Fig. 5.11, the interpolation filter $H_u(z)$ and the decimation filter $H_d(z)$ can be replaced by an equivalent single filter $H(z)$. This replacement is achievable since

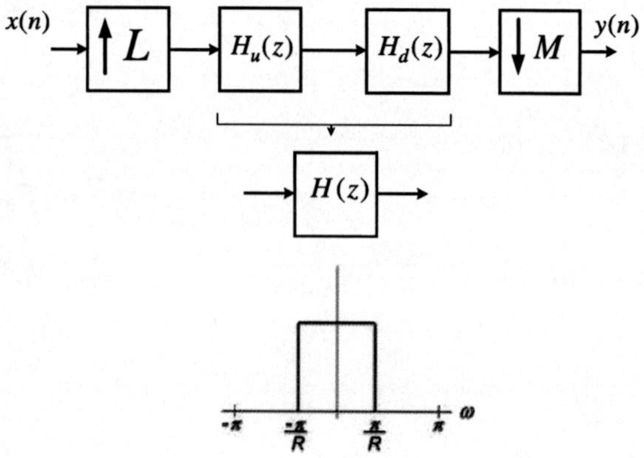


**Fig. 5.11** Rational-rate ($L/M$) sampling rate conversion scheme; $R = \max(M, L)$

both $H_d(z)$ and $H_u(z)$ are operating with the same sampling rate. The $H(z)$ design parameters, though, depend on the exact change in sample rate required. The filter must be able to perform both the interpolation and decimation effectively. To achieve this it is necessary that the normalized cutoff frequency of $H(z)$ be $\pi/\max\{L, M\}$.

**MATLAB:** Sampling rate conversion of a sequence can be achieved using the following MATLAB functions:

- `decimate` and `interpolate`: to decimate and interpolate a sequence by an integer factor, respectively.
- `re-sample` and `upfirdn`: to decimate/interpolate a sequence by a rational ratio. These functions offer several options.

*Example 2* Three different sampling rates are employed in digital audio applications, specifically, 44.1 kHz for digital music CDs, 48 kHz for digital audio tape, and 32 kHz for broadcasting. Down-converting from 48 to 44.1 kHz is equivalent to a sample-rate conversion by a rational factor of $L/M = 44.1/48 = 147/160$. The scheme shown in Fig. 5.11 can be used for the conversion. In this case $M > L$ and therefore the Low-pass filter design specification is dominated by the decimation process. That is, after being up-sampled by a factor of 147, the signal should be bandwidth limited to $\pm\pi/160$ before down-sampling. Figure 5.12 compares the input and the output sequences.

It is worth noting that the complexity of the rational sampling rate convertor is dependent on the complexity of the ratio $L/M$. For instance, when transferring digitally recorded music from digital tape to CD, the interpolated signal bandwidth will be $147 \times 48$ kHz $= 7.056$ MHz. From a practical perspective, moving audio signals to the megahertz range is undesirable. The techniques presented in the next section provide a means to overcome this kind of problem.

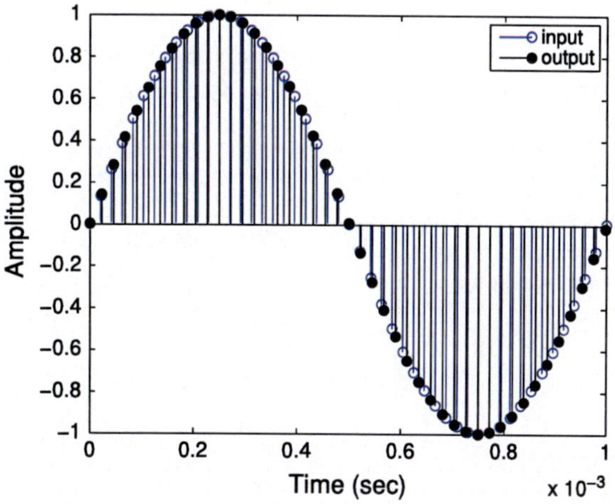


**Fig. 5.12** Down-converting from 48 to 44.1 kHz

## 5.4 Efficient Implementation of Multirate Systems

### *5.4.1 Noble Identities*

Two important identities are presented here for facilitating flexibility in the implementation of multirate systems. These identities relate to decimation and interpolation, and are shown in Fig. 5.13a and b, respectively. These identities are useful for simplifying the analysis and design of sophisticated multirate systems. They allow the different components of decimators/interpolators to be commuted as needed. As will be seen subsequently, these identities pave the way for significant computational savings to be obtained in multi-rate system implementations.

### *5.4.2 Polyphase Decomposition*

Consider the following $z$-transform domain input signal:

$$X(z) = 1 + 2z^{-1} + 3z^{-2} + 4z^{-3} + 5z^{-4} + 6z^{-5} + 7z^{-6}.$$

To perform a polyphase decomposition of this signal into two components, one just regroups the terms into even and odd powers of $z$:

$$X(z) = \underbrace{[1 + 3z^{-2} + 5z^{-4}]}_{X_0(z^2)} \quad + \quad z^{-1}\underbrace{[2 + 4z^{-2} + 6z^{-4}]}_{X_1(z^2)}.$$

or equivalently,

$$X(z) = \underbrace{[1 + 3z^{-2} + 5z^{-4}]}_{X_0(z^2)} \quad + \quad z\underbrace{[2z^{-2} + 4z^{-4} + 6z^{-6}]}_{X_1(z^2)}.$$

where $X_0(z^2)$ and $X_1(z^2)$ are the polyphase components of the original signal. These components are functions of $z^2$, and it easily be shown that $X(z) = X_0(z^2) + z^{-1}$ $X_1(z^2)$. This polyphase decomposition can be viewed as the decomposition of the

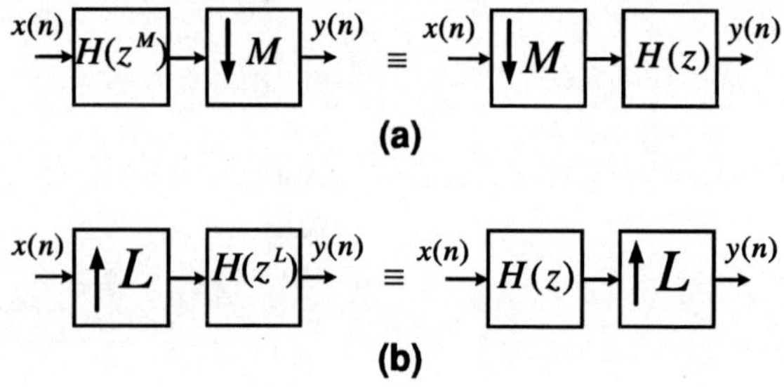


**Fig. 5.13** Useful identities: **a** decimation identity, **b** interpolation identity

original sequence into two subsequences, with each subsequence being a sequentially shifted version of the original sequence. Furthermore, because each of these two subsequences incorporates every second sample, they may be considered to be downsampled subsequences of the original input sequence.

More generally, for any sequence $\{x(n)\}$ with a $z$-transform given by

$$X(z) = \sum_{n=-\infty}^{\infty} x(n)z^{-n},$$

a re-arrangement into $M$ components can be achieved according to

$$X(z) = \sum_{k=0}^{M} z^{-k}X_k(z^M), \tag{5.22}$$

where

$$X_k(z) = \sum_{n=-\infty}^{\infty} x(Mn+k)z^{-n}, \quad 0 \le k \le M-1. \tag{5.23}$$

This decomposition of the input is illustrated in Fig. 5.14, where $\{x_k(n)\}$ is the $k$th subset of the parent sequence and the sub-sequences are related to each other according to

Fig. 5.14 $M$-band polyphase decomposition of a sequence

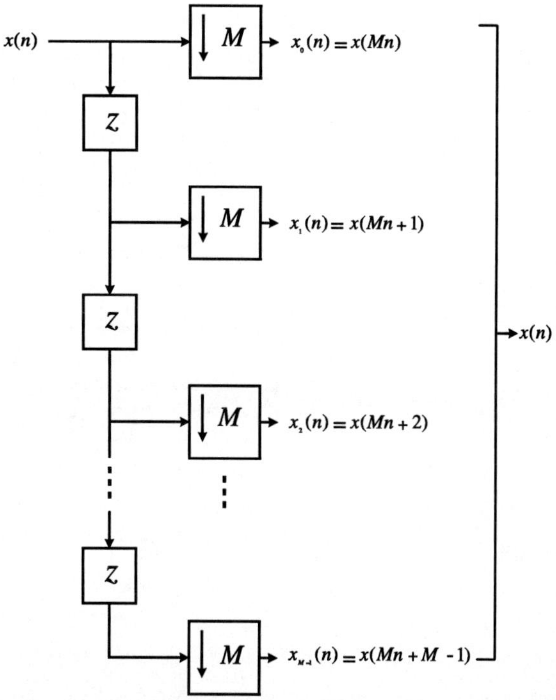

$$x_k(n) = x(Mn + k), \quad 0 \leq k \leq M - 1. \tag{5.24}$$

or,

$$x(n) = \sum_{k=0}^{M-1} x(Mn + k). \tag{5.25}$$

FIR based decimators can be implemented very efficiently by replacing the conventional low-pass FIR filter with its polyphase decomposition. To understand this more fully, recall the decimation structure shown in Fig. 5.8. The $M$-band polyphase decomposition of the filter transfer function $H(z)$ can be represented as

$$H(z) = \sum_{k=0}^{M-1} z^{-k} H_k(z^M), \tag{5.26}$$

and the direct realization of (5.26) is depicted in Fig. 5.15. According to the decimation noble identities (Fig. 5.13a), this $M$-band FIR polyphase filter can be replaced by its equivalent structure as shown in Fig. 5.16. As illustrated in Fig. 5.16, the FIR sub-band filter $H_0(z)$ is the $M$-fold decimated version of $H_0(z^M)$, and $H_1(z)$ is the $M$-fold decimated version of $H_1(z^M)$ and so on. By using this polyphase decomposition approach the computational requirements can be reduced by a factor of $M$, since the FIR filter can work at the lower (down-sampled) sampling rate $f_s/M$.

Similar to the case of decimation, the interpolator shown in Fig. 5.10 can be implemented using the polyphase decomposition technique. Utilizing the Noble identity for interpolation (see Fig. 5.13), an efficient $L$-band interpolation structure

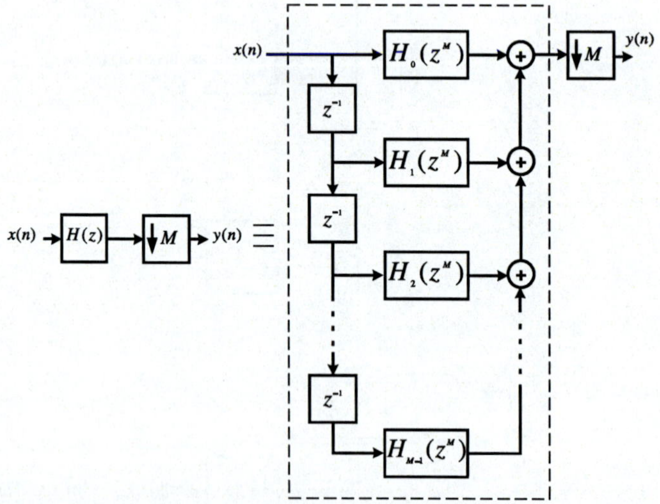


**Fig. 5.15** Polyphase decomposition of an FIR decimation filter

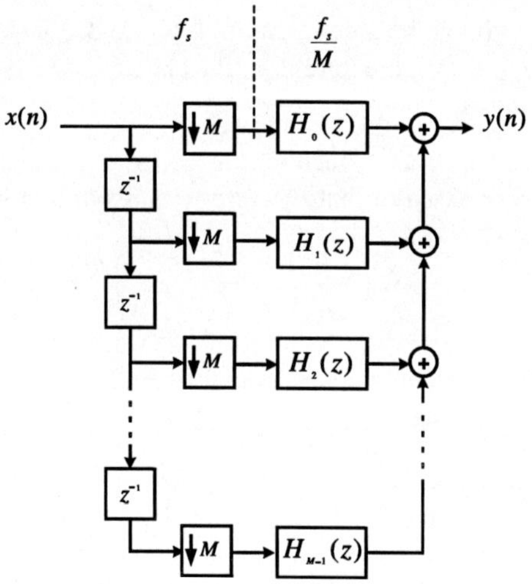


**Fig. 5.16** Efficient implementation of the polyphase decimation FIR filter shown in Fig. 5.15

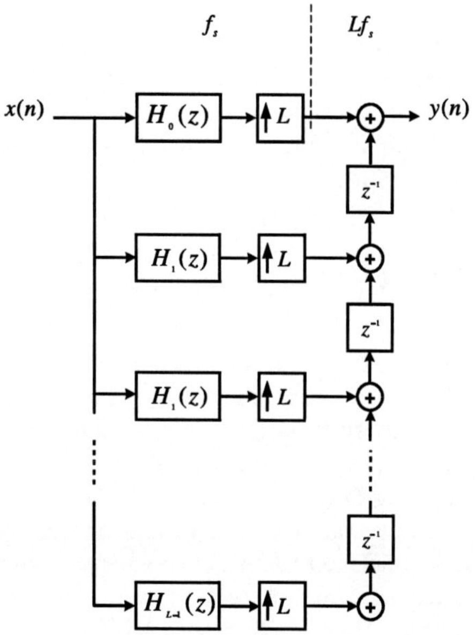


**Fig. 5.17** Efficient implementation of the polyphase interpolation FIR filter

can be realized as illustrated in Fig. 5.17. Generally, the computation can be reduced by a factor of $L$ compared to the traditional (single-rate) FIR filter realization. It is worth noting that in the case of linear-phase FIR filters, further reduction in the computation requirements can be gained due to the symmetry of the filter coefficients.

It can be shown that polyphase decomposition filters are, in fact, all-pass filters with variable phase-shifts [1] and this is the genesis of the term "polyphase" filters.

*Example 3* Consider the following linear-phase FIR low-pass filter of length $N = 6$. Note that the filter has a symmetric impulse response.

$$H(z) = h(0) + h(1)z^{-1} + h(2)z^{-2} + h(2)z^{-3} + h(1)z^{-4} + h(0)z^{-5}$$

Implement an $M = 3$-fold decimation filter using the polyphase decomposition technique.

*Solution:*

The three sub-filters can be obtained by inspection as:

$$H_0(z) = h(0) + h(2)z^{-3},$$
$$H_1(z) = h(1) + h(1)z^{-3},$$
$$H_2(z) = h(2) + h(0)z^{-3}.$$

Note that $H_0(z)$ and $H_2(z)$ have mirror image impulse responses while $H_1(z)$ has a symmetric impulse response. These relations can be utilized to further reduce the computational requirements.

These sub-filters operate on the input sequence according to the structure shown in Fig. 5.15.

### 5.4.3 Multistage Implementation

So far only single-stage decimation and interpolation structures have been considered. In many applications requiring large decimation/interpolation factors, however, single-stage implementation is often computationally too inefficient. In addition, for rational sampling rate conversion applications the intermediate sampling frequency after interpolation may be very high, making implementation problematical. This was evident in Example 3. To tackle these problems, multistage implementation can be utilized. As an illustration, Fig. 5.18 shows an $r$ stage implementation of a $M$-fold decimator. This multi-stage realization is achieved by splitting $M$ into factors $M = M_1 M_2 M_r$. Obviously, in order to do that, $M$ should be a composite integer. The filters $H_1(z)$, $H_2(z)$, and $H_3(z)$ should be designed such

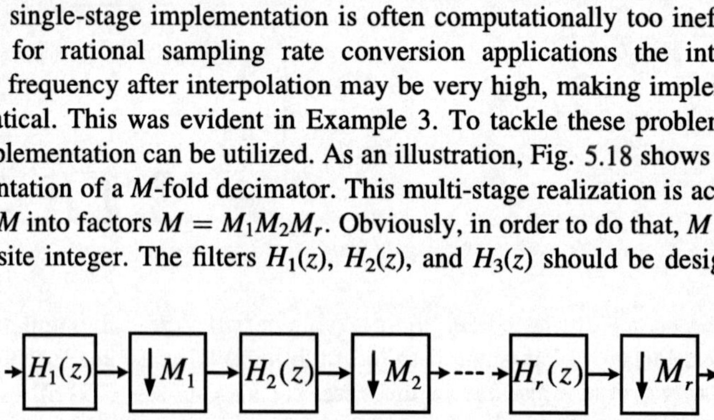


**Fig. 5.18** Multistage implementation of an $M$-fold decimator

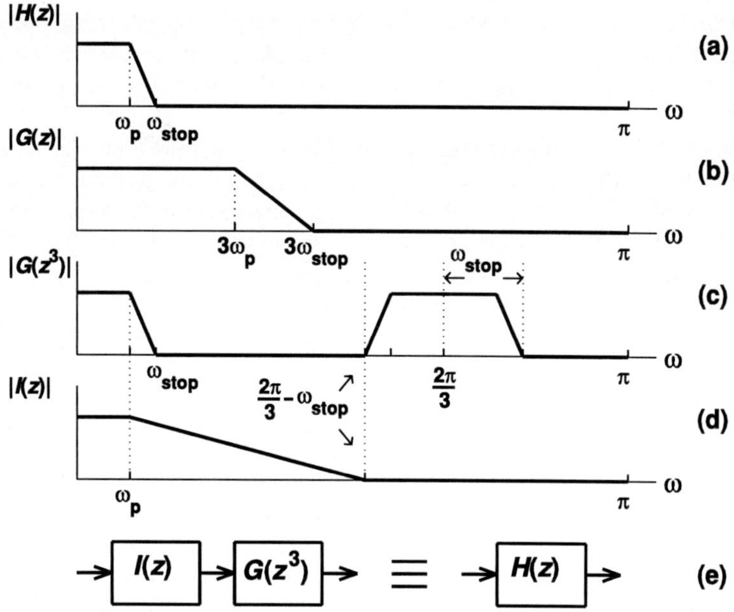


**Fig. 5.19** Multistage implementation of a threefold decimator using IFIR approach. **a** The desired frequency response, **b** the threefold stretched filter response $G(z)$, **c** the threefold upsampled version of $G(z)$, i.e., $G(z^3)$, **d** the required lowpass filter response $I(z)$, **e** the overall IFIR filter

that aliasing is avoided and overall pass-band and stop-band tolerances are satisfied.

### 5.4.3.1 Interpolated FIR Filter Design

Interpolated FIR (IFIR) design technique is a common approach to realise filters within multistage sampling rate converters. The rationale behind IFIR design is as follows. One normally requires many filter taps to obtain very sharp cutoffs with FIR filters. One can, however, use an elegant "slight of hand" to get sharp cutoffs without a large number of taps. To achieve this one first up-samples—recall that after up-sampling by a factor $M$ there are $M$ times as many images. Because these images must fit in a relatively small spectral area (i.e., they are compressed replicas), they tend to have sharp cutoffs. The second stage in the IFIR approach is to filter out the additional images using a low order filter. Then only the single transfer function image (with a sharp cutoff) is left.

To illustrate the IFIR approach, consider an example in which a $M = 3$-fold single-stage decimation filter $H(z)$ is required. The desired frequency response of this filter is shown in Fig. (5.19a), where $\omega_p$ and $\omega_{\text{stop}}$ are the pass-band and stop-band normalized frequency edges, respectively. In this case, the transition band will be $\omega_{stop} - \omega_p$. The equivalent IFIR design is begun by assuming a threefold "pre-stretched" filter $G(z)$ as shown in Fig. (5.19b). Then, $G(z)$ is upsampled by the same factor $M = 3$. This yields the filter $G(z^3)$, which has the desired magnitude response ($H(z)$) along with $M - 1$ image replicas, as shown in Fig. (5.19c). To eliminate the unwanted image replicas, another low-pass filter $I(z)$ should be cascaded with $G(z^3)$. The pass-band of $I(z)$ should be the same as the original one (with a cutoff of $\omega_p$), while the stop-band edge would extend from $\omega_p$ to the edge of the adjacent image, that is at $\frac{2\pi}{3} - \omega_{stop}$ (see Fig. (5.19d)). This second stage filter therefore has very relaxed cutoff requirements and can be implemented without a great deal of computation.

For a given filter specification, the overall order of IFIR designed filters tends to be significantly lower than for conventional filter designs.

**MATLAB:** The MATLAB function `ifir` can be used to design both $G(z^L)$ (where $L$ is the interpolation factor) and the image-suppressor filter $I(z)$.

*Example 4* The conventional decimator shown in Fig. 5.20a is to be implemented as a cascaded structure using the IFIR technique. The input signal $x(n)$ is originally sampled at 210 kHz and needs to be down-sampled to 7 kHz. The pass-band and stop-band frequencies are 3 and 3.5 kHz, respectively. The anti-aliasing filter $H(z)$ needs to be designed as an equiripple linear-phase FIR filter with the following specifications: The pass-band ripple is $\delta_p = 0.02$ and stop-band ripple is $\delta_s = 0.001$. Compare the computation complexity of the conventional and cascaded implementations.

*Solution:*

The MATLAB function `firpmord` (Parks-Maclellan FIR order estimator) is used to obtain the order of the FIR filters. According to the given specifications,

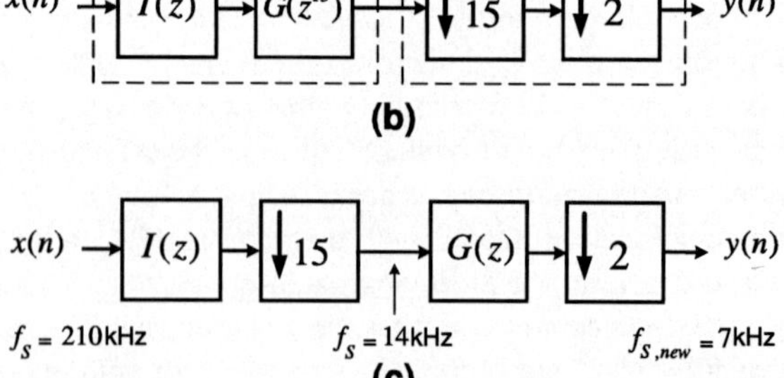


**Fig. 5.20** Multistage implementation of the decimator in Example 4 using IFIR design technique

**Table 5.1** Specifications of the IFIR filters in Example 4

| Filter | $f_p$ (kHz) | $f_{stop}$ (kHz) | $\delta_p$ | $\delta_{stop}$ | Order |
|---|---|---|---|---|---|
| $G(z)$ | 45 | 52.5 | 0.01 | 0.001 | 91 |
| $I(z)$ | 3 | 10.5 | 0.01 | 0.001 | 91 |

the order of the conventional single-stage anti-aliasing filter $H(e^{j\omega})$ is $N_{single} = 1279$.

Now the conventional single-stage filter $H(e^{j\omega})$ is to be replaced by the cascade implementation of $G(e^{j15\omega})I(e^{j\omega})$ using the IFIR approach as explained above. To meet the desired specifications, the pass-band ripple of $G(e^{j15\omega})$ and $I(e^{j\omega})$ should sum up at most to that of $H(e^{j\omega})$. Then, one may assume that the peak pass-band ripple is taken as $\delta_p/2$. On the other hand, to be conservative, the stop-band ripples of $G(e^{j15\omega})$ and $I(e^{j\omega})$ are assumed to be equal to that of $H(e^{j\omega})$. In this case, the specifications of the IFIR cascaded filters are as given in Table 5.1.

The order of the pre-stretched low-pass filter $G(e^{j\omega})$ is found to be $N_G = 91$. Thus, the order of its interpolated version is multiplied by a factor of 15, i.e., $G(e^{j15\omega})$, will be $N_G^{15} = 15 \times 19 = 1365$. Then, the order of the anti-imaging filter $I(e^{j\omega})$ is $N_I = 91$. This new IFIR equivalent combination $I(e^{j\omega})$ $G(e^{j15\omega})$, as illustrated in Fig. 5.20b, has a total of $N_G^{15} + N_I = 1365 + 91 = 1456$ coefficients. This, obviously, seems an increase in the order as compared to the single-stage one $N_H = 1279$. However, if one utilizes the decimation Noble identity, as shown in Fig. 5.13a, then the filter $G(e^{j15\omega})$ and the the subsequent 15-fold down-sampler can be replaced by the same down-sampler followed by the filter $G(e^{j\omega})$. This is shown in Fig. 5.20c. By doing that, the overall filter order reduces to $N_{overall} = N_G + N_I + 91 + 91 = 182$. Thus a reduction in the overall filter order of $N_H/N_{overall} = 1279/182 \simeq 7$ is achieved.

The IFIR technique leads to the design of a cascaded implementation ($G(z)$ and $I(z)$) which meets the required specifications using a much lower order. If a multiple stage approach is preferred, one may repeat this approach on the multiple stages. The different stages are obtained by factorizing $M = M_1M_2M_N$.

Note that the multistage implementation approach can be applied for interpolation filters as well.

## Reference

1. Vaidyanathan, P.P.: Multirate Systems and Filter Banks. Prentice Hall, USA (1993)

# Appendix A: Tutorials

## *Tutorial 1*

**Q**: Plot the unit step function $u(t)$. Also plot the functions $u(-t)$, $u(t-3)$, $u(-t-3)$ $u(t+5)$, $u(-t+5)$, and $y(t) = u(t) - u(t-1)$.

**Solution**:

$$u(t) = \begin{Bmatrix} 1, & t \geq 0 \\ 0, & t < 0 \end{Bmatrix}, \Rightarrow \quad \therefore \quad u(-t) = \begin{Bmatrix} 1, & -t \geq 0 \\ 0, & -t < 0 \end{Bmatrix} = \begin{Bmatrix} 1, & t \leq 0 \\ 0, & t > 0 \end{Bmatrix}$$

Generally we have:

$$u(t-a) = \begin{Bmatrix} 1, & t-a \geq 0 \\ 0, & t-a < 0 \end{Bmatrix} = \begin{Bmatrix} 1, & t \geq a \\ 0, & t < a \end{Bmatrix},$$

$$u(-t-a) = \begin{Bmatrix} 1, & -t-a \geq 0 \\ 0, & -t-a < 0 \end{Bmatrix} = \begin{Bmatrix} 1, & t \leq -a \\ 0, & t > -a \end{Bmatrix}$$

**Shift rules:**

If $g(t)$ is a function and $a > 0$, then the direction of horizontal shift from original location is found as follows:

| Function | Shift direction |
|---|---|
| $g(t+a)$ | – (Left) |
| $g(t-a)$ | + (Right) |
| $g(-t+a)$ | + (Right) |
| $g(-t-a)$ | – (Left) |

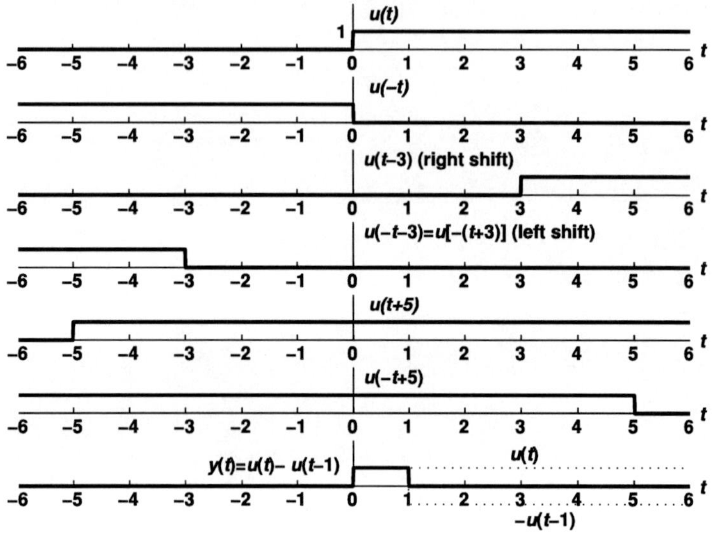


## Tutorial 2

**Q1**: Find the I/O (input–output) relation of the system shown below.

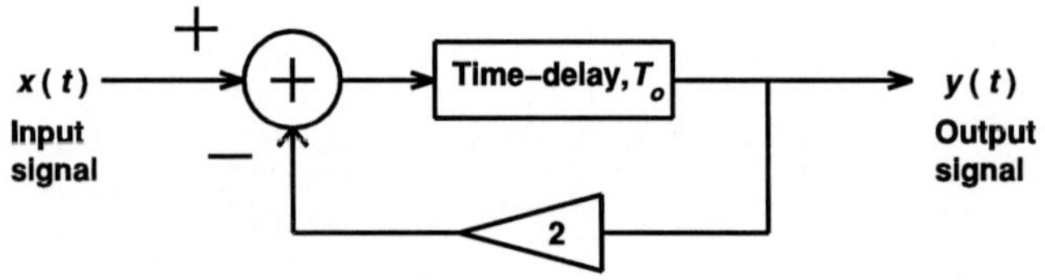


**Solution**: First, we should define the important internal points as shown below:

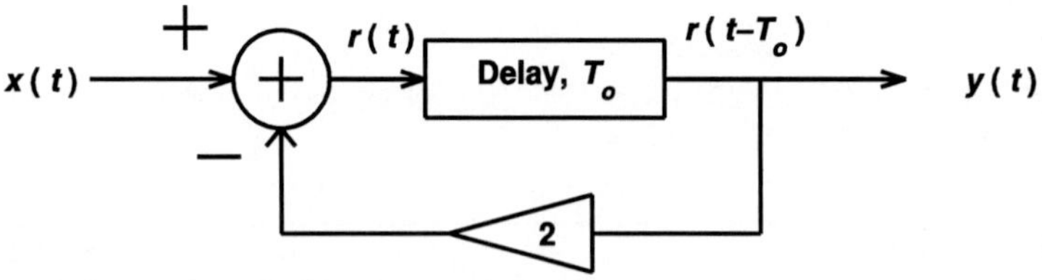


Now we write the system equations as follows:

$$y(t) = r(t - T_o); \quad r(t) = x(t) - 2y(t),$$

$$\therefore \quad y(t) = x(t - T_o) - 2y(t - T_o).$$

This is a ***feedback system*** since the I/O equation includes a delayed version of the output itself.

**Q2**: Find the I/O (input–output) relation of the system shown below.

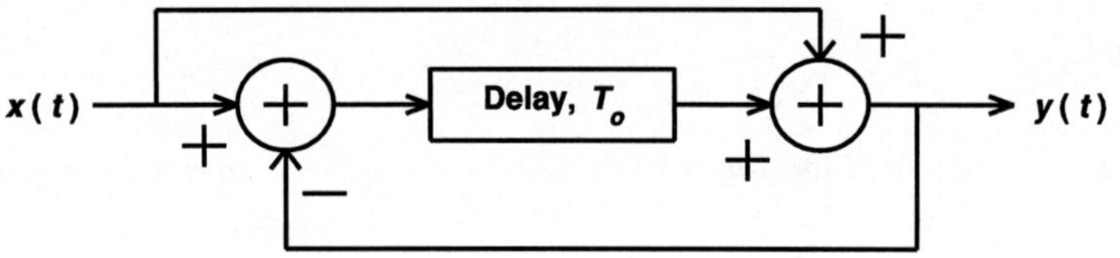


Ans. $y(t) = x(t) + x(t - T_o) - y(t - T_o)$.

## *Tutorial 3*

**Q**: Determine whether the analog time-delay $T_o$ ($T_o$ is constant) is:
1. Memoryless, 2. causal, 3. linear, 4. time-invariant, 5. stable.

**Solution**:

1. Since $y(t) = x(t - T_0)$, the output equals the input at a past time instant, $t - T_o$, hence it is a memory system.
2. The output $y(t)$ is not a function of $x(t + t_0)$, $t_0 > 0$, hence it is causal (does not depend on future values of the input).
3. Let **T** represents the system transformation.
   For the input $x(t) = ap(t) + br(t)$ we have:

$$y(t) = \mathbf{T}\{x(t)\} = x(t - T_0) = a \cdot p(t - T_0) + b \cdot r(t - T_0)$$
$$= a \cdot \mathbf{T}\{p(t)\} + b \cdot \mathbf{T}\{r(t)\}$$

   Hence, the system is linear.

4.
$$\mathbf{T}\{x(t - t_0)\} = x(t - t_0 - T_0) \quad (1)$$

   We have: $y(t) = x(t - T_0)$, hence,

$$y(t - t_0) = x(t - t_0 - T_0) \quad (2)$$

   From Eqs. 1 and 2 we get:

$$y(t - t_0) = \mathbf{T}\{x(t - t_0)\}.$$

   Hence, **T** is time-invariant.

5. If $|x(t)| \leq c \ \forall t$ (bounded), then we have $|x(t - T_0)| \leq c \ \forall t$.
Hence, $|y(t)| \leq c \ \forall t$ and the system is BIBO-stable.

## *Tutorial 4*

**Q**: Find the output of the following linear time-invariant system for a unit-step input (i.e., find the step response). Verify that the impulse response is the time derivative of the step response.

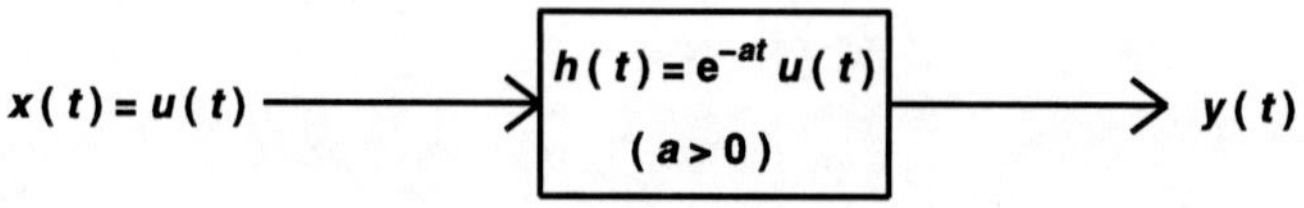


**Solution**: We have: $y(t) = u(t) * h(t) = \int_{-\infty}^{\infty} u(\lambda)h(t - \lambda)d\lambda$.

*Step 1*: Formulas of functions using $t$:

$$u(t) = \begin{Bmatrix} 1, & t \geq 0 \\ 0, & t < 0 \end{Bmatrix} \quad \text{and} \quad h(t) = e^{-at}u(t) = \begin{Bmatrix} e^{-at}, & t \geq 0 \\ 0, & t < 0 \end{Bmatrix}.$$

*Step 2*: Formulas of functions using $\lambda$:

$$u(\lambda) = \begin{Bmatrix} 1, & \lambda \geq 0 \\ 0, & \lambda < 0 \end{Bmatrix} \quad \text{and}$$

$$h(t - \lambda) = \begin{Bmatrix} e^{-a(t-\lambda)}, & t - \lambda \geq 0 \rightarrow \lambda \leq t \\ 0, & \lambda > t \end{Bmatrix}.$$

*Step 3*: Integrate to find the convolution. The range of integration can be found graphically as the zone of non-zero overlap between the two functions. This can be done by moving $h(t - \lambda)$ from left to right (i.e., by varying $t$ from $-\infty$ to $\infty$) while keeping $u(\lambda)$ fixed (see the plot below).

*Case 1*: $t > 0$

$$y(t) = \int_0^t \exp[-a(t - \lambda)]d\lambda$$

$$= (1/a)[\exp\{-a(t - \lambda)\}]_0^t = (1/a)[1 - \exp(-at)].$$

*Case 2*: $t < 0$

No overlap $\Rightarrow u(\lambda)h(t - \lambda) \equiv 0 \ \forall \lambda \Rightarrow y(t) \equiv 0$.

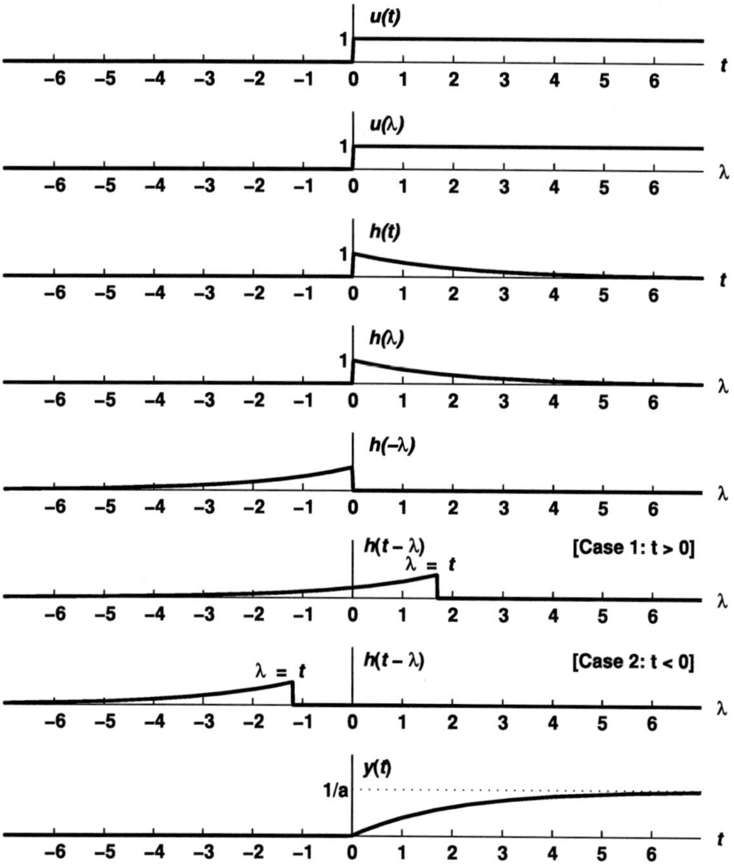


## *Tutorial 5*

**Q**: Evaluate $y(t) = x(t) * h(t)$, where $x(t)$ and $h(t)$ are as shown.

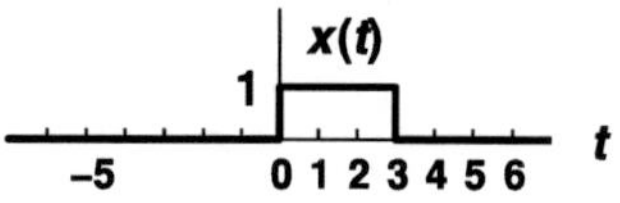


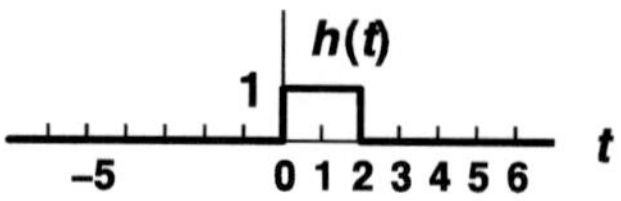


**Solution**: The convolution is given by $y(t) = x(t) * h(t) = \int_{-\infty}^{\infty} x(\lambda)h(t - \lambda)d\lambda$.

*Step 1*: Equations of $x(t)$ and $h(t)$:

$$x(t) = \begin{Bmatrix} 1, & 0 \le t \le 3 \\ 0, & \text{elsewhere} \end{Bmatrix} \quad \text{and} \quad h(t) = \begin{Bmatrix} 1, & 0 \le t \le 2 \\ 0, & \text{elsewhere} \end{Bmatrix}.$$

*Step 2*: Equations in terms of $\lambda$:

$$x(\lambda) = \begin{Bmatrix} 1, & 0 \le \lambda \le 3 \\ 0, & \text{elsewhere} \end{Bmatrix} \quad \text{and}$$

$$h(t-\lambda) = \begin{Bmatrix} 1, & 0 \le t-\lambda \le 2 \\ 0, & \text{elsewhere} \end{Bmatrix} = \begin{Bmatrix} 1, & \underbrace{t-2}_{\text{start}} \le \lambda \le \underbrace{t}_{\text{end}} \\ 0, & \text{elsewhere} \end{Bmatrix}$$

*Step 3*: Integrate to find $y(t)$.

*Case 1*: $t < 0$ or $t > 5 \to y(t) \equiv 0$ (no overlap).
*Case 2*: $0 < t \le 2 \to y(t) = \int_0^t dt = t.$
*Case 3*: $2 < t \le 3 \to y(t) = \int_{t-2}^t dt = 2.$
*Case 4*: $3 < t \le 5 \to y(t) = \int_{t-2}^3 dt = 5 - t.$

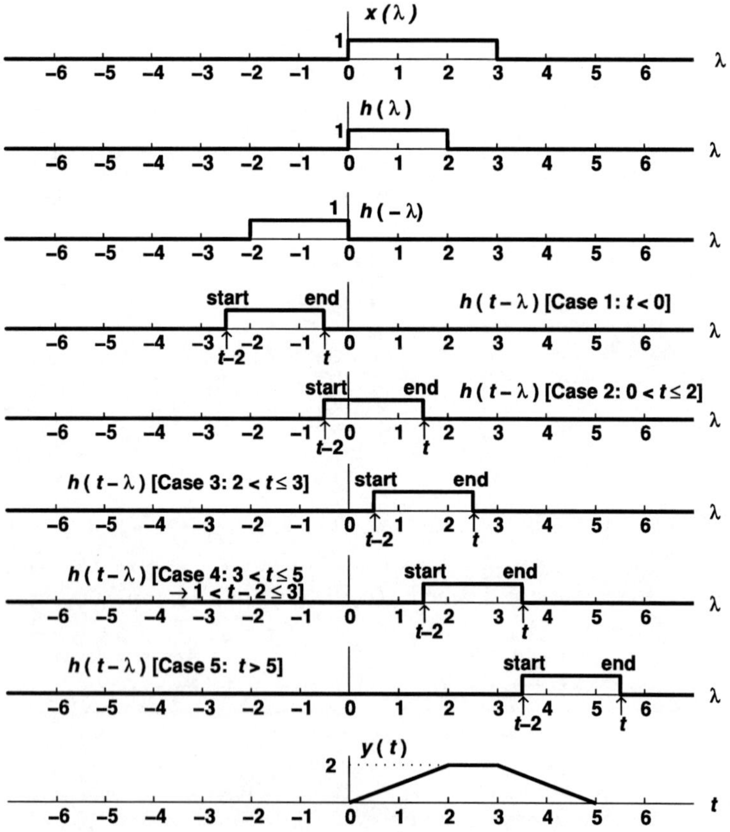

## *Tutorial 6*

**Q**: Find the convolution of the functions $x(t)$ and $h(t)$ as shown below.

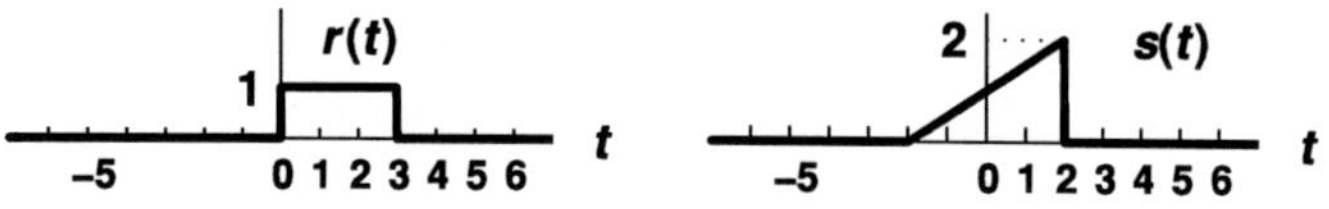


**Solution**: The convolution of $r(t)$ and $s(t)$ is given by:

$$y(t) = r(t) * s(t) = \int_{-\infty}^{\infty} r(\lambda)s(t-\lambda)d\lambda.$$

*Step 1*: Write the equations of $r(t)$ and $s(t)$ as follows:

$$r(t) = \begin{Bmatrix} 2, & 0 \le t \le 2 \\ 0, & \text{elsewhere} \end{Bmatrix}.$$

To find the equation of $s(t)$, we consider 2 points (as it is a straight line): $(t_1, s_1) = (-2, 0)$ and $(t_2, s_2) = (2, 2)$.

$$\therefore \quad \frac{s - s_1}{t - t_1} = \frac{s_2 - s_1}{t_2 - t_1} \Rightarrow s = \frac{t}{2} + 1 \Rightarrow s(t) = \begin{Bmatrix} t/2 + 1, & -2 \le t \le 2 \\ 0, & \text{elsewhere} \end{Bmatrix}$$

*Step 2*: Write the above equations in terms of $\lambda$, then sketch $r(\lambda)$ and $s(t - \lambda)$:

$$r(\lambda) = \begin{Bmatrix} 2, & 0 \le \lambda \le 2 \\ 0, & \text{elsewhere} \end{Bmatrix} \quad \text{and}$$

$$s(t-\lambda) = \begin{Bmatrix} (t-\lambda)/2 + 1, & -2 \le t - \lambda \le 2 \\ 0, & \text{elsewhere} \end{Bmatrix}.$$

$$= \begin{Bmatrix} (t-\lambda)/2 + 1, & \underbrace{t-2}_{\text{start}} \le \lambda \le \underbrace{t+2}_{\text{end}} \\ 0, & \text{elsewhere} \end{Bmatrix}$$

[*Note*: $2 \le t - \lambda \le 2 \Rightarrow -2 - t \le -\lambda \le 2 - t \Rightarrow t - 2 \le \lambda \le t + 2$].

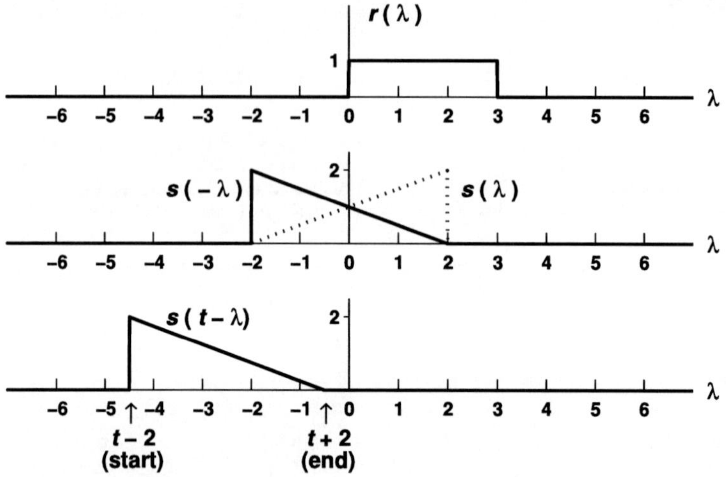


*Step 3*: Integrate to find $y(t)$. Consider the intervals of overlap between $r(\lambda)$ and $s(t-\lambda)$. These intervals can be defined using the endpoints of $r(\lambda)$ (which is fixed here) and $s(t-\lambda)$ (which is moving according to the shift $t$, noting that positive $t$ gives positive shift).

*Case 1*: End of $s(t-\lambda) <$ Start of $r(\lambda) \Rightarrow t+2 < 0 \Rightarrow t < -2$. No overlap, hence, $y(t) = 0$.

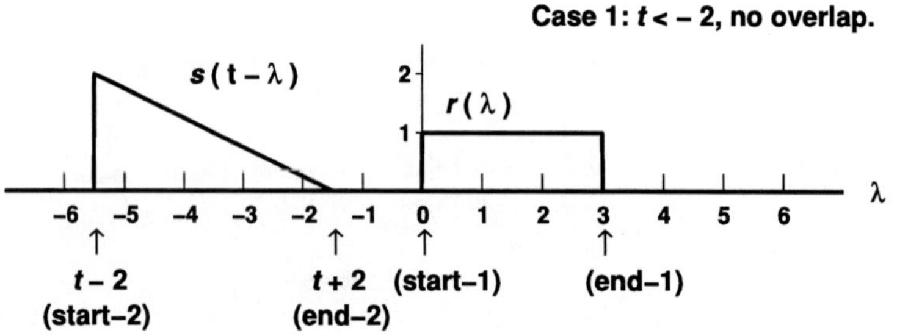


*Case 2*: As the end of $s(t-\lambda)$ [i.e., $\lambda = t+2$] moves right (when $t$ increases), $s(t-\lambda)$ moves right as well. The first overlap with $r(\lambda)$ occurs when the ***end*** of $s(t-\lambda)$ [i.e., $\lambda = t+2$] exceeds the ***start*** of $r(\lambda)$ [i.e., $\lambda = 0$]. Now consider that the ***end*** of $s(t-\lambda)$ is inside $r(\lambda)$, hence, $0 \le t+2 \le 3$, or equivalently $-2 \le t \le 1$. In this case we have:

$$y = \int_0^{t+2} \underbrace{[1]}_{r(\lambda)} \underbrace{[(t-\lambda)/2+1]}_{s(t-\lambda)} d\lambda = t^2/4 + t + 1$$

(where $-2 \le t \le 1$, as shown above).

**Case 2: – 2 ≤ *t* ≤ 1.**

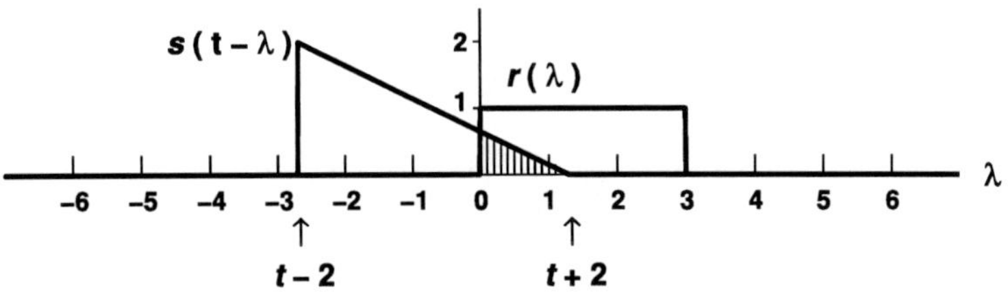


*Case 3*: As the ***end*** of $s(t - \lambda)$ [i.e., $\lambda = t + 2$] exceeds the ***end*** of $r(\lambda)$ [i.e., $\lambda = 3$], $s(t - \lambda)$ will fully overlap $r(\lambda)$ until the ***start*** of $s(t - \lambda)$ [i.e., $\lambda = t - 2$] reaches the ***start*** of $r(\lambda)$ [i.e., $\lambda = 0$]. Hence, we consider two conditions: $t + 2 > 3$ and $t - 2 < 0$, which give $t > 1$ and $t < 2$, or equivalently $1 < t < 2$. In this case we have:

$$y = \int_0^3 \underbrace{[1]}_{r(\lambda)} \underbrace{[(t-\lambda)/2 + 1]}_{s(t-\lambda)} d\lambda = (3/2)t + 3/4$$

(where $1 < t < 2$, as shown above).

**Case 3: 1 < *t* < 2.**

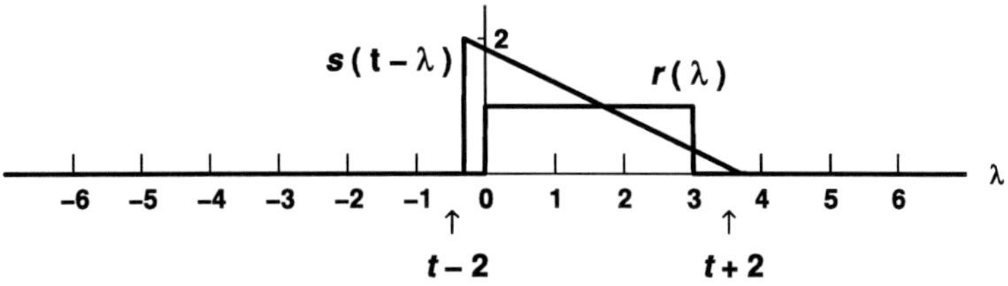


*Case 4*: As the ***start*** of $s(t - \lambda)$ [i.e., $\lambda = t - 2$] moves beyond the ***start*** of $r(\lambda)$ [i.e., $\lambda = 0$], overlap takes place only between $\lambda = t - 2$ [***start*** of $s(t - \lambda)$] and $\lambda = 3$ [***end*** of $r(\lambda)$]. Hence, we consider the condition $0 \le t - 2 \le 3$, or equivalently, $2 \le t \le 5$. Here we have:

$$y = \int_{t-2}^3 \underbrace{[1]}_{r(\lambda)} \underbrace{[(t-\lambda)/2 + 1]}_{s(t-\lambda)} d\lambda = 15/4 + t/2 - t^2/2$$

where $2 \le t \le 5$, as shown above).

**Case 4: $2 \leq t \leq 5$.**

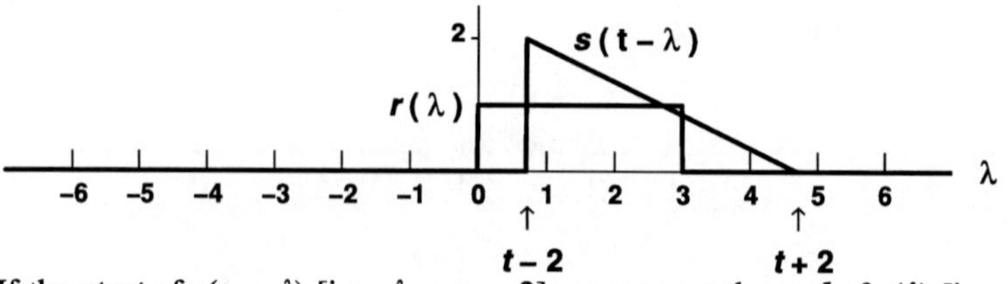


*Case 5*: If the ***start*** of $s(t-\lambda)$ [i.e., $\lambda = t-2$] moves past the ***end*** of $r(\lambda)$ [i.e., $\lambda = 3$], ***no overlap*** occurs. The condition here is written as $t-2>3$, or equivalently $t>5$. Here $y(t)=0$.

**Case 5: $t \geq 5$, no overlap.**

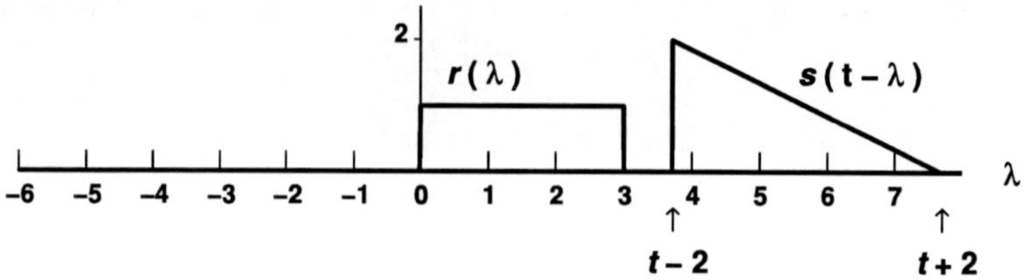


**Summary of Results**

$$y(t) = \left\{ \begin{array}{cc} t^2/4+t+1, & -2 \leq t \leq 1 \\ (3/2)t+3/4, & 1<t<2 \\ (15/4)+t/2-t^2/4, & 2 \leq t \leq 5 \\ 0 & \text{elsewhere} \end{array} \right\}$$

**Q**: Solve the above problem as $y(t) = s(t) * r(t) = \int_{-\infty}^{\infty} s(\lambda) r(t-\lambda) d\lambda$. You should get the same answer, since convolution is commutative.

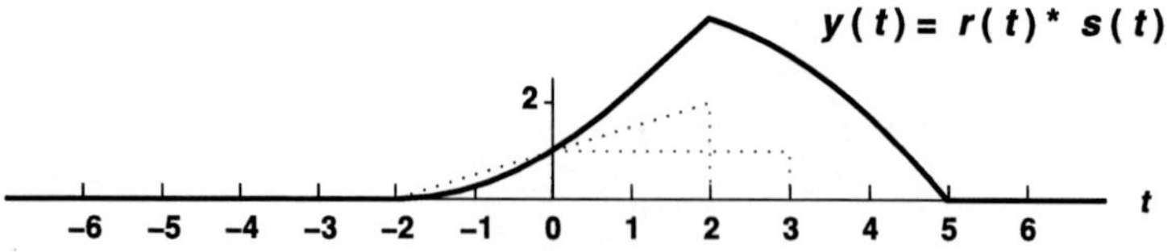

## Tutorial 7

**Q**: Show that $\delta(t) = \lim_{a \to 0} \frac{1}{2a} \Pi_{2a}(t)$ $[a > 0]$. What is the importance of this result?

**Solution**: We should prove that the function $g(t) = \frac{1}{2a} \Pi_{2a}(t)$ (shown below) satisfies the definition and properties of the delta function as $a \to 0$.

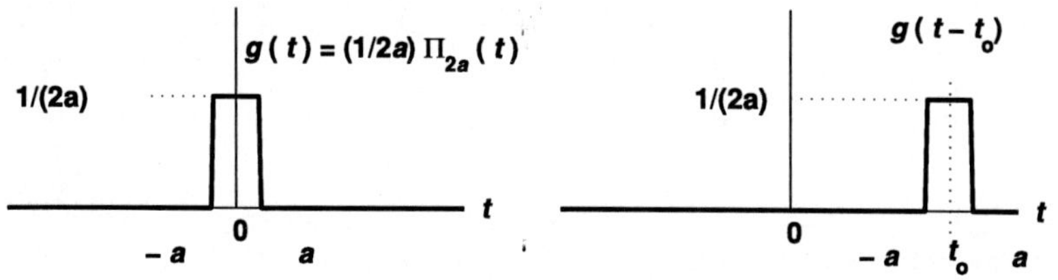


The delta function is defined as $\int_{-\infty}^{\infty} x(t)\delta(t - t_o)dt = x(t_o)$ for any continuous function $x(t)$ [see Tables]. Applying the same integral to $g(t)$ we get:

$$\int_{-\infty}^{\infty} x(t)g(t - t_o)dt = \frac{1}{2a} \int_{t_o - a}^{t_o + a} x(t)dt \tag{1}$$

The ***Mean Value Theorem for integrals*** states that for any continuous function $s(t)$ we have:

$$\int_{c}^{d} s(t)dt = [d - c]s(t_m),$$

where $t_m$ is a number between $c$ and $d$ ($c < t_m < d$). Applying this theorem to Eq. 1 above we get:

$$\int_{-\infty}^{\infty} x(t)g(t - t_o)dt = \frac{1}{2a} \int_{t_o - a}^{t_o + a} x(t)dt = \frac{1}{2a}[(t_o + a) - (t_o - a)]x(t_m) = x(t_m),$$

where $t_o - a < t_m < t_o + a$. As $a \to 0$, we have $t_m \to t_o$; and hence

$$\int_{-\infty}^{\infty} x(t) \lim_{a \to 0}\{g(t - t_o)\}dt = \lim_{a \to 0} \int_{-\infty}^{\infty} x(t)g(t - t_o)dt = \lim_{a \to 0}\{x(t_m)\} = x(t_o).$$

The distinguishing ***Properties*** of the delta function are evenness, unit area, and spiky shape (Tables). These properties are satisfied by the above function:

*Evenness*: we have $g(-t) = g(t)$.
*Unit area*: $\int_{-\infty}^{\infty} g(t)dt = (1/2a)(2a) = 1$.
*Spiky shape*: $\lim_{a\to 0} g(t) = \lim_{a\to 0}\left\{\begin{matrix} \frac{1}{2a} & |t|<a \\ 0 & \text{elsewhere} \end{matrix}\right\} = \left\{\begin{matrix} \infty & t=0 \\ 0 & \text{elsewhere} \end{matrix}\right\}.$

This result justifies the use of narrow pulse to approximate the delta function in practical applications, noting that it is impossible to generate an exact delta function.

## *Tutorial 8*

**Q**: Show that the unit-step response $\rho(t)$ of any LTI system is related to its impulse response $h(t)$ by: $h(t) = \frac{d\rho(t)}{dt}$.
**Solution**: The I/O relation for LTI systems gives: $\rho(t) = h(t) * u(t) = \int_{-\infty}^{\infty} h(\lambda)$ $u(t-\lambda)d\lambda$.
The unit-step function is given by:

$$u(\lambda) = \left\{\begin{matrix} 1, & \lambda > 0 \\ 0, & \text{elsewhere} \end{matrix}\right\}$$

Hence, $u(t-\lambda) = \left\{\begin{matrix} 1, & t-\lambda > 0 \Rightarrow \lambda < \underbrace{t}_{\text{(end)}} \\ 0, & \text{elsewhere} \end{matrix}\right\}.$

Now we have $\rho(t) = \int_{-\infty}^{t} h(\lambda)d\lambda$.
Therefore, $\rho'(t) = h(t)$ [see Tables].

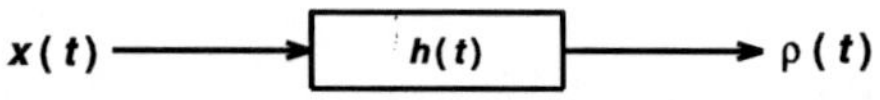


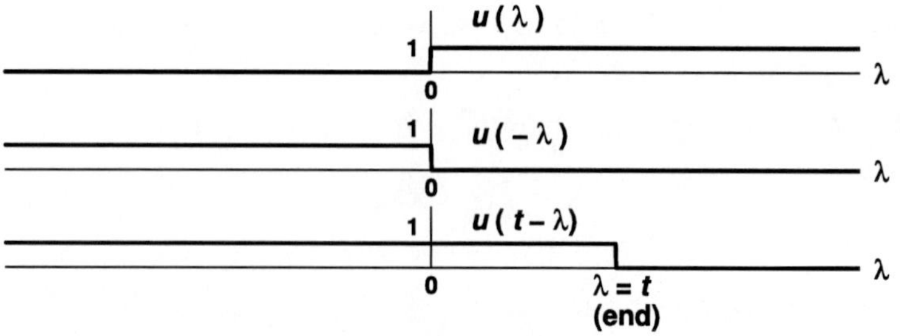

## *Tutorial 9*

**Q**: For the periodic square wave $x(t)$ shown below, find:
(A) The complex Fourier series, (B) the trigonometric Fourier series.

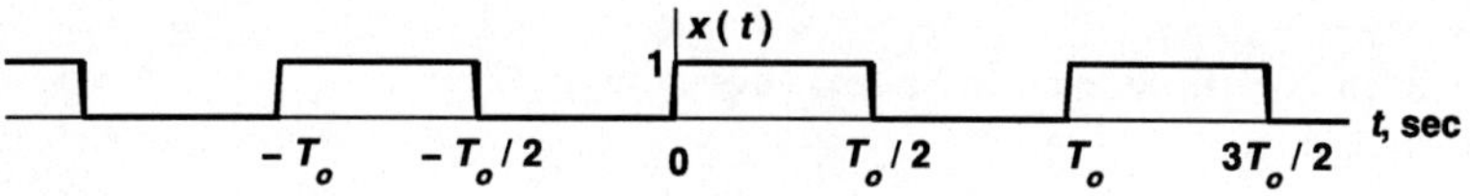


**Solution**:
(A) Since $x(t)$ is periodic, it has Fourier series expansion [Tables]:

$$x(t) = \sum_{k=-\infty}^{\infty} X_k e^{+j2\pi k f_0 t}$$

From the above figure, the signal fundamental frequency is $f_o = 1/T_o$ Hz. The Fourier coefficients (for $k \neq 0$) are given by [see Tables]:

$$X_k = \frac{1}{T_o} \int_0^{\frac{T_o}{2}} x(t) e^{-j2\pi k f_o t} dt = \frac{1}{T_o} \int_0^{\frac{T_o}{2}} e^{-j2\pi k f_o t} dt$$
$$= \left(\frac{1}{T_o}\right)\left(\frac{1}{-j2\pi k f_o}\right)\left[e^{-j2\pi k f_o t}\right]_0^{\frac{T_o}{2}} = \frac{1 - e^{-jk\pi}}{j2\pi k} = \frac{1-(-1)^k}{j2\pi k}$$

[Note that we used $f_o\, T_o = 1$ and $e^{-j\pi} = \cos(\pi) - j\sin(\pi) = -1$].
For $k = 0$ we have: $X_o = \frac{1}{T_o}\int_0^{\frac{T_o}{2}} x(t)dt = \frac{1}{2}$.
Now:

1. If $k$ is even, i.e., $k = 2n$, then $X_k = X_{2n} = 0$ (if $n \neq 0$)
2. If $k$ is odd, i.e., $k = 2n + 1$, then $X_k = X_{2n+1} = \frac{1}{j\pi(2n+1)}$.

$$\therefore \quad x(t) = \frac{1}{2} + \frac{1}{j\pi} \sum_{n=-\infty}^{\infty} \frac{1}{2n+1} e^{j(2n+1)2\pi f_0 t}$$

(B) From above we have:

$$x(t) = \sum_{k=-\infty}^{\infty} X_k e^{+j2\pi k f_0 t} = X_0 + \sum_{k=1}^{\infty} \left[X_k e^{j2\pi k f_0 t} + X_{-k} e^{-j2\pi k f_0 t}\right]$$
$$= a_0 + \sum_{k=1}^{\infty} \{a_k \cos(2\pi k f_0 t) + b_k \sin(2\pi k f_0 t)\}$$

where we used Euler's formula to get:

$$a_0 = X_0, \quad a_k = X_k + X_{-k}, \quad b_k = j(X_k - X_{-k}).$$

Hence, $a_o = X_0 = 1/2$.
If $k$ is even ($k \neq 0$), then $a_k = b_k = 0$ (see above).
If $k$ is odd, then:

$$a_k = a_{2n+1} = X_{(2n+1)} + X_{-(2n+1)} = 0 \quad \text{and} \quad b_k = b_{2n+1} = \frac{2}{(2n+1)\pi}.$$

Therefore, we can write the signal $x(t)$ as follows:

$$\begin{aligned} x(t) &= \frac{1}{2} + \sum_{n=0}^{\infty} b_{2n+1} \sin(2\pi(2n+1)f_0 t) \\ &= \frac{1}{2} + \frac{2}{\pi}\left[\sin(\omega_0 t) + \frac{1}{3}\sin(3\omega_0 t) + \frac{1}{5}\sin(5\omega_0 t) + \cdots\right]. \end{aligned}$$

**Q**: If $x(t)$ above is passed through the system shown below, find the output $y(t)$ and the average power in $x(t)$ and $y(t)$.
**Ans.** $P_x = 0.5$; $P_y = 0.47$.

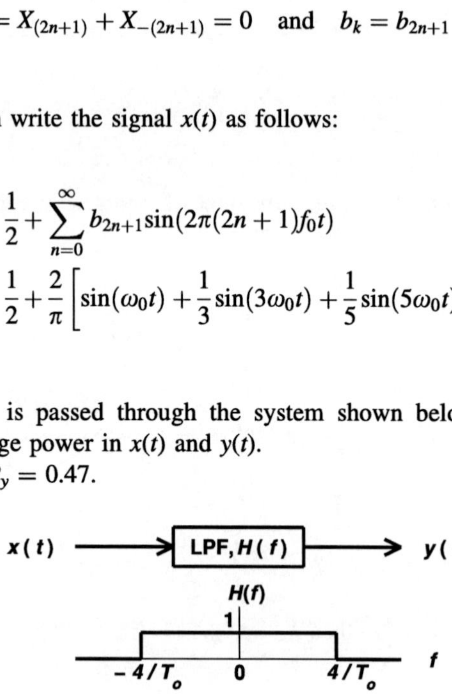


## *Tutorial 10*

**Q1**: Show that $x(t)\delta(t - t_o) = x(t_o)\delta(t - t_o)$.
**Solution**: Let $g(t)$ be any function that is continuous at $t = t_o$, and let $\phi(t) = g(t)x(t)$. Then using Tables we get:

$$\int_{-\infty}^{\infty} g(t)[x(t)\delta(t - t_o)]dt = \int_{-\infty}^{\infty} [g(t)x(t)]\delta(t - t_o)dt = \int_{-\infty}^{\infty} \phi(t)\delta(t - t_o)dt = \phi(t_o).$$

Now: $\phi(t_o) = g(t_o)x(t_o) = x(t)\int_{-\infty}^{\infty} g(t)\delta(t - t_o)dt = \int_{-\infty}^{\infty} g(t)[x(t_o)\delta(t - t_o)]dt.$

Hence we have: $\int_{-\infty}^{\infty} g(t)\underbrace{[x(t)\delta(t - t_o)]}_{s(t)}dt = \int_{-\infty}^{\infty} g(t)\underbrace{[x(t_o)\delta(t - t_o)]}_{r(t)}dt$

Since $g(t)$ is arbitrary, the above equation implies that:

$$s(t) = r(t), \quad \text{or} \quad x(t)\delta(t - t_o) = x(t_o)\delta(t - t_o).$$

**Q2:** Consider the cascaded and parallel systems shown in Figs. a and b. Find the equivalent impulse responses for the equivalent system shown in Fig. c.

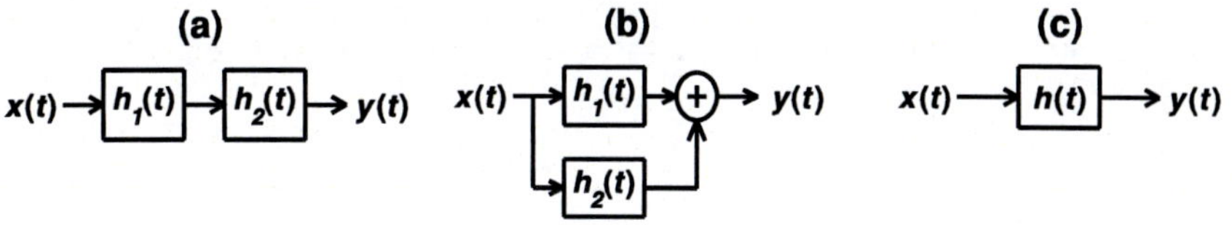


**Solution:**

(a) Let $y_1(t) = x(t) * h_1(t)$.

$\therefore$

$$y(t) = y_1(t) * h_2(t) = [x(t) * h_1(t)] * h_2(t) = x(t) * \underbrace{[h_1(t) * h_2(t)]}_{h(t)} \Rightarrow h(t)$$

$$= h_1(t) * h_2(t)$$

where we used the *associative* property of convolution.

(b) Let $y_1(t) = x(t) * h_1(t)$, $y_2(t) = x(t) * h_2(t)$.

$\therefore$

$$y(t) = y_1(t) + y_2(t) = x(t) * h_1(t) + x(t) * h_2(t) = x(t) * \underbrace{[h_1(t) + h_2(t)]}_{h(t)} \Rightarrow h(t)$$

$$= h_1(t) + h_2(t)$$

where we used the *distributive* property of convolution.

## *Tutorial 11*

Q: For the periodic pulse train $p(t) = \sum_{k=-\infty}^{\infty} \delta(t - kT)$ find:
(A) Complex Fourier series (FS), (B) trigonometric FS, (C) Fourier Transform (FT).

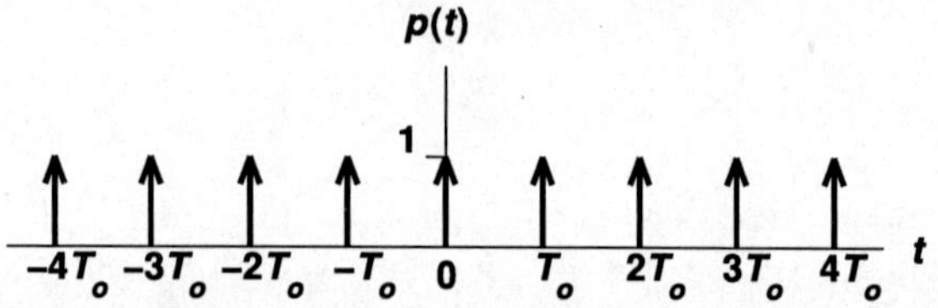


**Solution**:

(A) $p(t) = \sum_{k=-\infty}^{\infty} P_k e^{+j2\pi k f_o t}$, where $\omega_o = 2\pi f_o = \frac{2\pi}{T_o}$ is the fundamental frequency, $T_o$ being the fundamental period, and the Fourier coefficients are given by [see Tables]:

$$P_o = \frac{1}{T_o}\int_{-T_o/2}^{T_o/2} p(t)dt = \frac{1}{T_o}\int_{-T_o/2}^{T_o/2} \delta(t)dt = \frac{1}{T_o}$$

$$P_k = \frac{1}{T_o}\int_{-T_o/2}^{T_o/2} p(t)e^{-j2\pi k f_o t}dt = \frac{1}{T_o}\int_{-T_o/2}^{T_o/2} \delta(t)e^{-j2\pi k f_o t}dt = \frac{1}{T_o}$$

$$\therefore \; p(t) = \frac{1}{T_o}\sum_{-\infty}^{\infty} e^{j2\pi k f_o t}$$

(B) $p(t) = a_o + \sum_{k=1}^{\infty}[a_k\cos(k\omega_o t) + b_k\sin(k\omega_o t)]$, where [see Tables]:

$$a_o = P_o = \frac{1}{T_o}, \quad a_k = P_k + P_{-k} = \frac{2}{T_o}, \quad b_k = j(P_k - P_{-k}) = 0.$$

$$\therefore \; p(t) = \frac{1}{T_o} + \frac{2}{T_o}\sum_{k=1}^{\infty}\cos(k\omega_o t).$$

(C) Using part (A) we have [using Tables]:

$$\mathcal{F}\{p(t)\} = \mathcal{F}\left\{\frac{1}{T_o}\sum_{k=-\infty}^{\infty} e^{j2\pi k f_o t}\right\} = \frac{1}{T_o}\sum_{k=-\infty}^{\infty}\mathcal{F}\{e^{j2\pi k f_o t}\}$$

$$= \frac{1}{T_o}\sum_{k-\infty}^{\infty}\delta(f - kf_o) = f_o\sum_{k-\infty}^{\infty}\delta(f - kf_o)$$

Hence, a ***pulse train*** in the ***time domain*** is Fourier-transformed to ***a pulse train*** in the ***frequency domain***.

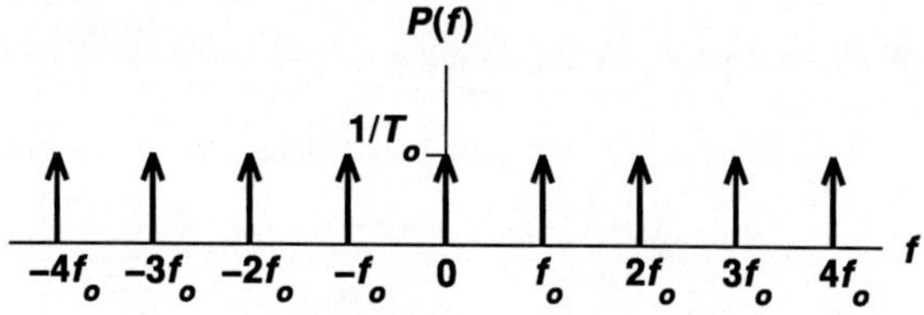


## Tutorial 12

**Q1**: Show that $x(t)\cos\omega_0 t \overset{\mathcal{F}}{\longleftrightarrow} \frac{1}{2}X(f-f_0)+\frac{1}{2}X(f+f_0)$.

**Solution**: From Tables we have:

$$\cos\omega_0 t \overset{\mathcal{F}}{\longleftrightarrow} \frac{1}{2}\delta(f-f_0)+\frac{1}{2}\delta(f+f_0).$$

Let $c(t)=\cos\omega_0 t$, $C(f)=\mathcal{F}\{\cos\omega_0 t\}$. Using Tables we have:

$$F\{x(t)\cdot c(t)\}=X(f)*C(f)=X(f)*\left[\frac{1}{2}\delta(f-f_0)+\frac{1}{2}\delta(f+f_0)\right]$$
$$=\frac{1}{2}X(f-f_0)+\frac{1}{2}X(f+f_0)]$$

where we used the relation $X(f) * \delta(f-f_0)=X(f-f_o)$ from Tables.

**Q2**: (A) Find $\mathcal{F}\{x(t)=e^{-a|t|}, a>0\}$. (B) Using part (A), find $\mathcal{F}\left\{\frac{2}{1+t^2}\right\}$.

**Solution**:
(A)

$$X(f)=\int_{-\infty}^{\infty} x(t)e^{-j2\pi ft}dt=\int_{-\infty}^{\infty} e^{-a|t|}e^{-j2\pi ft}dt$$
$$=\int_{-\infty}^{0} e^{at}e^{-j2\pi ft}dt=\int_{0}^{\infty} e^{-at}e^{-j2\pi ft}dt=\frac{1}{a-j2\pi f}+\frac{1}{a+j2\pi f}=\frac{2a}{a^2+(2\pi f)^2}.$$

(B)

$$x(t)\overset{\mathcal{F}}{\longleftrightarrow}X(f)\Rightarrow X(t)\overset{\mathcal{F}}{\longleftrightarrow}x(-f)\quad\text{[duality (Tables)]}$$
$$\Rightarrow X(t)\overset{\mathcal{F}}{\longleftrightarrow}x(f)\quad\text{[if } x(t) \text{ is even]}$$

Using duality of FT and part (A) above we get: $\mathcal{F}\left\{\frac{2}{1+(2\pi t)^2}\right\} = e^{-|f|}$.
From Tables (scaling property) we have: $s(kt) \leftrightarrow \frac{1}{|k|}S(\frac{f}{k})$.
Applying this to the above result we get:

$$\mathcal{F}\left\{\frac{2}{1+(t)^2}\right\} = 2\pi e^{-2\pi|f|}$$

## *Tutorial 13*

**Q**: Determine the impulse response of the ideal bandstop filter whose frequency response is shown below.

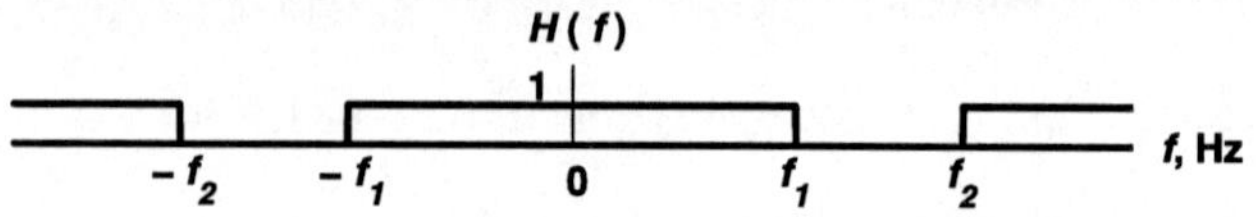


**Solution**: It is better that we arrange *H(f)* in terms of well-known functions. In this question it can be written as follows:

$$H(f) = 1 - G(f),$$

where *G(f)* is shown below.
To find an explicit formula for *G(f)*, let $f_0 = \frac{f_1+f_2}{2}, B = f_2 - f_1$.

$$\therefore \quad G(f) = \prod_B(f - f_0) + \prod_B(f + f_0) = X(f - f_0) + X(f + f_0),$$

where $X(f) = \prod_B(f)$.
Now from Tables we have:

1. $\text{sinc}(Lt) \overset{\mathcal{F}}{\longleftrightarrow} \frac{1}{L}\prod_L(f) \Rightarrow L \cdot \text{sinc}(Lt) \overset{\mathcal{F}}{\longleftrightarrow} \prod_L(f)$ [multiply both sides by *L*]

   Hence, for this question we have: $\underbrace{B\text{sinc}(Bt)}_{x(t)} \overset{\mathcal{F}}{\longleftrightarrow} \underbrace{\prod_B(f)}_{X(f)}$.

2. $x(t)\cos(2\pi f_0 t) \overset{\mathcal{F}}{\longleftrightarrow} \frac{1}{2}X(f - f_0) + \frac{1}{2}X(f + f_0)$,

$$\therefore \quad \underbrace{2x(t)\cos(2\pi f_0 t)}_{g(t)=2B\,\text{sinc}(Bt)\cos(2\pi f_0 t)} \overset{\mathcal{F}}{\longleftrightarrow} \underbrace{X(f - f_0) + X(f + f_0)}_{G(f)}$$

$$\therefore \quad h(t) = F^{-1}\{1 - G(f)\} = F^{-1}\{1\} - F^{-1}\{G(f)\} = \delta(t) - g(t)$$
$$= \delta(t) - 2B\,\text{sinc}(Bt)\cos(2\pi f_0 t).$$

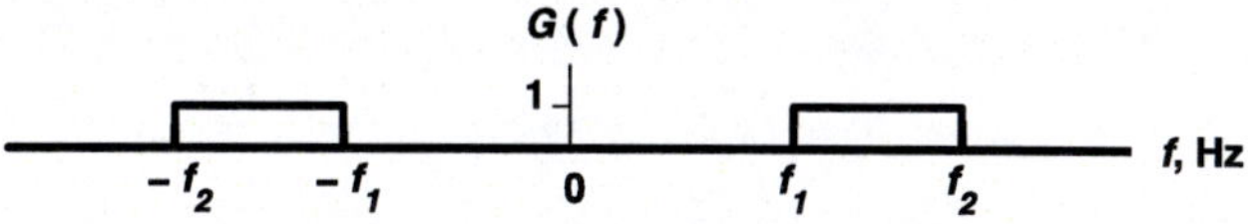


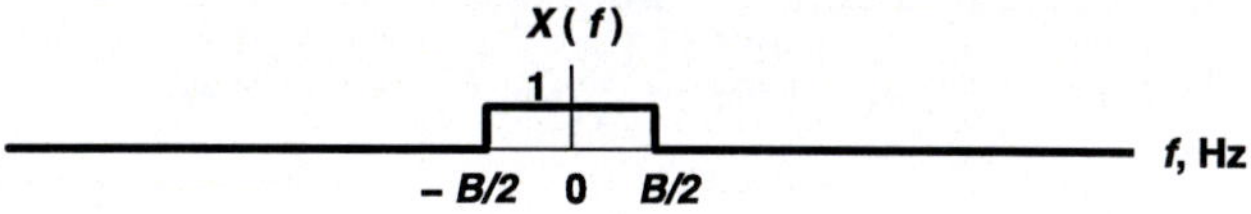


## Tutorial 14

**Q**: (A) Determine the impulse response of the ideal band-stop filter whose frequency response is shown in Fig. 1 below.
(B) Determine the impulse response of the ideal band-stop filter whose frequency response is shown in Fig. 2 below.

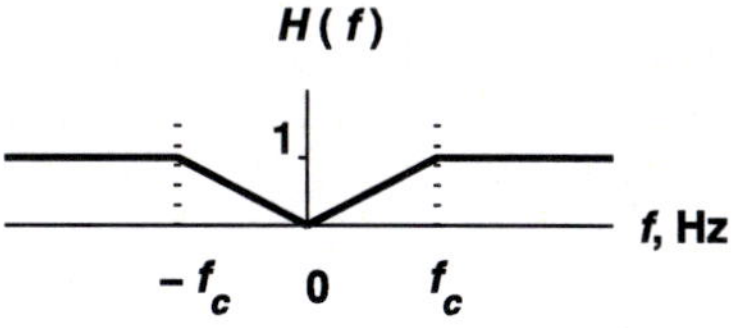


**(1)**

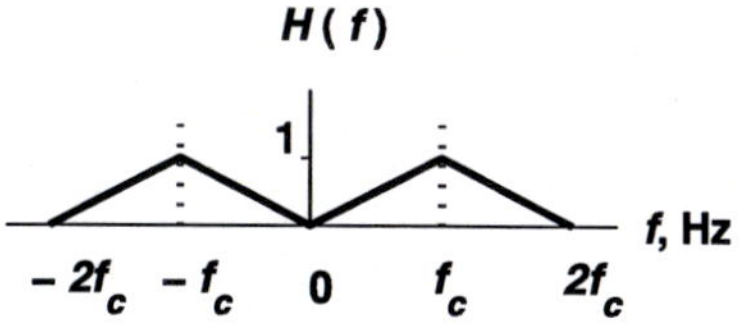


**(2)**

**Solution**:
(A) The frequency response $H(f)$ can be written as:
$H(f) = 1 - G(f)$, where $G(f) = \Lambda_{2f_c}(f)$ [see the figure below].

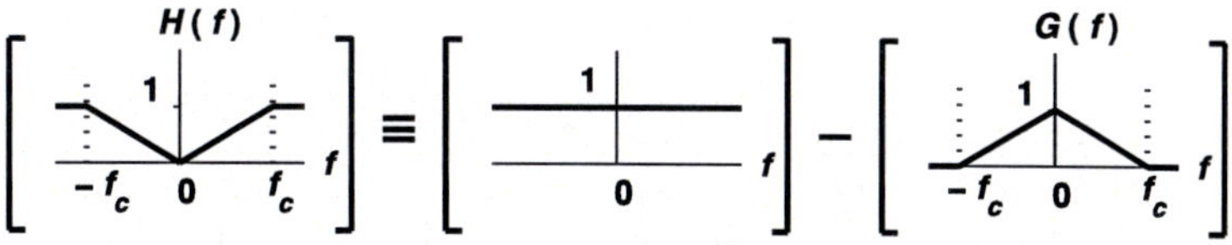


From Tables we have: $\Lambda_T(t) \overset{\mathcal{F}}{\longleftrightarrow} \frac{T}{2}\text{sinc}^2(\frac{T}{2}f)$, and:

$$x(t) \overset{\mathcal{F}}{\longleftrightarrow} X(f) \Rightarrow X(t) \overset{\mathcal{F}}{\longleftrightarrow} x(-f) \quad [\text{duality}(\textit{Tables})]$$

$$\Rightarrow X(t) \overset{\mathcal{F}}{\longleftrightarrow} x(f) [\text{If } x(t) \text{ is even}]$$

$$\therefore \quad \frac{B}{2}\text{sinc}^2\left(\frac{B}{2}t\right) \overset{\mathcal{F}}{\longleftrightarrow} \Lambda_B(f) \Rightarrow \underbrace{\frac{2f_c}{2}\text{sinc}^2(\frac{2f_c}{2}t)}_{g(t)} \overset{\mathcal{F}}{\longleftrightarrow} \underbrace{\Lambda_{2f_c}(f)}_{G(f)}$$

$$\therefore \quad h(t) = \mathcal{F}^{-1}\{H(f)\} = \mathcal{F}^{-1}\{1 - G(f)\} = \delta(t) - f_c\,\text{sinc}^2(f_c t).$$

(B) In this part, the frequency response reminds us of the modulation property. It can be written as:

$$H(f) = G(f - f_c) + G(f + f_c).$$

Hence, using Tables or Tutorial 13 we get:

$$h(t) = 2g(t)\cos(2\pi f_c t) = 2f_c \mathrm{sinc}^2(f_c t) \cdot \cos(2\pi f_c t).$$

## *Tutorial 15*

**Q**: Show that:

(A) $x(t) \cdot y(t) \overset{\mathcal{F}}{\longleftrightarrow} X(f) * Y(f)$

(B) $x(t) * y(t) \overset{\mathcal{F}}{\longleftrightarrow} X(f) \cdot Y(f)$

**Solution**:

(A)

$$\begin{aligned}
\mathcal{F}\{x(t)y(t)\} &= \int_{-\infty}^{\infty} x(t)y(t)e^{-j2\pi ft}dt \\
&= \int_{-\infty}^{\infty} \mathcal{F}^{-1}\{X(f)\}y(t)e^{-j2\pi ft}dt \\
&= \int_{-\infty}^{\infty} \left[ \int_{-\infty}^{\infty} X(\lambda)e^{+j2\pi\lambda t}d\lambda \right] y(t)e^{-j2\pi ft}dt \\
&= \int_{-\infty}^{\infty} X(\lambda) \underbrace{\left[ \int_{-\infty}^{\infty} y(t)e^{-j2\pi(f-\lambda)t}dt \right]}_{Y(f-\lambda)} d\lambda \\
&= \int_{-\infty}^{\infty} X(\lambda)Y(f-\lambda)d\lambda = X(f) * Y(f).
\end{aligned}$$

(B)

$$\mathcal{F}\{x(t)*y(t)\} = \int_{-\infty}^{\infty}\left[\int_{-\infty}^{\infty} x(\lambda)y(t-\lambda)d\lambda\right]e^{-j2\pi ft}dt$$

$$= \int_{-\infty}^{\infty} x(\lambda)\underbrace{\left[\int_{-\infty}^{\infty} y(t-\lambda)e^{-j2\pi ft}dt\right]}_{\text{Let } v=t-\lambda} d\lambda$$

$$= \int_{-\infty}^{\infty} x(\lambda)\left[\int_{-\infty}^{\infty} y(v)e^{-j2\pi f(v+\lambda)}dv\right] d\lambda$$

$$= \left(\int_{-\infty}^{\infty} x(\lambda)e^{-j2\pi f\lambda}d\lambda\right)\left(\int_{-\infty}^{\infty} y(v)e^{-j2\pi fv}dv\right)$$

[since $\lambda$ and $v$ are independent]

$$= \mathcal{F}\{x(t)\}\cdot\mathcal{F}\{y(t)\} = X(f)\cdot Y(f)$$

## *Tutorial 16*

**Q**: Find the s-domain *voltage* transfer function and the *voltage* impulse response for the analog systems shown below. Assume zero initial conditions (IC's).

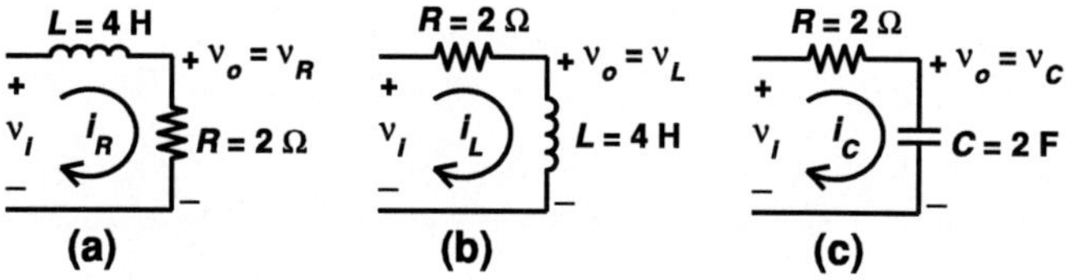


**Solution**:

$$\underbrace{v_L(t) = Ldi(t)/dt}_{\text{voltage across L}} \underset{\text{Tables}}{\overset{\mathcal{L}}{\longleftrightarrow}} V_L(s) = L[sI(s) - i(0^-)] = Ls\cdot I(s) \quad [\text{zero IC's}].$$

Hence, inductor impedance is represented in the complex-frequency domain as:

$$X_L(s) = V_L(s)/I(s) = sL \quad [\text{zero IC's}].$$

$$\text{Now } \underbrace{v_R(t) = R \cdot i(t)}_{\text{voltage across R}} \overset{\mathcal{L}}{\longleftrightarrow} V_R(s) = RI(s) \Rightarrow X_R(s) = V_R(s)/I(s) = R.$$

$$\underbrace{i_C(t) = Cdv_C(t)/dt}_{\text{current through C}} \underset{\text{Tables}}{\overset{\mathcal{L}}{\longleftrightarrow}} I_C(s) = C \cdot [sV_C(s) - v_C(0^-)] = Cs \cdot V_C(s) \quad \text{[zero IC's]}$$

$$\Rightarrow X_C(s) = V_C(s)/I_C(s) = 1/sC.$$

To summarize impedance transformations to the s-domain (zero IC's):
$R \Rightarrow R$; $L \Rightarrow sL$; $C \Rightarrow 1/(sC)$.
Now we can transform the above circuits to the s-domain using $\mathcal{L}$-Tables:

(a) $H(s) = V_o(s)/V_i(s) = 2/[4s + 2] \Rightarrow h(t) = \mathcal{L}^{-1}\{(1/2)[1/(s + 0.5)]\} = 0.5e^{-0.5t}$

(b) $H(s) = \frac{4s}{4s+2} = \frac{4s+2-2}{4s+2} = 1 - \frac{(1/2)}{s+(1/2)} \Rightarrow h(t) = \delta(t) - (1/2)e^{-\frac{1}{2}t}$.

(c) $H(s) = \frac{1/(2s)}{(1/2s)+2} = \frac{1}{1+4s} = \frac{1}{4}\frac{1}{s+(1/4)} \Rightarrow h(t) = \frac{1}{4}e^{-\frac{1}{4}t}$.

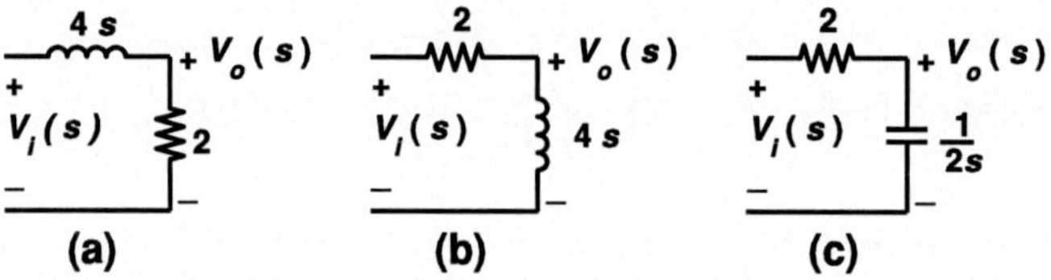


## *Tutorial 17*

**Q**: Find the current and voltage across the capacitor below after the switch is closed at $t = 0$.

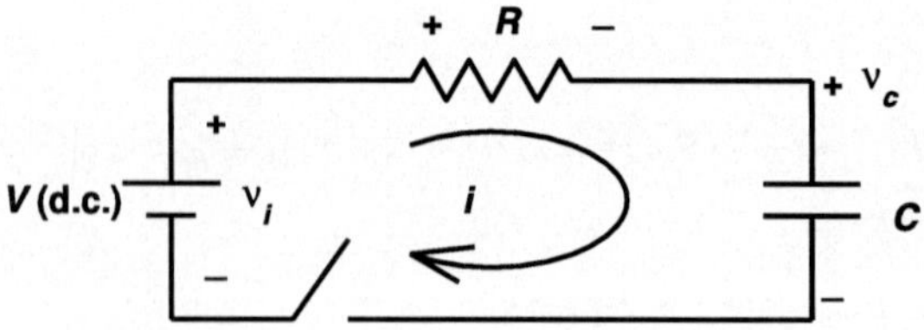

**Solution**: Let the capacitor initial voltage be $V_o$ (volts). After the switch is closed, a d.c. voltage of $V$ volts will suddenly be applied, hence, $v_i\,(t) = Vu(t)$ volts.

$$\underbrace{i(t) = i_C(t) = C\frac{dv_C(t)}{dt}}_{\text{current through C}} \underset{\text{Tables}}{\overset{\mathcal{L}}{\longleftrightarrow}} I(s) = C \cdot [sV_C(s) - v_C(0^-)] = C[sV_C(s) - V_o],$$

or simply:

$$I = C \cdot [sV_C - V_o] \tag{1}$$

$$v_c = v_i - iR \overset{\mathcal{L}}{\longleftrightarrow} V_C = V_i - IR = V/s - IR \tag{2}$$

From (1), (2) we get:

$$I = \frac{1}{R}\frac{V - V_o}{s + 1/RC} = I_o\frac{1}{s + 1/RC} \tag{3}$$

[where $I_o = (V - V_o)/R$.]
From Tables: $i(t) = I_o\, e^{-t/RC}\, u(t)$ amperes.
Substituting (3) in (2) we get: $V_C = V[\frac{1}{s} - \frac{1}{s+1/RC}] + \frac{V_o}{s+1/RC}$
From Tables: $v_c\,(t) = [V(1 - e^{-t/RC}) + V_o\, e^{-t/RC}]u(t)$ volts.

## *Tutorial 18*

**Q**: The switch S in Fig. 1 below has been closed for a long time. Find the current across the capacitor after the switch is opened at $t = 0$.

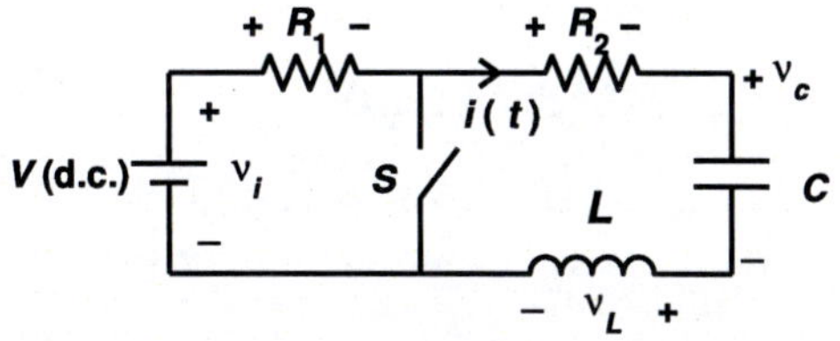


(1)

**Solution**: As the switch was closed for a long time, both $v_C\,(0^-)$ and $i_L\,(0^-)$ are zero. Even if there was an initial voltage $V_o$ on $C$ and an initial current $I_o$ in $L$ when S was closed, the final current and voltages in the right loop will all be zero. This is due to the consumption of energy by $R_2$. To prove this, consider the right loop where the switch has been closed as shown in Fig. 2.

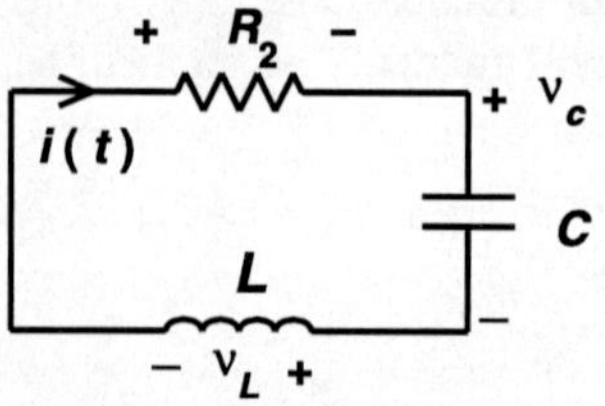


(2)

Let $R = R_2$. Now $V_R = IR$.

$$i(t) = C\frac{dv_C}{dt} \overset{\mathcal{L}}{\longleftrightarrow} I(s) = C[sV_C(s) - v_C(0^-)] \rightarrow I = C[sV_C - V_o]$$

$$\rightarrow V_C = \frac{I}{Cs} + \frac{V_o}{s}$$

$$v_L(t) = L\frac{di(t)}{dt} \overset{\mathcal{L}}{\longleftrightarrow} V_L(s) = L[sI(s) - i(0^-)] \rightarrow V_L = L[sI - I_o].$$

$$V_C + V_L + V_R = 0 \rightarrow \frac{I}{Cs} + \frac{V_o}{s} + sLI - LI_o + IR = 0$$

$$\rightarrow I = \frac{LCI_o s - CV_o}{LCs^2 + RCs + 1}$$

$$I = I_o\frac{s - \frac{V_o}{LI_o}}{s^2 + \frac{R}{L}s + \frac{1}{LC}} = I_o\frac{s - \frac{V_o}{LI_o} + (\frac{R}{2L} - \frac{R}{2L})}{(s + \frac{R}{2L})^2 + (\frac{1}{LC} - \frac{R^2}{4L^2})} = I_o\left[\frac{s+a}{F(s)} - \xi\frac{\omega_o}{F(s)}\right]$$

where

$$a = \frac{R}{2L}, \quad \omega_o^2 = \frac{1}{LC} - \frac{R^2}{4L^2}, \quad F(s) = (s+a)^2 + \omega_o^2, \quad \xi = \left(\frac{V_o}{LI_o} + \frac{R}{2L}\right)/\omega_o.$$

$\therefore \quad i(t) = I_o e^{-at}[\cos(\omega_o t) - \xi\sin(\omega_o t)]u(t).$

If $I_0$, then $i(t) = -(V_o/L\omega_o)e^{-at}\sin(\omega_o t)u(t)$.

If $D = \dfrac{R^2}{L^2} - \dfrac{4}{LC} > 0$, we use partial fractions to get :

$i(t) = \left[k_1 e^{-b_1 t} + k_2 e^{-b_2 t}\right]u(t)$, with:

while $\quad k_1 = -\lambda/$

$$b_1, b_2 = \frac{R}{2L} \mp \frac{1}{2}\sqrt{\frac{R^2}{L^2} - \frac{4}{LC}} > 0,$$

$k_1 = -I_o(k - b_1)/b \quad$ (where $k = V_o/(LI_o), b = b_2 - b_1$)

$k_2 = I_o(k - b_2)/b \quad$ for $I_o \neq 0$,

$b$ (where $\lambda = V_o/L$) and $k_2 = \lambda/b$ for $I_o = 0$.

Unless $R = 0$, all above expressions for $i(t)$ tend to zero after a long time, so are the expressions for $v_L(t) = Ldi(t)/dt$ and $v_R = i(t)R$. Hence, $v_C = -(v_L + v_R) \rightarrow 0$ as well.

**Now back to the main question.** Let $R = R_1 + R_2$. We have $v_i(t) = Vu(t) \rightarrow V_i(s) = 1/s$. Using $\mathcal{L}$ rules with zero IC's we get:

$$V_C = I/(Cs), V_L = LsI.$$
$$v_i(t) = v_R + v_C + v_L \rightarrow 1/s = IR + I/(Cs) + LsI$$
$$\rightarrow I = \frac{(1/L)}{s^2 + (R/L)s + 1/(LC)}$$

If $D < 0$, we get: $I = (1/\omega_o L)[\omega_o/F(s)] \rightarrow i(t) = (1/\omega_o L)e^{-at}\sin(\omega_o t)u(t)$.
If $D > 0$, we get $I = (1/L)[a_1/(s + b_1) + a_2/(s + b_2)]$ where $a_1 = 1/b$, $a_2 = -1/b$, hence

$$i(t) = (1/bL)[e^{-b_1 t} - e^{-b_2 t}]u(t).$$

## *Tutorial 19*

**Q1**: Find the transfer function of the feedback system shown below. All signals are voltages.

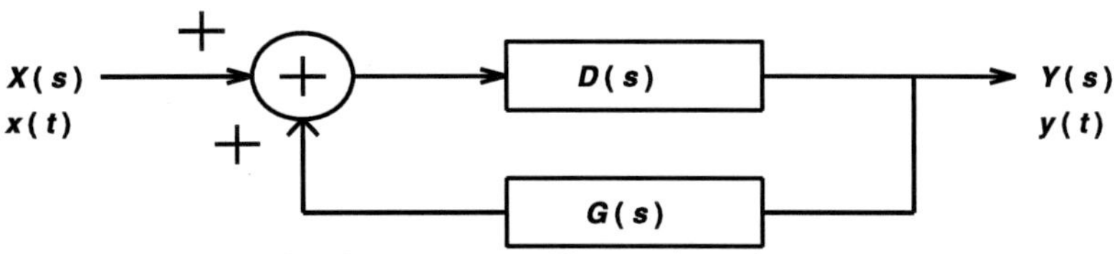


**Solution**: First, we define important auxiliary points [i.e., $e(t)$ and $r(t)$] as shown below.

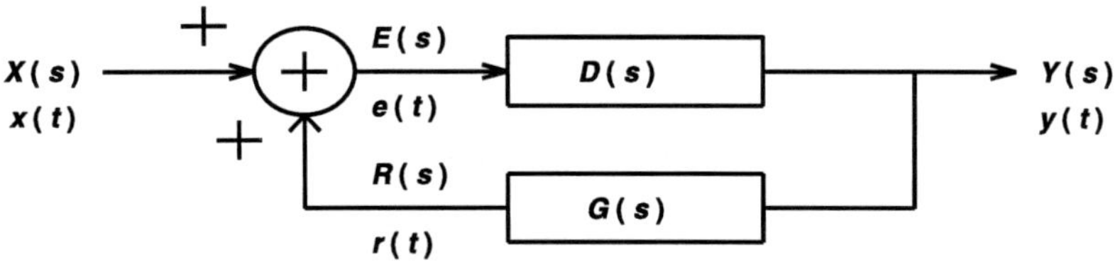


Now we write the system equations as follows:

$$e(t) = x(t) + r(t) \quad (1)$$

This equation is in the time-domain. It is easier to transform to the s-domain using $\mathcal{L}$, since the convolution in the $t$-domain would be transformed into a simple multiplication in the $s$-domain. Taking the $\mathcal{L}$ of both sides of (1) we get:

$$E(s) = X(s) + R(s) \quad (2)$$

Similarly,

$$R(s) = G(s)Y(s) \tag{3}$$

$$Y(s) = D(s)E(s) \tag{4}$$

Using (2), (3), and (4) we get: $Y(s)[1 - D(s)G(s)] = D(s)X(s)$

$$\therefore \quad H(s) = \frac{Y(s)}{X(s)} = \frac{D(s)}{1 - D(s)G(s)}.$$

**Q2**: Find the transfer function of the following feedback system.

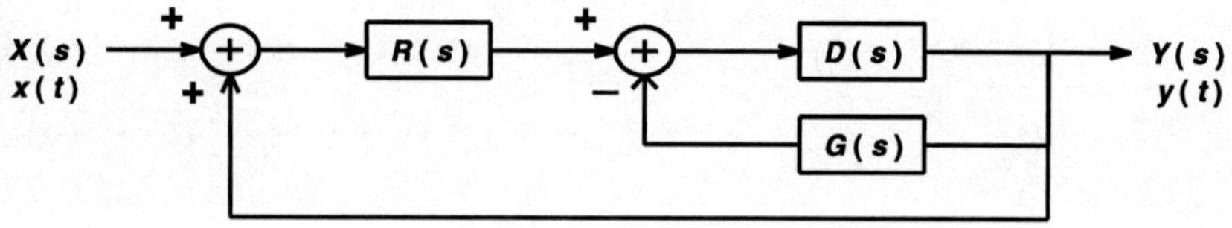


## *Tutorial 20*

**Q**: The relationship between $\mathcal{F}\{x(t)\}$ and $\mathcal{L}\{x(t)\}$ is given by:

$$F\{x(t)\} = L\{x(t)\}|_{s=j\omega}$$
$$\text{or}: X(\omega) = X(s)\big|_{s=j\omega}$$

where $\omega = 2\pi f$. This holds on the condition that:

$$\int_{-\infty}^{\infty} |x(t)|dt < \infty \quad [\text{i.e.}, x(t) \text{ is } \textbf{\textit{absolutely integrable}}].$$

Now decide whether $X(\omega)$ can be found from $X(s)$ for the functions:
1. $\delta(t)$, 2. $u(t)$, 3. $t^n\ u(t)$, $n$ is integer $\geq 0$.

**Solution:**

1. $\int_{-\infty}^{\infty} |x(t)|dt = \int_{-\infty}^{\infty} \delta(t)dt = 1 < infty$. Hence, $X(\omega) = X(s)|_{s=j\omega}$.
   From Tables: $\Delta(s) = L\{\delta(t)\} = 1$. $\therefore$ $\Delta(\omega) = \Delta(s)\big|_{s=j\omega} = 1$.
2. $\int_{-\infty}^{\infty} |u(t)|dt = \int_{0}^{\infty} u(t)dt = \int_{0}^{\infty} 1.dt \to \infty$.
   Hence, we **cannot** use $X(\omega) = X(s)|_{s=j\omega}$.
   Note that the Laplace transform of $u(t)$ is: $U(s) = \frac{1}{s}$ (Tables), and the Fourier transform of $u(t)$ is: $U(f) = \frac{1}{2}\delta(f) + \frac{1}{j2\pi f}$ (Tables).

Therefore, $U(\omega) = \pi\delta(\omega) + \frac{1}{j\omega}$ [Using the relation $\delta(2\pi f) = \frac{1}{2\pi}\delta(f)$ from Tables.]
It is apparent that $U(\omega) \neq U(s)|_{s=j\omega}$.

3. The function $t^n u(t)$ has **no** Fourier transform, **but it has** Laplace transform given by $\frac{n!}{s^{n+1}}$.

## *Tutorial 21*

Q: Expand the following expression using ***partial fraction expansion***, then find $y(t) : Y(s) = \frac{2s^2+13s+12}{(s+1)(s^2+5s+6)}$.

**Solution**: To find the *inverse* Laplace transform ($\mathcal{L}^{-1}$) for an expression of the form:

$$Y(s) = \frac{N(s)}{D(s)} = \frac{a_0 + a_1 s + \cdots + a_m s^m}{(s-p_1)(s-p_2)\cdots(s-p_n)},$$

we apply partial fraction expansion rules as follows:

$$Y(s) = \frac{c_1}{s-p_1} + \frac{c_2}{s-p_2} + \cdots + \frac{c_n}{s-p_n}.$$

***Notes:***

1. If $m$ = degree $[N(s)] \geq n$ = degree $[D(s)]$, perform long division first.
2. For complex conjugate poles $p_{k+1} = p_k^*$ we have $c_{k+1} = c_k^*$.
3. For a pole of multiplicity $r$ at $s = p_k$ there will be $r$ terms: $c_{k1}/(s-p_k) + c_{k2}/(s-p_k)^2 + \cdots + c_{kr}/(s-p_k)^r$ in the expansion.

**Now back to the main question.**

$$Y(s) = \frac{2s^2+13s+12}{(s+1)(s-p_1)(s-p_2)} \Rightarrow s^2+5s+6=0 \Rightarrow p_{1,2} = \frac{-5 \pm \sqrt{5^2 - 4\times 6}}{2}$$
$$= -2, -3.$$

Hence, $Y(s) = \frac{2s^2+13s+12}{(s+1)(s+2)(s+3)}$. Let $Y(s) = \frac{a}{s+1} + \frac{b}{s+2} + \frac{c}{s+3}$.

$$\therefore \quad \frac{2s^2+13s+12}{(s+1)(s+2)(s+3)} = \frac{a}{s+1} + \frac{b}{s+2} + \frac{c}{s+3}$$

To find $a$ do the following:

1. Multiply by $(s+1)$ : $\frac{2s^2+13s+12}{(s+1)(s+2)(s+3)} = a + \frac{b(s+1)}{s+2} + \frac{c(s+1)}{s+3}$.
2. Put $s = -1$: $\frac{2-13+12}{(-1+2)(-1+3)} = a + 0 + 0 \rightarrow a = \frac{1}{2}$.

Similarly we get: $b = 6$ and $c = -9/2$.
From **Tables** we find $\mathcal{L}^{-1}$ of $Y(s)$ as follows:

$$y(t) = ae^{-t} + be^{-2t} + ce^{-3t} = \frac{1}{2}e^{-t} + 6e^{-2t} - \frac{9}{2}e^{-3t}.$$

**MATLAB**: The above question can be solved on MATLAB® as follows:

$$\mathtt{A} = [2\,13\,12]; \quad \mathtt{B} = [1\,6\,11\,6]; \quad [\mathtt{R},\mathtt{P},\mathtt{K}] = \mathtt{residue(A,B)}$$

where `R` gives the coefficients $c_i$, `P` gives the poles $p_i$, and `K` is empty if degree(`A`) < degree(`B`). Note that `B` can be obtained using `B = conv([1 1],[1 5 6])` {multiplication of two polynomials}.

## *Tutorial 22*

**Q**: A linear electrical system is described by the second-order differential equation:

$$y''(t) + 5y'(t) + 6y(t) = x(t) \tag{1}$$

with initial conditions:

$$y(0) = 2; y'(0) = 1.$$

If the input voltage is given by $x(t) = e^{-t}\,u(t)$, find the output $y(t)$.
**Solution**: From $\mathcal{L}$ Tables we have:

$$y'(t) \overset{\mathcal{L}}{\longleftrightarrow} sY(s) - y(0^-) = sY(s) - 2$$

$$y''(t) \overset{\mathcal{L}}{\longleftrightarrow} s[sY(s) - y(0^-)] - y'(0^-) = s^2Y(s) - 2s - 1$$

$$x(t) = e^{-t}u(t) \overset{\mathcal{L}}{\longleftrightarrow} X(s) = \frac{1}{s+1}$$

Taking $\mathcal{L}$ of both sides of Eq. 1, we get:

$$[s^2Y(s) - 2s - 1] + 5[sY(s) - 2] + 6Y(s) = \frac{1}{s+1}$$

$$Y(s)[s^2 + 5s + 6] = 2s + 11 + \frac{1}{s+1} = \frac{2s^2 + 13s + 12}{s+1}$$

$$\therefore \quad Y(s) = \frac{2s^2 + 13s + 12}{(s+1)(s^2 + 5s + 6)}$$

$$\therefore \quad y(t) = \frac{1}{2}e^{-t} + 6e^{-2t} - \frac{9}{2}e^{-3t} \quad \text{(From Tutorial 21)}$$

## *Tutorial 23*

**Q**: Find the autocorrelation function of $x(t) = \prod_1(t-0.5)$.
Note that $x(t)$ is a ***deterministic*** signal.
**Solution**: From Tables: $R(\tau) = \int_{-\infty}^{\infty} x(\lambda)x(\tau + \lambda)d\lambda$. Note that $R(\tau) = x(\tau) * x(-\tau)$.
We follow the same steps that are necessary to find the convolution of two signals.

From Tables: $\prod_T(t) = \begin{Bmatrix} 1, & -T/2 \le t \le T/2 \\ 0, & \text{elsewhere} \end{Bmatrix}$

$$\therefore \quad x(\lambda) = \prod\nolimits_1(\lambda - 0.5) = \begin{Bmatrix} 1, & -0.5 \le \lambda - 0.5 \le 0.5 \\ 0, & \text{elsewhere} \end{Bmatrix} = \begin{Bmatrix} 1, & 0 \le \lambda \le 1 \\ 0, & \text{elsewhere} \end{Bmatrix}$$

$$\therefore \quad x(\tau + \lambda) = \begin{Bmatrix} 1, & 0 \le \tau + \lambda \le 1 \\ 0, & \text{elsewhere} \end{Bmatrix} = \begin{Bmatrix} 1, & \underbrace{-\tau}_{\text{start}} \le \lambda \le \underbrace{1-\tau}_{\text{end}} \\ 0, & \text{elsewhere} \end{Bmatrix}$$

Note that since we have $x(\tau + \lambda)$ inside the integral, a positive $\lambda$-shift $\tau$ will give negative shift to the function w.r.t the y-axis (***contrary to the case of convolution***). Now we move the function $x(\tau + \lambda)$ from left to right while $x(\lambda)$ is fixed. We get the following cases:

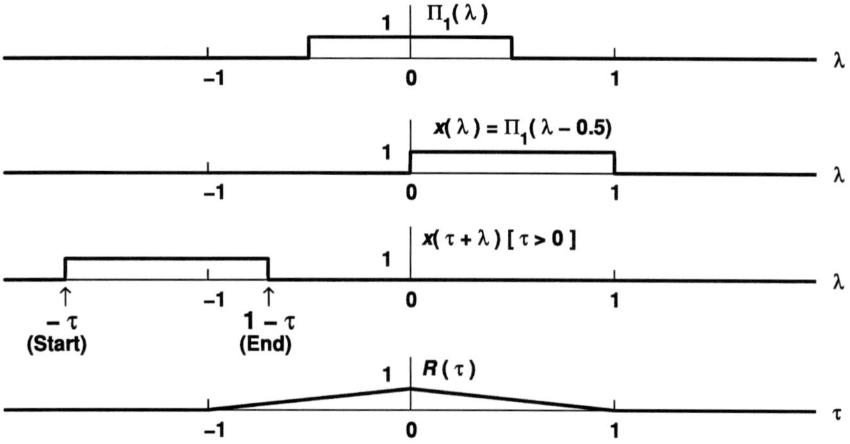


*Case 1*: $1 - \tau < 0$ (i.e., $\tau > 1$) $\Rightarrow R(\tau) = 0$ [no overlap].
*Case 2*: $0 \le 1 - \tau < 1$ (i.e., $0 < \tau \le 1$) $\Rightarrow R(\tau) = \int_0^{1-\tau} d\lambda = 1 - \tau$.
*Case 3*: $0 \le -\tau \le 1$ (i.e., $-1 \le \tau \le 0$) $\Rightarrow R(\tau) = \int_{-\tau}^{1} d\lambda = 1 + \tau$.
*Case 4*: $-\tau > 1$ (i.e., $\tau < -1$) $\Rightarrow R(\tau) = 0$ [no overlap].

Notice that $R(\tau)$ is ***Case even***, with an ***Case absolute maximum at*** $\tau = 0$.
**MATLAB**: The above question can be solved on MATLAB® as follows:

```
x = stepfun(t,0) - stepfun(t,1);   y = xcorr(x) * Ts;
```

## *Tutorial 24*

**Q**: A white noise signal *n*(*t*) is passed through a *RC* circuit as shown. Find the autocorrelation function of the output noise.

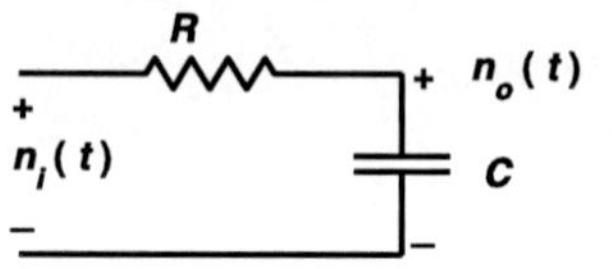


**Solution**:

$$H(f) = \frac{X_C}{R + X_C} = \frac{\frac{1}{j\omega C}}{R + \frac{1}{j\omega C}} = \frac{1}{1 + j2\pi fCR},$$

$$G_n(f) = \eta/2 \underset{WKT}{\overset{\mathcal{F}}{\longleftrightarrow}} R_n(\tau) = \frac{\eta}{2}\delta(\tau),$$

$$G_{n_0}(f) = |H(f)|^2 G_n(f) = \frac{\eta/2}{1 + 4\pi^2 f^2 R^2 C^2} = \frac{\eta}{2}\frac{\frac{1}{R^2C^2}}{\frac{1}{R^2C^2} + (2\pi f)^2}$$

$$= \frac{\eta}{4RC}\frac{2\frac{1}{RC}}{\left(\frac{1}{RC}\right)^2 + (2\pi f)^2}.$$

$\therefore\ R_{n_0}(\tau) = F^{-1}\{G_{n_0}(f)\} = \frac{\eta}{4RC}e^{-\frac{|\tau|}{RC}}$ [Tables; Wiener–Kinchin Theorem].
Note that $R_n(\tau) = \mathcal{E}\{n(t)n(t+\tau)\}$, where $\mathcal{E}$ is the statistical expectation functional [Compare with the deterministic signal in Tutorial 22].

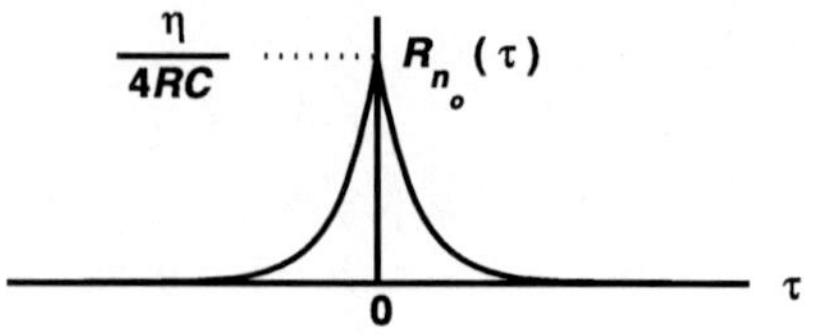

## Tutorial 25

**Q1**: A random telegraph signal *x*(*t*) with autocorrelation function $R_x(\tau) = e^{-a|\tau|}$, *a* being a constant, is applied to an RC circuit as shown. Find the power spectral density (PSD) of the output random signal *y*(*t*).

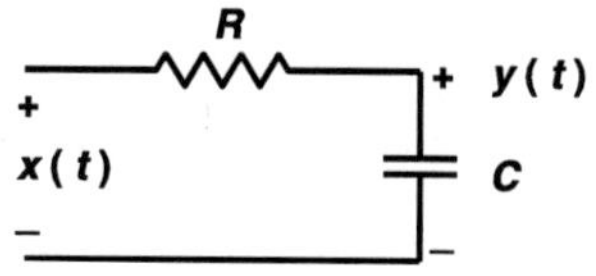


**Solution**: We have: $H(f) = \frac{1}{1+j2\pi fRC}$

$$\therefore \quad G_x(f) = F\{R_x(\tau)\} = F\{e^{-a|\tau|}\} = \frac{2a}{a^2 + (2\pi f)^2} \quad \text{[Tables; using WKT]}.$$

$$\therefore \quad G_y(f) = |H(f)|^2 G_x(f) = \left\{\frac{1}{1 + (2\pi f)^2 R^2 C^2}\right\}\left\{\frac{2a}{a^2 + (2\pi f)^2}\right\}.$$

**Q2**: A random noise current signal *x*(*t*) has a ***Case double-sided*** PSD normalized to unit resistance given by:

$$G_x(f) = e^{-0.01|f|} \quad \text{W/Hz/}\Omega \ (\text{Amp}^2/\text{Hz}).$$

The signal is passed through an ideal BPF with lower cutoff frequency at 1 kHz and upper cutoff frequency at 5 kHz as shown. Determine the noise power at the output if the load resistance is 100 Ω.

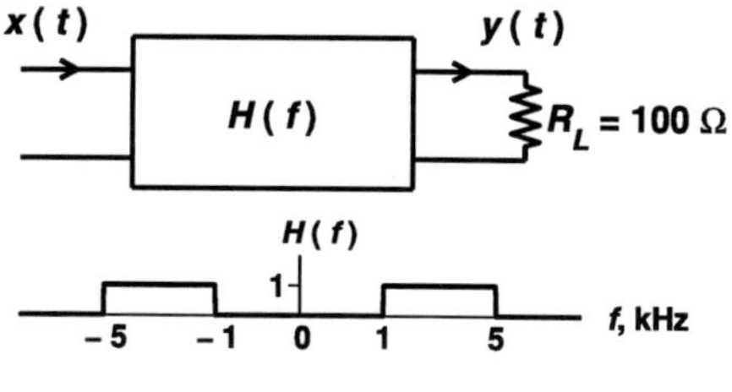


**Solution**: The normalized output power is:

$$P_y = \int_{-\infty}^{\infty} G_y(f)df = \int_{-\infty}^{\infty} G_x(f)|H(f)|^2 df = 2\int_{1k}^{5k} e^{-0.01f} df = 0.008 \text{ W}/\Omega.$$

The total output noise power is 100 $P_y = 0.8$ W.

## Tutorial 26

**Q**: The autocorrelation function of a signal $x(t)$ is defined as follows:

$$R_x(\tau) = \begin{cases} \int_{-\infty}^{\infty} x(\lambda)x(\tau+\lambda)d\lambda, & \text{If } x \text{ is an energy signal} \\ \frac{1}{T_o}\int_{-T_o/2}^{T_o/2} x(\lambda)x(\tau+\lambda)d\lambda, & \text{If } x \text{ is a periodic power signal} \\ \mathcal{E}\{x(t)x(t+\tau)\}, & \text{If } x \text{ is WSS random signal} \\ \{= \lim_{T\to\infty}\frac{1}{T}\int_{-T/2}^{T/2} x(\lambda)x(\tau+\lambda)d\lambda & \text{when } x \text{ is ergodic}\} \end{cases}$$

with the following ***properties***:

1. $R_x(-\tau) = R_x(\tau)$ [i.e., it is an ***even*** function].

$$2.\ R_x(0) = \begin{cases} E(\text{signal energy}), & \text{If } x \text{ is an energy signal} \\ P(\text{average signal power}), & \text{If } x \text{ is a periodic power signal} \\ P\ (\text{average signal power}) = \mathcal{E}\{x^2(t)\}, & \text{If } x \text{ is WSS random signal} \\ \overset{\text{ergodic}}{\longrightarrow} \lim_{T\to\infty}\int_{-T/2}^{T/2} x^2(t)dt & \end{cases}$$

$R_x(\tau)$ has an absolute maximum at $\tau = 0$, i.e., $|R_x(\tau)| \le R_x(0)$
Wiener–Kinchin Relations:

$$R_x(\tau) \overset{\mathcal{F}}{\longleftrightarrow} \begin{cases} \underbrace{|X(f)|^2}_{ESD} & \text{If } x \text{ is an energy signal} \\ \underbrace{|X_k|^2}_{PSD} & \text{If } x \text{ is a periodic power signal} \\ \underbrace{G_x(f)}_{PSD} & \text{If } x \text{ is WSS random signal} \end{cases}$$

***Prove properties 1, 2, and 3 for energy signals.***

**Solution**:

1. $R_x(-\tau) = \int_{-\infty}^{\infty} x(\lambda)x(-\tau+\lambda)d\lambda \overset{v=-\tau+\lambda}{\longrightarrow} \int_{-\infty}^{\infty} x(\tau+v)x(v)dv = R_x(\tau)$.
2. $R_x(0) = \int_{-\infty}^{\infty} x^2(\lambda)d\lambda = E$ (always positive).
3. $|R_x(\tau)|^2 \le \int_{-\infty}^{\infty}|x(\lambda)|^2 d\lambda \int_{-\infty}^{\infty}|x(\tau+\lambda)|^2 d\lambda = [\int_{-\infty}^{\infty}|x(\lambda)|^2 d\lambda]^2$ [using Schwarz's inequality, Tables.]

For real signals: $|x(t)|^2 = x^2(\lambda) \to R_x^2(\tau) \le R_x^2(0) \to |R_x(\tau)| \le R_x(0)$.

## *Tutorial 27*

**Q**: A random current signal $x(t)$ has a two-sided, normalized (to unit resistance) power spectral density (PSD) given by:

$$G_x(f) = \left\{ \begin{array}{ll} 0, & 0 < |f| < 1 \text{ Hz} \\ \frac{0.01}{|f|}, & 1 < |f| < \infty \text{ Hz} \end{array} \right\} \text{Amp}^2/\text{Hz}$$

The signal is passed through a filter whose transfer function is $H(s) = \frac{1}{s}$.

1. Determine the magnitude transfer function of the filter.
2. Determine the power transfer function of the filter.
3. Determine the average power (normalized to 1 Ω) of the output signal $y(t)$.

**Solution**:

1. $|H(\omega)| = |H(s)|_{s=j\omega} = \frac{1}{|j\omega|} = \frac{1}{\omega}$ or $|H(f)| = \frac{1}{2\pi f}$.
2. $|H(f)|^2 = \frac{1}{4\pi^2 f^2}$.
3. $P = \int_{-\infty}^{\infty} G_x(f)|H(f)|^2 df = 2\int_1^{\infty} \frac{0.01}{f} \frac{1}{4\pi^2 f^2} df = 0.000253\ \text{W}/\Omega$.

## *Tutorial 28*

**Q**: A signal $x(t) = 4\cos(\omega_0\, t)$ is transmitted through two additive white Gaussian noise (AWGN) channels and received as $s_1(t) = x(t) + n_1(t)$ and $s_2(t) = x(t) + n_2(t)$, where $\text{SNR}_1 = 10$ dB and $\text{SNR}_2 = 0$ dB. Find:

1. The expected values of $s_1(t)$ and $s_2(t)$.
2. Noise power in both cases.
3. Roughly plot the pdf of $n_1(t)$ and $n_2(t)$. Which one has more spread around the mean?

**Solution**:

1. Since we have AWGN, then $\mathcal{E}\{n_1(t)\} = \mathcal{E}\{n_2(t)\} = 0$ [at any time $t$].
   Note that $\mathcal{E}\{n(t)\}$ = ***statistical mean*** of $n(t) = m = \int_{-\infty}^{\infty} np(n)dn$, while the ***time mean*** is $m_t = \frac{1}{T}\int_0^T n(t)dt$, $T$ being the total time. If $t$ is discrete, then $m_t = \frac{1}{N}\sum_{k=0}^{N-1} n(k)$. If $n(t)$ is ***ergodic***, then $m = m_t$.

$$\therefore\quad \mathcal{E}\{s_1(t)\} = \mathcal{E}\{x(t) + n_1(t)\} = \mathcal{E}\{x(t)\} + \overbrace{\mathcal{E}\{n_1(t)\}}^{0} = x(t).$$

Similarly: $\mathcal{E}\{s_2(t)\} = x(t)$.

2.
$$\text{SNR}_1\,\text{dB} = 10\log_{10}(\text{SNR}_1) \rightarrow \text{SNR}_1 = 10^{\text{SNR}_1(\text{dB})/10} = 10.$$
$$\text{SNR}_1 = P_x/P_{n_1} \Rightarrow P_{n_1} = P_x/\text{SNR}_1 = P_x/10.$$
$$P_x = 4^2/2 = 8 \Rightarrow P_{n_1} = 8/10 = 0.8 \equiv -00.96\,\text{dB}.$$
$$\text{Similarly: } P_{n_2} = 8 \equiv 9\,\text{dB}.$$

3. Gaussian pdf is given by: $P(n) = \frac{1}{\sqrt{2\pi}\sigma} e^{-(n-m)^2/2\sigma^2}$ where $m$ = mean, $\sigma^2$ = variance.
Both $n_1$ and $n_2$ have zero means. Their variances are:

$$\sigma_1^2 = P_{n_1} = 0.8 \rightarrow \sigma_1 \approx 0.9 \quad \text{and} \quad \sigma_2^2 = P_{n_2} = 8 \rightarrow \sigma_2 \approx 2.8.$$

Since $n_2$ has more power, i.e., more variance, we expect that it has wider spread as shown.

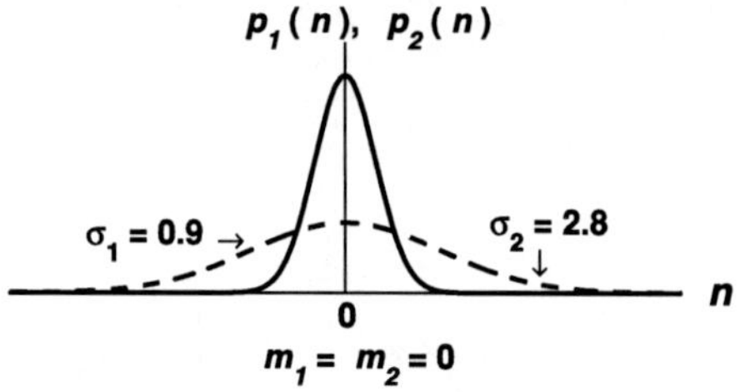


## *Tutorial 29*

Q: ***Range Estimation by Radar:*** A narrow pulse $s(t)$ is transmitted by the radar station towards the airplane. The pulse will hit the plane and reflect back to the radar, where a matched filter (correlator) is utilized to estimate the distance between the plane and the radar. Explain how the correlator is used for range estimation.

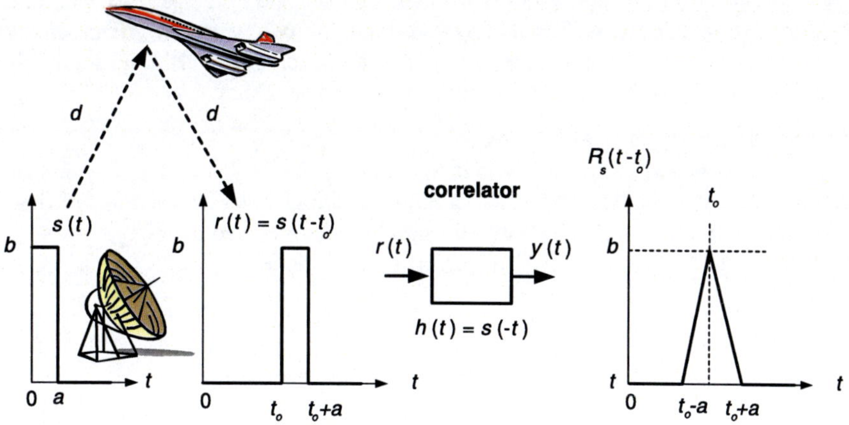


**Solution**: The received signal $r(t)$ is a delayed version of the transmitted signal $s(t)$, which is also corrupted by noise. It is given by: $r(t) = s(t - t_0) + n(t)$. The ***correlator is a matched filter*** with impulse response given by:

$$h(t) = s(-t) \text{ [A reflected version of } s(t) \text{ around vertical axis].}$$

$$y(t) = r(t) * h(t) = \int_{-\infty}^{\infty} r(\lambda)h(t-\lambda)d\lambda = \int_{-\infty}^{\infty} r(\lambda)s(\lambda - t)d\lambda$$

$$= \int_{-\infty}^{\infty} [s(\lambda - t_o) + n(\lambda)]s(\lambda - t)d\lambda$$

$$\overset{v=\lambda-t}{\longrightarrow} = \int_{-\infty}^{\infty} s(v + t - t_o)s(v)dv + \int_{-\infty}^{\infty} n(v+t)s(v)dv = R_s(t - t_o) + R_{ns}(t).$$

*Note*: since $h(t) = s(-t)$, the system ***convolution*** is essentially a ***correlation***. Since the correlation is a measure of similarity, $R_{ns}(t) \approx 0$.
From Tutorial 21, the autocorrelation $R_s(t - t_o)$ of a square pulse $s(t)$ is a triangular function with a maximum at $t - t_o = 0$, or $t = t_o$. Hence, if the pulse width $a$ is very small, we have just a pulse at the time instant $t = t_o$ at the receiver, from which we can decide the time delay $t_o$.
Now $2d = ct_o$ (where $c$ = velocity of signal propagation $\approx 10^8$ m/s).
Hence, the distance to the airplane is $d = ct_o/2$.

## *Tutorial 30*

**Q**: In ***binary baseband communication***, binary data (a sequence of 0's and 1's) are transmitted through a channel using two ***orthogonal signals***, $s_0(t)$ [to represent logic 0] and $s_1$ $(t)$ [to represent logic 1]. If the data rate is $R$ bps, then the time duration of $s_0$ and $s_1$ is $T = 1/R$ (s). A bank of two ***matched filters*** and a comparator are used at the receiver for optimal detection of signals in noise. In this question assume noise-free transmission, with $s_0$ and $s_1$ as shown, and $R = 1$ kbps.

1. Find the energy of $s_0$ and $s_1$ over $T$ s.
2. Show that $s_0$ and $s_1$ are orthogonal.
3. Find the outputs of the two matched filters at the receiver as functions of time over (0, $T$), knowing logic 0 was transmitted and received as $r(t)$.
4. What is the decision of the receiver in step (3) above at $t = T$?

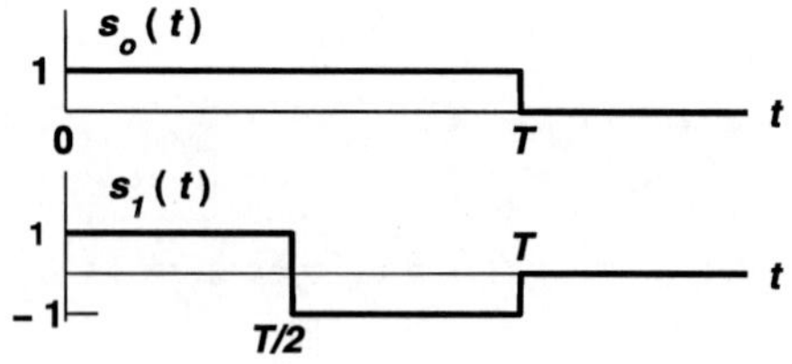


**Solution**:

1. $E_o = \int_0^T s_0^2(t)dt = T = 0.001$ J; $E_1 = \int_0^T s_1^2\,(t)dt = 0.001$ J.
2. $\int_0^T s_0(t)s_1(t)dt = \int_0^{T/2} dt - \int_{T/2}^{T} dt = 0 \rightarrow s_o$ and $s_1$ are orthogonal.
3

$$r_0 = \text{Output of the 1st correlator} = \int_0^t r(t)s_o(t)dt = \int_0^t s_o^2(t)dt = t$$

$$r_1 = \text{Output of the 2nd correlator} = \int_0^t r(t)s_1(t)dt = \int_0^t s_o(t)s_1(t)dt$$

$$= \begin{Bmatrix} \int_0^t dt = t, & 0<t\leq T/2 \\ \int_0^{T/2} dt - \int_{T/2}^{t} dt = T-t, & T/2<t\leq T \end{Bmatrix}$$

4. At $t = T$, the receiver makes a decision based on the output of the comparator. Since $r_o\,(T) = T > r_1\,(T) = 0$, the decision is that logic 0 was transmitted. No error is expected since we considered a noise-free condition.

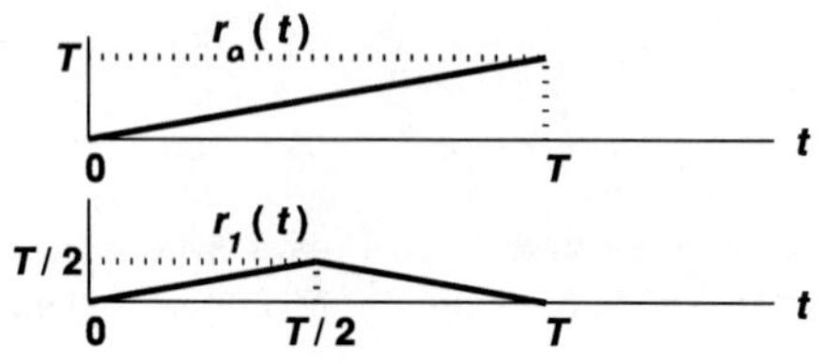

## *Tutorial 31*

**Q**: A random signal frequently encountered in application is a sinusoid with random phase as follows: $s(t) = A\cos(\omega_o t + \theta)$ where $\theta$ is a random variable *uniformly distributed* over $(0, 2\pi)$. Find the autocorrelation function of $s(t)$.

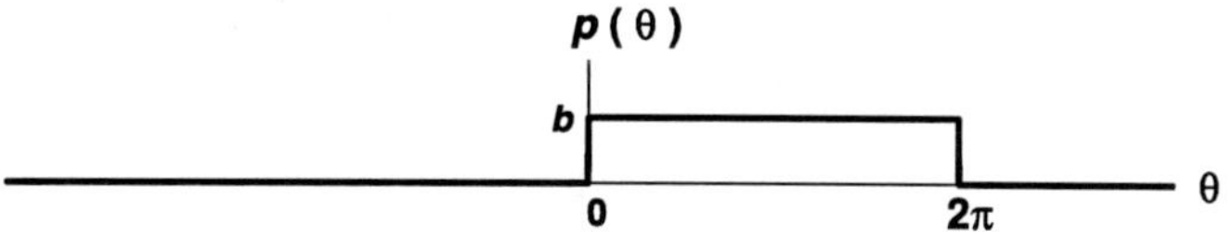


**Solution**: Uniform distribution implies that: $2\pi b = 1$; hence, $b = 1/(2\pi)$. The pdf of $\theta$ is given by:

$$p(\theta) = \begin{Bmatrix} 1/(2\pi), & 0 \le \theta < 2\pi \\ 0, & \text{elsewhere} \end{Bmatrix}.$$

Since $s(t)$ is a random signal, its autocorrelation is given by:

$$R_s(\tau) = \mathcal{E}\{s(t)s(t+\tau)\} = \mathcal{E}\left\{ A^2\cos\underbrace{(\omega_o + \theta)}_{x}\cos\underbrace{[\omega_o(t+\tau)+\theta]}_{y} \right\}$$

$$= \mathcal{E}\left\{ \frac{A^2}{2}[\cos(\underbrace{2\omega_o t + \omega_o\tau + 2\theta}_{x+y}) + \cos(\underbrace{\omega_o\tau}_{y-x})] \right\} \quad \text{(using Tables)}$$

$$= \frac{A^2}{2}\mathcal{E}\{\cos(2\omega_o t + \omega_o\tau + 2\theta)\} + \frac{A^2}{2}\mathcal{E}\{\cos(\omega_o\tau)\}$$

$$= \frac{A^2}{2}\int_0^{2\pi} \cos(2\omega_o t + \omega_o\tau + 2\theta)\underbrace{\frac{1}{2\pi}}_{p(\theta)}\,d\theta + \frac{A^2}{2}\cos(\omega_o\tau)$$

$$= \frac{A^2}{2}\cos(\omega_o\tau)$$

*Note*: If we sample $s(t)$ with a sampling rate of $f_s = 1/T_s = 1/(T_o/4) = 4f_o$, the resulting samples are uncorrelated since $R_s\,(T_s) = R_s\,(T_o/4) = 0$.

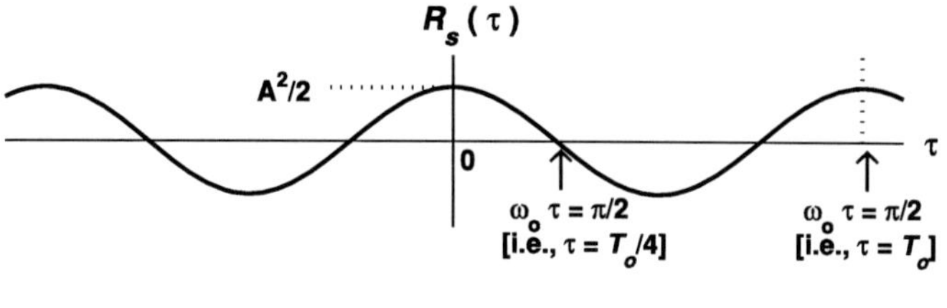

## Tutorial 32

**Q**: The pulse $x(t)$ shown in Fig. a below is sent along a communication line. At the receiver, a matched filter $h(t) = x(-t)$ is used for optimal detection.

1. Assume that the noise at the receiver has a flat spectrum, but is bandlimited to 10 kHz. The variance of noise is $\sigma^2$. What is the maximum signal power to noise power ratio (SNR) at the input of the matched filter (i.e., what is $\mathrm{SNR}_i$ at the time of optimal detection)?
2. What is the maximum SNR at the output of the matched filter, $\mathrm{SNR}_o$?
3. What is the ratio $(\mathrm{SNR}_o|_{\max})/(\mathrm{SNR}_i|_{\max})$?
4. Repeat the above steps for $s(t)$ in Fig. b.

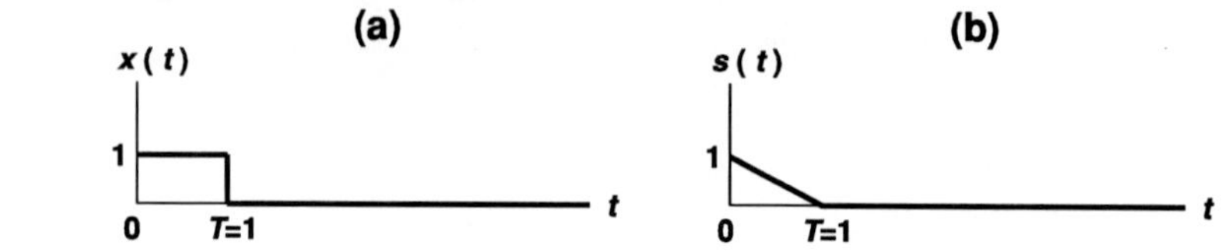


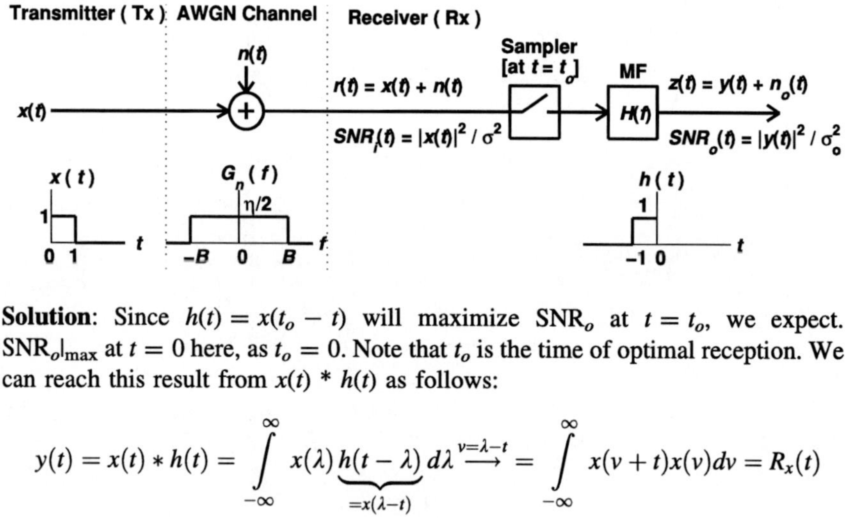


**Solution**: Since $h(t) = x(t_o - t)$ will maximize $\mathrm{SNR}_o$ at $t = t_o$, we expect. $\mathrm{SNR}_o|_{\max}$ at $t = 0$ here, as $t_o = 0$. Note that $t_o$ is the time of optimal reception. We can reach this result from $x(t) * h(t)$ as follows:

$$y(t) = x(t) * h(t) = \int_{-\infty}^{\infty} x(\lambda)\underbrace{h(t-\lambda)}_{=x(\lambda-t)}d\lambda \overset{v=\lambda-t}{\longrightarrow} = \int_{-\infty}^{\infty} x(v+t)x(v)dv = R_x(t)$$

Following Tutorial 23, we find $y(t)$ as shown below.

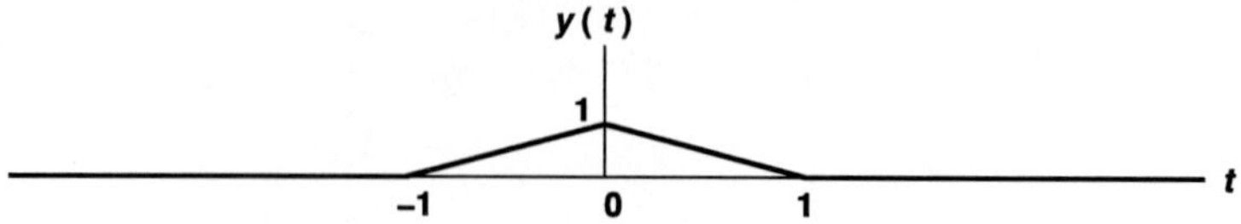

1.

$$\mathrm{SNR}_i(t) = |x(t)|^2/\sigma^2 = 1/\sigma^2 \quad \forall t \in (0, T) \tag{1}$$

2.

$$\mathrm{SNR}_o(t)|_{\max} = |y(0)|^2/\sigma_o^2 = 1/\sigma_o^2 \quad \text{(using the figure above)} \tag{2}$$

Now we find $\sigma_0^2$ as follows:

$$P_n = \sigma^2 = \int_{-\infty}^{\infty} G_n(f)df = \int_{-B}^{B} \frac{\eta}{2} df = \eta B \tag{3}$$

$$P_{n_0} = \sigma_0^2 = \int_{-\infty}^{\infty} G_{n_0}(f)df = \int_{-\infty}^{\infty} G_n(f)|H(f)|^2 df = \frac{\eta}{2} \int_{-B}^{B} |H(f)|^2 df \tag{4}$$

Since $h(t) = x(-\,t)$, we have

$$H(f) = X(-f) \quad [\mathit{Tables}] \tag{5}$$

Since $x(t)$ is real, we have

$$X(-f) = X^*(f) \tag{6}$$

$$\therefore \quad |H(f)|^2 = H(f)H^*(f) = X^*(f)X(f) = |X(f)|^2 \tag{7}$$

From (4) and (7) we have:

$$\sigma_o^2 = \frac{\eta}{2} \int_{-B}^{B} |X(f)|^2 df \tag{8}$$

Since $|X(f)|$ is a sinc function [Tables], then $|X(f)|^2$ decays rapidly and we have:

$$\sigma_0^2 \approx \frac{\eta}{2} \int_{-\infty}^{\infty} |X(f)|^2 df \tag{9}$$

Using Parseval's theorem we have:

$$\sigma_o^2 \approx \frac{\eta}{2} \int_{-\infty}^{\infty} |X(f)|^2 df = \frac{\eta}{2} \int_{-\infty}^{\infty} |x(t)|^2 dt = \frac{\eta}{2} \int_{0}^{1} 1 \cdot dt = \frac{\eta}{2} \tag{10}$$

From (2) and (10) we have:

$$\mathrm{SNR}_o(t)|_{\max} = 1/\sigma_o^2 = 2/\eta \tag{11}$$

3. From (1), (2), (3) and (11) we have:

$$\frac{\mathrm{SNR_o(t)}|_{\max}}{\mathrm{SNR}_i(t)|_{\max}} = \frac{1/\sigma_0^2}{1/\sigma^2} = \frac{\sigma^2}{\sigma_0^2} = \frac{\eta \cdot B}{\eta/2} = 2B = 20000$$

## *Tutorial 33*

**Q1**: Find the magnitude and phase response of the constant time-delay system.

**Solution**: The time-delay I/O equation is $y(t) = x(t - T)$.
Taking the Fourier transform we get: $Y(f) = X(f)e^{-j2\pi fT}$ [Tables].

$$\therefore \quad H(f) = Y(f)/X(f) = e^{-j2\pi fT} \quad [\therefore \quad h(t) = \delta(t - T), \text{using } \mathit{Tables}].$$

Hence, the magnitude response is $|H(f)| = 1$ (constant), i.e., the time-delay is an ***all pass*** filter. The phase response is $\phi(f) = \angle H(f) = -2\pi \cdot fT$. Note that the phase is *frequency dependent* in a linear relationship (hence the name *linear phase*).

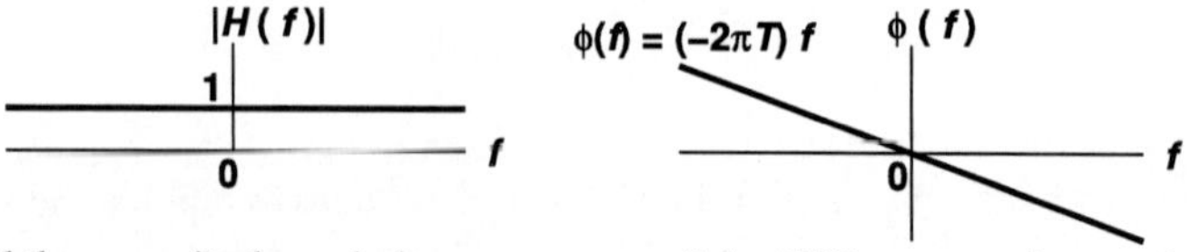


**Q2**: Find the magnitude and phase response of the *Hilbert transformer* defined by:

$$H(f) = \begin{Bmatrix} -j, & f \geq 0 \\ j, & f < 0 \end{Bmatrix}.$$

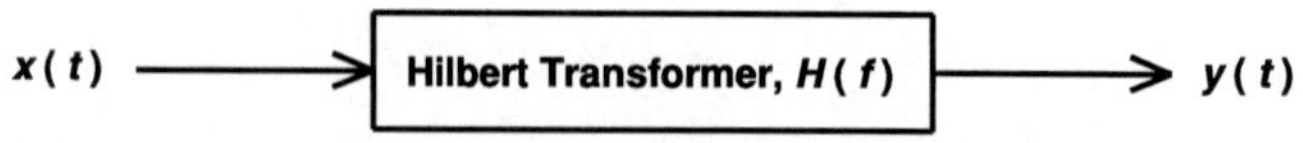


**Solution**: $H(f) = -j\mathrm{sgn}(f) \to h(t) = 1/(\pi \cdot)$ [Tables, Duality of $\mathcal{F}$].
$\therefore \quad |H(f)| = |-j\mathrm{sgn}(f)| = 1$ (constant), i.e., Hilbert transformer (HT) is an ***all-pass*** filter.

$$\phi(f) = \angle H(f) = \angle[0 - j\mathrm{sgn}(f)] = \left\{ \begin{array}{ll} \tan^{-1}(\frac{-1}{0}) = \tan^{-1}(-\infty) = -\frac{\pi}{2}, & f \geq 0 \\ \tan^{-1}(\frac{1}{0}) = \tan^{-1}(+\infty) = \frac{\pi}{2}, & f < 0 \end{array} \right\}.$$

Hence $\phi(f)$ is ***constant*** for physical frequencies $f \geq 0$.

**Comparison**: A constant time-delay $T$ gives a constant (frequency-independent) time-delay and a frequency-dependent phase shift $\phi(f)$; HT gives frequency-independent (constant) phase shift (90°) and frequency-dependent time-delay $[t_d = (\pi/2)/\omega]$.

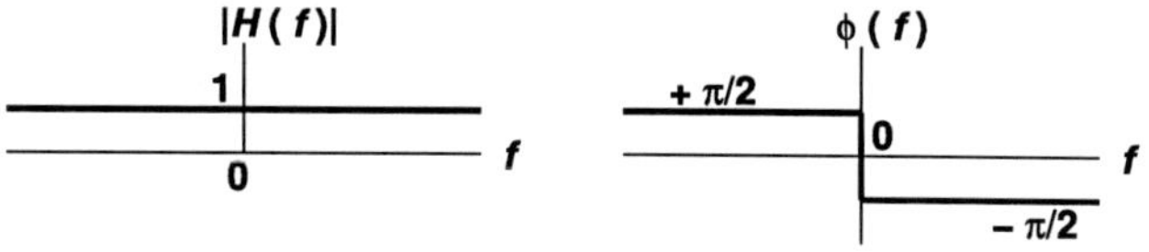


## *Tutorial 34*

**Q1**: Design a BP Chebychev-I 1 dB ripple filter with center frequency $f_0 = 10$ kHz, bandwidth BW = 1 kHz, maximum gain $G_m = 1$, and stopband gain $\leq -10$ dB for $f \leq f_1 = 7$ kHz and $f \geq f_2 = 13$ kHz. True load resistance is $R_L = 10\,\Omega$.

**Solution**:

$$\omega_u = \omega_0 + \omega_b/2 = 2\pi(10\text{k}) + 2\pi(1\text{k})/2 = 21\pi\text{k rad/s}$$

where $\omega_b = BW = \omega_u - \omega_l$.

$$\omega_l = \omega_0 - \omega_b/2 = 2\pi(10\text{k}) - 2\pi(1\text{k})/2 = 19\pi\text{k rad/s}.$$

$$\omega_g = \sqrt{\omega_l\omega_u} = \sqrt{(21\pi\text{k})(19\pi\text{k})} = 19.97\pi\text{k rad/s}.$$

Using LP → BP transformation (Tables): $\omega_N = \frac{\omega^2 - \omega_g^2}{\omega\omega_b}$, we find the normalized LP frequencies that correspond to $f_1$ and $f_2$ as follows:
$\omega_{1N} = -7.25$ and $\omega_{2N} = 5.32$.
We check the order $n$ that gives gain $\leq -10$ dB for $|\omega_{2N}| \leq \omega_N \leq |\omega_{1N}|$; or $5.32 \leq \omega_N \leq 7.25$. From Chebychev-I curves (Tables), $n = 1$ gives gain $\leq -10$ dB for $\omega_N = 7.25$, but not for $\omega_N = 5.32$. Hence, $n = 2$ is the suitable choice. From Tables we obtain the *normalized* transfer function as follows:

$$H(s_N) = \frac{a_0}{1.1 + 1.09 s_N + s_N^2} \Rightarrow G_{\mathrm{dc}} = \frac{G_M}{\sqrt{1 + \varepsilon^2}} = \frac{1}{\sqrt{1 + 0.25}} = \frac{a_0}{1.1} \Rightarrow a_0 = 0.98.$$

Using the transformation LPN → BP (Tables), we obtain the final transfer function:

$$H(s) = \frac{0.98}{1.1 + 1.09\left(\frac{s^2+\omega_g^2}{s\omega_b}\right) + \left(\frac{s^2+\omega_g^2}{s\omega_b}\right)^2}$$

which is a 4th-order filter. Note that since $s_N \to \frac{s^2+\omega_g^2}{\omega_b s}$ *is quadratic, number of poles is* **doubled** *by using this transformation.*
***Circuit Design***: Using Tables, the scaling factors $Z = 10$ and $F = \omega_b = 2\pi\text{k}$, and the relation $\omega_g^2 = 1/(LC)$, we design the BPF circuit as shown below.

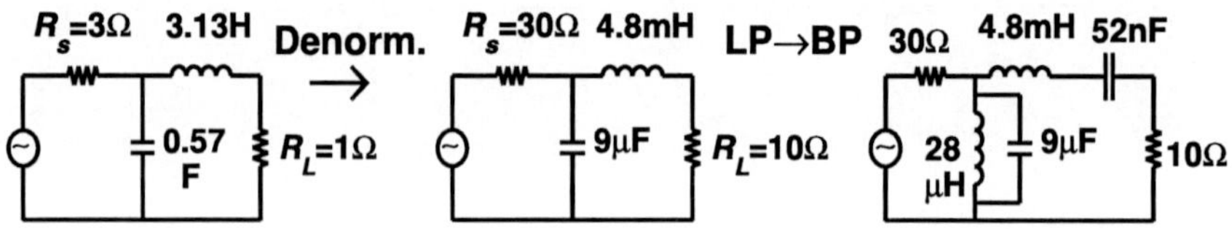


## *Tutorial 35*

**Q**: Find the values of $k$ for which the analog system shown below is stable ($k$ is real).

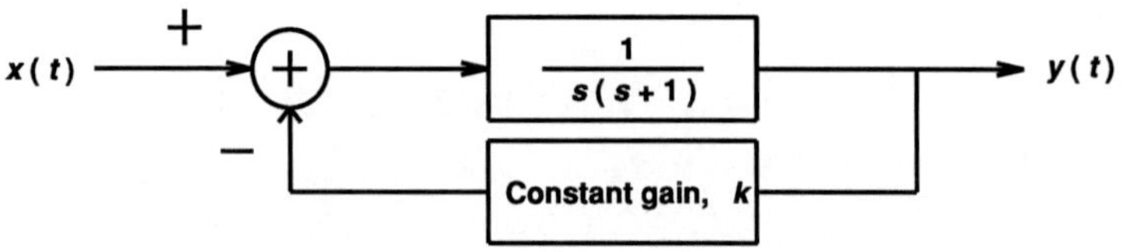


**Solution**: Using same approach as in Tutorial 19, we find the transfer function of the system as follows:

$$H(s) = \frac{1}{s^2+s+k} = \frac{1}{(s-p_1)(s-p_2)}$$

$$p_{1,2} = \frac{-1 \pm \sqrt{1-4k}}{2} = -\frac{1}{2} \pm \frac{1}{2}\sqrt{1-4k} = -\frac{1}{2} \pm \sqrt{\frac{1}{4} - k}$$

For the system to be stable, its poles $p_1$ and $p_2$ should be in the ***left half of the s-plane***.

Since $k$ is real, $\sqrt{\frac{1}{4}-k}$ is either real (if $\frac{1}{4}\geq k$) or pure imaginary (***but not complex***) (if $\frac{1}{4}<k$). If $\sqrt{\frac{1}{4}-k}$ is imaginary, then both $p_1$ and $p_2$ will be in the left half of the $s$-plane and the system is always stable.
Now assume $\sqrt{\frac{1}{4}-k}$ is real, i.e., $k\leq\frac{1}{4}$. For stability we need the following condition:

$$-\frac{1}{2}+\sqrt{\frac{1}{4}-k}<0 \quad \text{(to keep } p_1<0\text{)}.$$

$$\therefore \quad \frac{1}{4}-k<\frac{1}{4}\rightarrow k>0.$$

Combining $k\leq\frac{1}{4}$ and $k>0$ gives $0<k\leq\frac{1}{4}$ as the overall condition in this case.
Now we ***summarize*** the real and imaginary cases above:

1. Case 1 (the root is imaginary): $k>\frac{1}{4}$
2. Case 2 (the root is real): $0<k\leq\frac{1}{4}$

The combination of the two cases gives the final condition $k>0$.

## *Tutorial 36*

**Q**: Design a passive bandpass Butterworth filter with the following specifications:

1. 4th-order
2. Center frequency = 10 kHz
3. Bandwidth = 3 kHz
4. True load impedance = 600 Ω
5. Maximum gain = 1.

**Solution**:

$$\omega_l=\omega_0-\frac{\omega_b}{2}=17\text{k}\pi;\quad \omega_g=\sqrt{\omega_u\omega_l}=19.77\text{k}\pi$$

The scaling factors are: ISF = Z = 600 and FSF = $F=\omega_b=6k\pi$.
Since LP → BP transformation is quadratic (Tables), we need a 2nd-order LPF as the LP-prototype. Its transfer function is found from Tables as follows:

$$H(s_N)=\frac{a_o}{b_o+b_1s_N+s_N^2}\qquad \text{where}\quad b_o=1\quad\text{and}\quad b_1=1.41.$$

$$G_{\text{d.c}}=G_m=1=\frac{a_o}{b_o}\rightarrow a_o=b_o=1.$$

Using the LP → BP transformation $s_N = \frac{s^2+\omega_g^2}{s\omega_b}$ (from Tables), we get the transfer function of the BPF as follows:

$$H(s) = \frac{1}{1 + 1.41\frac{s^2+\omega_g^2}{s\omega_b} + \left(\frac{s^2+\omega_g^2}{s\omega_b}\right)^2}$$

**Circuit Design:** Using Tables, the scaling factors $Z = 600$ and $F = \omega_b = 6\pi$k, and the relation $\omega_g^2 = 1/(LC)$, we design the BPF circuit as shown below.

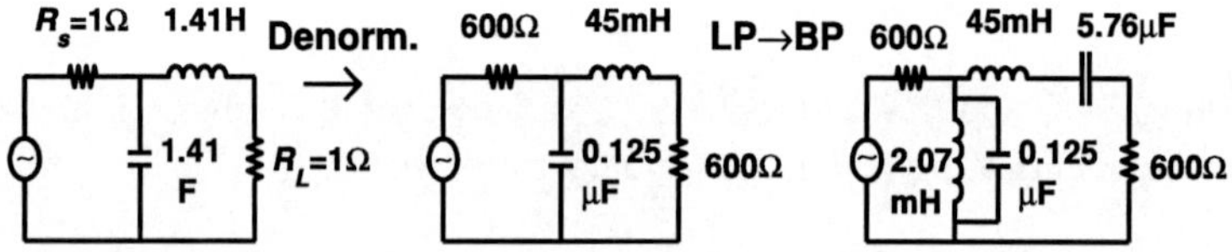


## *Tutorial 37*

**Q**: Find the transfer function of the following active filter and explain its function.

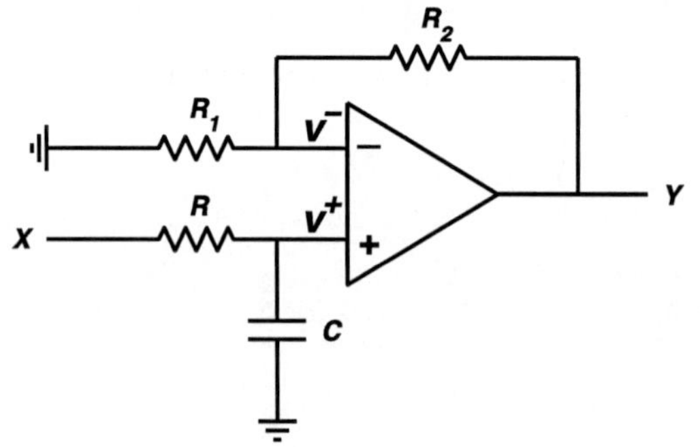


**Solution**:

$$\frac{Y}{V^+} = \left(1 + \frac{R_2}{R_1}\right),$$

$$V^+ = X\frac{X_c}{R + X_c} = X\frac{1/(j\omega C)}{R + 1/(j\omega C)} = X\frac{1}{1 + j\omega RC}$$

$$\therefore \quad \frac{Y}{X} = \frac{Y}{V^+}\frac{V^+}{X} = \left(1 + \frac{R_2}{R_1}\right)\frac{1}{1 + j\omega RC} = \frac{G_m}{1 + j\omega RC}$$

where $G_m = 1 + R_2/R_1$ is the gain of the filter.

The magnitude response is:

$$|H(\omega)| = \frac{G_m}{\sqrt{1 + (\frac{\omega}{\omega_o})^2}}$$

where $\omega_o = \frac{1}{\sqrt{RC}}$.
Now we have:

$$H(0) = G_m;$$
$$H(\infty) = 0;$$
$$|H(\omega_o)| = G_m/\sqrt{2}.$$

Hence, the system is a LPF with cutoff frequency $\omega_o = \frac{1}{\sqrt{RC}}$.

## Tutorial 38

**Q**: Determine and plot the spectrum at all points shown in the signal processing system below. Consider only the frequency range 0–40 Hz.

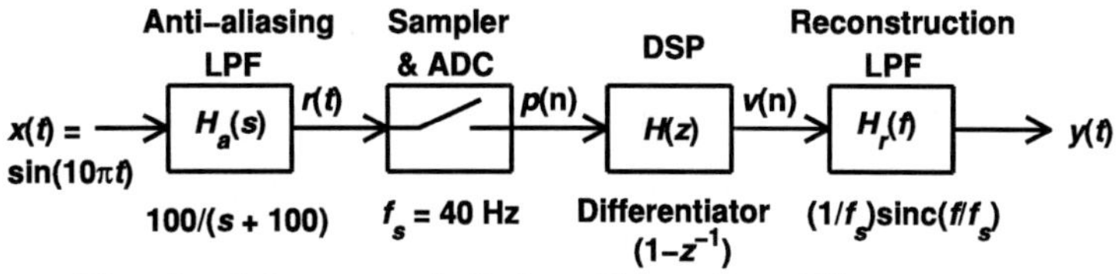


**Solution**: The signal frequency is $2\pi f_o = 10\pi \rightarrow f_o = 5\,\text{Hz}$.

$$X(f) = \frac{1}{2j}\delta(f - f_o) - \frac{1}{2j}\delta(f + f_o) \text{ (Tables)} \Rightarrow |X(f)| = \frac{1}{2}\delta(f - f_o) + \frac{1}{2}\delta(f + f_o).$$

For $|R(f)|$ we have: $A = 0.5|H_a(j2\pi f_o)| = 0.5\left|\frac{100}{j10\pi + 100}\right| = 0.47$ (see the figure below).
For $|V(f)|$ we have: $B_1 = 18.8\left|H(e^{j2\pi f_0 T_s})\right| = 18.8\left|1 - e^{-j2\pi 5/40}\right| = 14.4$, and $B_2 = 18.8\left|H(e^{j2\pi(f_s - f_o)T_s})\right| = 18.8\left|H(e^{j(2\pi - 2\pi f_o T_s)})\right| = 18.8\left|H(e^{-j2\pi f_o T_s})\right| = B_1$.
For $|Y(f)|$ we have:

$$C = B_1|H_r(f_0)| = 14.4\left|\frac{1}{40}\,\text{sinc}\left(\frac{5}{40}\right)\right| = 0.35 \quad \text{and} \quad D = B_2|H_r(35)| = 0.05.$$

*Note: no need here to divide ($1 - z^{-1}$) by $T_s$ for accurate differentiation, since the sampler performs this function.*

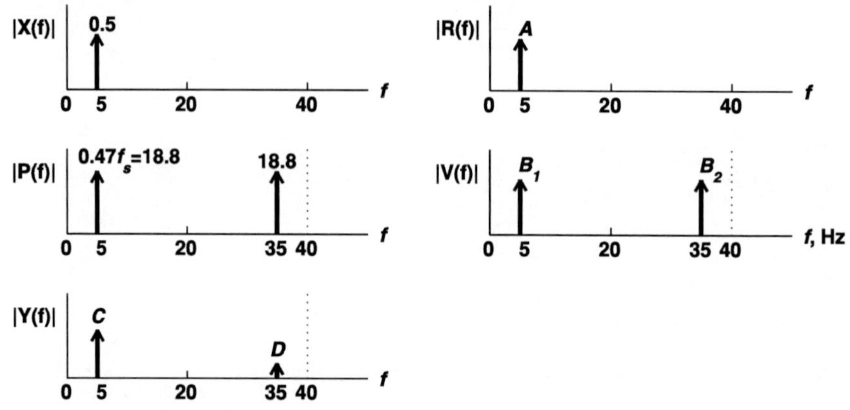


## *Tutorial 39*

**Q**: If the interest rate $r$ is fixed and there are no account fees, then a savings account in a bank represents an LTI system. Let $x(n)$ denote the amount of money deposited (or withdrawn) in the $n$th day, and $y(n)$ be the total amount of money in the account at the end of the $n$th day. Assume $r$ is compounded daily.

1. Find the impulse response $h(n)$ of the system using a time-domain approach.
2. Find the system output [using $h(n)$] at the $n$th day.
3. Find the system difference equation, the transfer function $H(z)$, and the impulse response $h(n)$. Can we use the fft to find the output $y(n)$?

**Solution**:

1. We need an input which is a delta function $\delta(n)$. Put \$1 in the opening day ($n = 0$) and \$0 afterwards. Now we have:

   $$y(0) = x(0) = \delta(0) = 1; \quad y(1) = 1 + r;$$
   $$y(2) = y(1) + ry(1) = y(1)(1 + r) = (1 + r)^2. \text{ Similarly: } y(n) = (1 + r)^n.$$

   This is the impulse response of the system, $\{h(n)\}$.
2. $y(n) = x(n) * h(n) = \sum_{k=0}^{n} x(k)h(n-k) = \sum_{k=0}^{n} x(k)[1+r]^{n-k}$.
   For example, if $r = 0.01\% = 0.0001$ and we put \$100 in the opening day, \$50 in the next day, \$50 in the 3rd day, and nothing afterwards, then at the end of the year we have: $y(365) = x(0)(1+r)^{365} + x(1)(1+r)^{364} + x(1)(1+r)^{363} =$ \$207.41.
3. New balance = old + $r \cdot$ old + new deposit

$$\therefore \quad y(n) = y(n-1) + ry(n-1) + x(n) = (1+r)y(n-1) + x(n)$$

$$\therefore \quad Y(z) = (1+r)z^{-1}Y(z) + X(z) \Rightarrow Y(z)[1-(1+r)z^{-1}] = X(z)$$

$$\therefore \quad H(z) = \frac{Y(z)}{X(z)} = \frac{1}{1-(1+r)z^{-1}} = \frac{z}{z-(1+r)} \rightarrow h(n) = (1+r)^n \text{ (Tables)}.$$

This is an **IIR** filter, with a pole at $z = 1 + r$ and a zero at $z = 0$. It is unstable. We can use the FFT if $\{x(n)\}$ has finite length and we are interested in $y(n)$ for a finite range of $n$, i.e., if $0 \le n \le M$, where $M$ is a finite number.

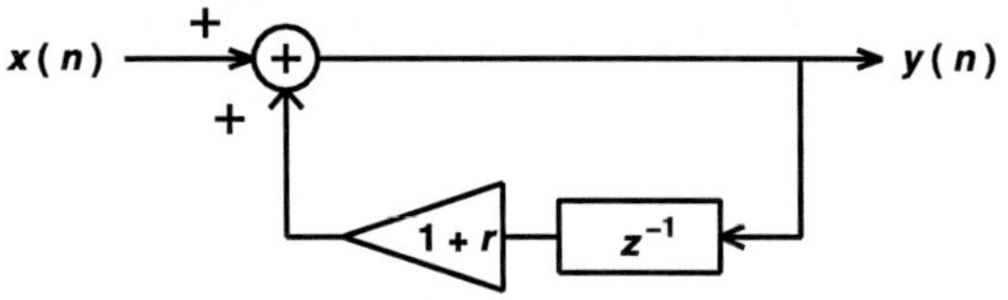


## *Tutorial 40*

**Q**: An engineer wants to make a decision about the trend of a fluctuating stock price. Based on a study of the stock market, he found that a period of 3 days is sufficient for a temporary decision, where the importance (weight) of the price at any day is twice that at the previous day. How can this engineer use the fft algorithm to estimate the weighted average of the stock price over the last week, knowing that the last seven prices were {12, 9, 10, 13, 8, 6, 9} dollars?
**Solution**: We need a 3-tap moving average FIR filter ($N = 3$). We should find its impulse response first.
Assume that the percentage weight of today's price is $a \Rightarrow h(0) = a$.
The weight of yesterday's price will be $a/2 \Rightarrow h(1) = a/2$.
The weight of the price 2 days ago will be $(a/2)/2 = a/4 \Rightarrow h(2) = a/4$.
Now we have $1 = a + \frac{a}{2} + \frac{a}{4}$, from which $a = 0.57$. Hence, the tap weights are: $h_0 = 0.57$, $h_1 = 0.28$, and $h_2 = 0.14$. The average price suitable for a decision will be the output of the FIR filter:

$$y(n) = h(n) * x(n) = \sum_{k=0}^{N-1} h(k)x(n-k); \quad \text{where } h = [h_o\ h_1\ h_2] \quad \text{and}$$

$$x = \begin{matrix} [12 & 9 & 10 & 13 & 8 & 6 & 9] \\ 0 & 1 & 2 & 3 & 4 & 5 & 6 \end{matrix}$$

The length of $\{y(n)\}$ sequence is $L = N_h + N_x - 1 = 3 + 7 - 1 = 9$.
For efficient computation (especially in real applications, where $N_h$ and $N_x$ can be very large), we should use the fft as follows:

1. Zero-pad $h$ to get $h_z = [h_o\ h_1\ h_2\ 0\ 0\ 0\ 0\ 0\ 0]$ and $x$ to get $x_z = [12\ 9\ 10\ 13\ 8\ 6\ 9\ 0\ 0]$.
2. Compute: $H_z = \text{fft}(h_z)$, $X_z = \text{fft}(x_z)$, and $Y_z = X_z\,H_z$.
3. Find $y_z = \text{ifft}(Y_z)$ to get

| | | | | | | | | | |
|---|---|---|---|---|---|---|---|---|---|
| $y_z = y =$ | [6.8 | 8.5 | 10 | 11.5 | 9.7 | 7.5 | 8 | 3.4 | 1.2] |
| | 0 | 1 | 2 | 3 | 4 | 5 | 6 | 7 | 8 |

Consider only the central 5 values of $\{y(n)\}$, i.e., items no. 2, 3, …, 6.
*Note*: the coefficients $\{h(n)\}$ can be adaptive based on change of market variables.

## *Tutorial 41*

**Q**: Find $x(n)$ if

(A) $X(z) = \frac{3}{z-2}$,
(B) $X(z) = \frac{1}{z^2-3z+2}$.

**Solution**:

(A) $X(z) = \frac{3}{z-2} = 3z^{-1}\frac{z}{z-2} = 3z^{-1}R(z)$.
From Tables we find: $r(n) = 2^n\ u(n)$ and $z^{-1}R(z) \overset{ZT}{\longleftrightarrow} r(n-1) = 2^{n-1}u(n-1)$.
Hence we have $x(n) = 3r(n-1) = 3\cdot 2^{n-1}\ u(n-1)$.

(B) $z^2 - 3z + 2 = 0 \Rightarrow p_{1,2} = \frac{-(-3)\pm\sqrt{(-3)^2-4(1)(2)}}{2(1)} = 1, 2.$
$X(z) = \frac{1}{(z-p_1)(z-p_2)} = \frac{1}{(z-1)(z-2)} = \frac{a}{z-1} + \frac{b}{z-2}$ (where we used ***partial fraction expansion***).

To find $a$:

1. Multiply by $z-1: \frac{z-1}{(z-1)(z-2)} = a + b\frac{z-1}{z-2}$
2. Put $z = 1: \frac{1}{(1-2)} = a + 0 \to a = -1.$

Similarly we find $b = 1$.

$$\therefore\quad X(z) = \frac{-1}{z-1} + \frac{1}{z-2} = (-1)z^{-1}\frac{z}{z-1} + z^{-1}\frac{z}{z-2}.$$

From Tables and (A) above we find:

$$x(n) = -u(n-1) + 2^{n-1}u(n-1) = [2^{n-1} - 1]u(n-1).$$

*Note:* Normally in such questions, it is better to expand $X(z)/z$ rather than $X(z)$. Applying this to Q2 above, we get $x(n) = \frac{1}{2}\delta(n) + [2^{n-1} - 1]u(n)$. Show that the two answers are equivalent.

## Tutorial 42

**Q**: Find the values of $b$ for which the system shown below is stable, knowing that $b$ is real.

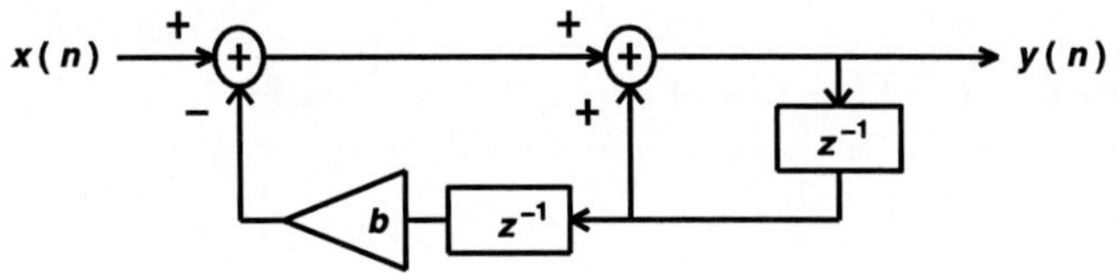


**Solution**: We first define intermediate signals $r(n)$, $g(n)$ as shown below.

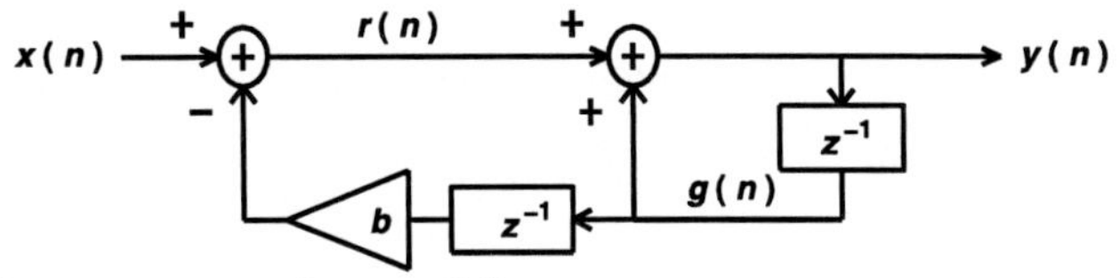


We write the system equations as follows:

$$y(n) = r(n) + g(n) \quad (1)$$

$$g(n) = y(n-1) \quad (2)$$

$$r(n) = x(n) - by(n-2) \quad (3)$$

From (1), (2), and (3) we get:

$$y(n) = x(n) + y(n-1) - by(n-2) \quad (4)$$

Taking the ZT of both sides of (4) and re-arranging terms we get:

$$Y(z)[1 - z^{-1} + bz^{-2}] = X(z) \Rightarrow H(z) = \frac{Y(z)}{X(z)} = \frac{1}{1 - z^{-1} + bz^{-2}} = \frac{z^2}{z^2 - z + b}$$

The system has two poles at $p_{1,2} = \frac{1 \pm \sqrt{1-4b}}{2}$.
Since $b$ is real, then $\sqrt{1-4b}$ is either real or pure imaginary, but not complex.

*Case 1*: If $1 > 4b$ (i.e., $\sqrt{1-4b}$ is real positive), then $|z_p| < 1 \Rightarrow -3 < \pm\sqrt{1-4b} < 1$. $\left.\begin{matrix} +\sqrt{1-4b} < 1 \rightarrow 0 < b \\ -3 < -\sqrt{1-4b} \rightarrow -2 < b \end{matrix}\right\} \Rightarrow 0 < b$ (intersection of $b > -2$ and $b > 0$.)

*Case 2:* If $1 \leq 4b$ (i.e., $\sqrt{1-4b}$ is pure imaginary), then

$$|z_p| < 1 \Rightarrow \left|1 \pm j\sqrt{4b-1}\right| < 2 \Rightarrow b < 1.$$

Hence, combining Case 1 and Case 2 we get $0 < b < 1$.

**MATLAB**: try ≫ `b = 0.8; r = roots([1 −1 b]); a = abs(r)`

## *Tutorial 43*

**Q**: Find the values of $b$ for which the system shown below is stable, knowing that $b$ is real.

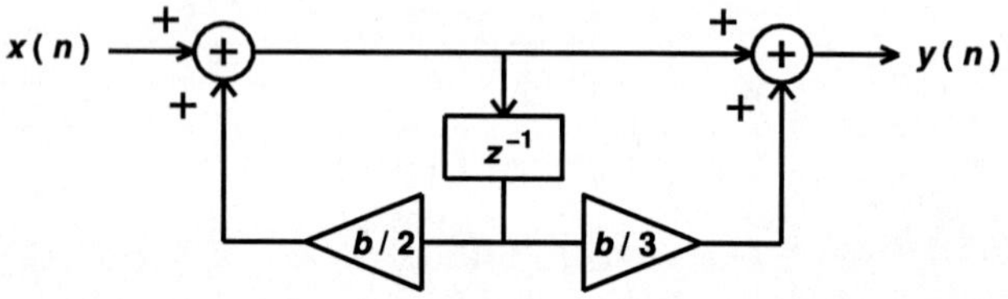


**Solution**: It is better to define some auxiliary points on such diagrams. In this question we define $r(n)$, hence the other side of $z^{-1}$ will be $r(n-1)$.

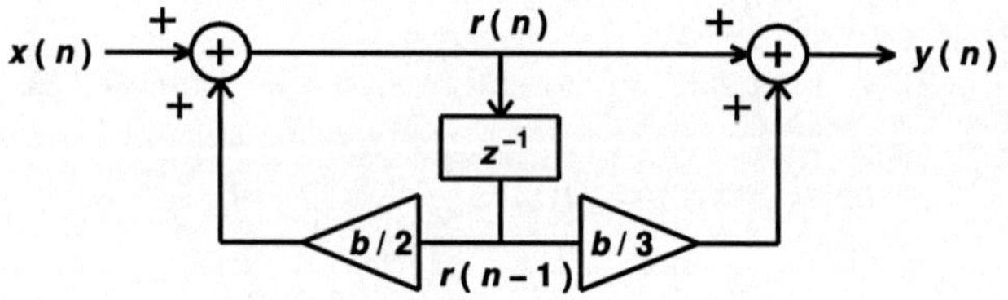


$$y(n) = r(n) + (b/3)r(n-1) \tag{1}$$

$$r(n) = x(n) + (b/2)r(n-1) \tag{2}$$

It is not easy to find the I/O relationship from the time domain equations. Hence, take the z-transform of (1) and (2) as follows:

$$Y(z) = R(z) + \frac{b}{3}z^{-1}R(z) = \left[1 + \frac{b}{3}z^{-1}\right]R(z) \tag{3}$$

$$R(z) = X(z) + \frac{b}{2}z^{-1}R(z) \Rightarrow X(z) = \left[1 - \frac{b}{2}z^{-1}\right]R(z) \tag{4}$$

$$\therefore \quad H(z) = \frac{Y(z)}{X(z)} = \frac{1 + (b/3)z^{-1}}{1 - (b/2)z^{-1}} = \frac{z + (b/3)}{z - (b/2)} \tag{5}$$

The system has a zero at $z = -b/3$ and a pole at $z = b/2$.
For stability we should have: $|b/2| < 1 \rightarrow \therefore |b| < 2$.

**Q**: From Eq. 5 above, find the impulse response $h(n)$ and the difference equation of the above system.

## Tutorial 44

**Q**: Using the impulse invariance method, design a digital Chebychev LPF with the following specifications:

1. $T_s = 0.01$ s (hence, $f_s = 100$ Hz),
2. $f_c = 10$ Hz ($\therefore \ \omega_c = 20\pi$ rad/s),
3. $G_m = 1$, gain $< 0.1$ (i.e., $-20$ dB) for $30 \le f \le f_s/2 = 50$ Hz),
4. 3 dB ripple is allowed in the passband.

**Solution**: We need an analog Chebychev filter with the above specifications, with gain less than $-20$ dB for normalized frequency $f_n \ge 30/10 = 3$ (normalized w.r.t $f_c$). From Tables we find the filter order $n = 2$, with normalized transfer function:

$$H_N(s) = \frac{a_o}{0.7079 + 0.6449 s_N + s_N^2}$$

Since $n$ is even, we have $G_{dc} = G_m/\sqrt{1+\varepsilon^2} = 1/\sqrt{1.9953} = 0.7079$.
Hence, $a_o/0.7079 = 0.7079 \Rightarrow a_o = 0.5$. The denormalized analog transfer function is obtained by the substitution $s_N = s/\omega_c = s/(20\pi)$ to get:

$$H_a(s) = \frac{1974}{2795 + 40.5s + s^2}$$

Using Partial Fraction Expansion (Tutorial 19), we can write $H_a(s)$ as follows:

$$H_a(s) = \frac{c_1}{s - p_1} + \frac{c_2}{s - p_2},$$

where $p_1 = -20.25 + 48.18i$, $p_2 = \overline{p_1} = -20.25 - 48.18i$, $c_1 = -20.2i$, $c_2 = \overline{c_1} = 20.2i$.
The z-domain poles are:

$$z_1 = \exp(p_1 T_s) = 0.68 + 0.45i, \quad z_2 = \exp(p_2 T_s) = \overline{z_1} = 0.68 - 0.45i.$$

Hence, the transfer function of the digital filter would be:

$$\begin{aligned} H(z) &= T_s\left[c_1\frac{z}{z-z_1}+c_2\frac{z}{z-z_2}\right] \\ &= T_s z\left[c_1\frac{1}{z-z_1}+\overline{c_1}\frac{1}{z-\overline{z_1}}\right] \\ &= T_s z\left[\frac{c_1(z-\overline{z_1})+\overline{c_1}(z-z_1)}{(z-z_1)(z-\overline{z_1})}\right] \\ &= T_s z\frac{(c_1+\overline{c_1})z-(c_1\overline{z_1}+\overline{c_1}z_1)}{z^2-(z_1+\overline{z_1})z+|z_1|^2} \\ &= \frac{0.15z}{z^2-1.36z+0.667}. \end{aligned}$$

## *Tutorial 45*

**Q**: Using the impulse invariance method, design a digital filter with sampling frequency 100 Hz and impulse response that matches the response of the following 3rd-order analog Butterworth filter:

$$H(s)=\frac{1}{(s+1)(s-p)(s-\overline{p})},$$

where $p = -0.5 + 0.866\text{i}$

**Solution**: Using Partial Fraction Expansion, we write the transfer function as follows:

$$H_a(s)=\frac{1}{s+1}+\frac{c}{s-p}+\frac{\overline{c}}{s-\overline{p}},$$

where $c = -0.5 - 0.28\text{i}$.
The z-domain poles are:

$z_1 = \exp\,(p_1\,T_s) = \exp\,(-\,T_s) = 0.99,$
$z_2 = \exp\,(pT_s) = 0.995 + 0.0086\text{i},$
$z_3 = \overline{z_2} = 0.995 - 0.0086\text{i}.$

$$\therefore\quad H(z)=T_s\left[\frac{z}{z-0.99}+c\frac{z}{z-z_2}+\overline{c}\frac{z}{z-\overline{z_2}}\right]$$

## Tutorial 46

**Q**: Using the ***bilinear transform***, design a 4th-order BP Butterworth digital filter with center frequency $\Omega_o = 1.5$, maximum gain $G_m = 1$, and bandwidth $\Omega_b = 0.4$.
**Solution**: Since the transformation LP → BP is quadratic (Tables), we need a 2nd-order prototype analog LPF. The transfer function of this filter is given by (Tables):

$$H(s_N) = \frac{a_o}{b_o + b_1 s_N + s_N^2}$$

where $b_o = 1$ and $b_1 = 1.41$.

$$G_{\text{d.c}} = G_m = 1 = \frac{a_o}{b_o} \Rightarrow a_o = b_o = 1. \Rightarrow H(s_N) = \frac{1}{1 + 1.41 s_N + s_N^2}.$$

$$\omega_u = \tan(\Omega_u/2) = \tan\left[\left(\Omega_o + \frac{\Omega_b}{2}\right)/2\right] = \tan(1.7/2) = 1.14$$

$$\omega_l = \tan(\Omega_l/2) = \tan\left[\left(\Omega_o - \frac{\Omega_b}{2}\right)/2\right] = \tan(1.3/2) = 0.76$$

$$\omega_b = \omega_u - \omega_l = 0.38; \omega_g = \sqrt{\omega_l \omega_u} = 0.93$$

Using the LP → BP transformation $s_N = \frac{s^2 + \omega_g^2}{s\omega_b}$ (from Tables), we get the transfer function of the analog BPF as follows:

$$H(s) = \frac{1}{1 + 1.41\frac{s^2+\omega_g^2}{s\omega_b} + \left(\frac{s^2+\omega_g^2}{s\omega_b}\right)^2}$$

Using the bilinear transformation $s = (z - 1)/(z + 1)$, we obtain the transfer function of the required BP digital filter as follows:

$$H(z) = \frac{1}{1 + 1.41\frac{[(z-1)/(z+1)]^2+\omega_g^2}{[(z-1)/(z+1)]\omega_b} + \left(\frac{[(z-1)/(z+1)]^2+\omega_g^2}{[(z-1)/(z+1)]\omega_b}\right)^2}$$

## Tutorial 47

**Q**: A radar station received the two signals at two different time intervals. A correlator is used to analyze these signals using their autocorrelations. The correlator outputs are as shown below. Comment on the structure of these signals.

**Solution**: Figure a shows approximately a delta function with some disturbance, hence the first signal is just a broadband noise. The pdf, mean, and variance (power) of this noise can be found from the input signal itself.
Figure b shows a symmetric shape with disturbance everywhere and a maximum at the origin, along with a narrow spike at the origin, hence the second signal is a piece of information corrupted by broadband noise.

## *Tutorial 48*

**Q**:
(A) Find the Hilbert transform of $x(t) = \sin(\omega_o t)$. signals.
(B) Find the Hilbert transform of $x(t) = \sin(\omega_o t) + \delta(t)$.
(C) Find the analytic signal $z(t)$ associated with the real signal $x(t) = \cos(\omega_o t)$.
(D) Compare the spectra of these signals and comment.

**Solution:**
(A) Let $\mathcal{H}[\sin(\omega_o t)] = y(t)$.

The transfer function of $\mathcal{H}$ is $H(f) = \begin{Bmatrix} e^{-j\pi/2} = -j, & f > 0 \\ e^{+j\pi/2} = j & f < 0 \end{Bmatrix} = -j\text{sgn}(f)$.

From Tables: $X(f) = \frac{1}{2j}\delta(f - f_o) - \frac{1}{2j}\delta(f + f_o)$. Hence we have:

$$Y(f) = H(f) \cdot X(f) = -j\text{sgn}(f)[\frac{1}{2j}\delta(f - f_o) - \frac{1}{2j}\delta(f + f_o)]$$

$$= -\frac{1}{2}\delta(f - f_o) - \frac{1}{2}\delta(f + f_o) \Rightarrow y(t) = -\cos(\omega_o t) \text{ (Tables)}$$

(B) Let $\mathcal{H}[\delta(t)] = d(t), \mathcal{H}[\sin(\omega_o t)] = y(t)$, and $\mathcal{H}[x(t)] = z(t)$.
From Tables, $\Delta(f) = FT\{\delta(t)\} = 1$

$$\therefore \quad D(f) = -j\text{sgn}(f) \cdot \Delta(f) = -j\text{sgn}(f)$$

From Tables we have: $d(t) = 1/(\pi \cdot t)$; an from (A) we have: $y(t) = -\cos(\omega_o t)$.
Since $\mathcal{H}$ is a ***linear*** system, we have: $z(t) = y(t) + d(t) = -\cos(\omega_o t) + 1/(\pi \cdot t)$.
(C) $z(t) = x(t) + j\mathcal{H}\{x(t)\}$.

Using an approach similar to that in (A) above, we have
$\mathcal{H}\{\cos(\omega_o t)\} = \sin(\omega_o t)$.
Hence, $z(t) = \cos(\omega_o t) + j\sin(\omega_o t) = e^{j\omega_o t}$.
Now from Tables we have:

$$X(f) = \frac{1}{2}\delta(f - f_o) + \frac{1}{2}\delta(f + f_o); \quad Z(f) = \delta(f - f_o).$$

(D) Using the analytic signal, the negative part of spectrum is removed, while the positive part is scaled by 2. This indicates that the use of the analytic signal will lead to spectrum economy.

## *Tutorial 49*

**Q**: Consider a first-order sinusoidal DPLL (SDPLL) under noise-free conditions, center frequency $f_o = 1$ Hz, and input signal $x(t) = \sin(6t + \pi/4)$.

1. Find the system equation in terms of the digital filter gain $G_1$.
2. Find the steady-state phase error $\phi_{ss}$ in terms of $G_1$.
3. Find the range of $G_1$ that ensures locking on the above incoming frequency.
4. Choose a value for $G_1$ inside the locking range and find the corresponding $\phi_{ss}$.
5. Assuming $t(0) = 0$, plot $x(t)$ with the first three DCO pulses for the abovevalue of $G_1$.

**Solution**:
1. From Tables we find the system equation as follows:

$$\phi(k+1) = \phi(k) - \omega.G_1\sin[\phi(k)] + (\omega - \omega_o)T_o \tag{1}$$

We have: $\omega = 6$ rad/s, $\omega_o = 2\pi$ rad/s, and $T_o = 1$ s.
Hence, the system equation will be given by:

$$\phi(k+1) = \phi(k) - 6G_1\sin[\phi(k)] - 0.28 \tag{2}$$

2. From (2) we have:

$$\phi_{ss} = \phi_{ss} - 6G_1\sin[\phi_{ss}] - 0.28 \Rightarrow \phi_{ss} = \sin^{-1}(-0.047/G_1) \tag{3}$$

3. Locking conditions are as follows:
*Condition 1*: From (3) we have:

$$|-0.047/G_1| < 1 \Rightarrow G_1 > 0.047 \tag{4}$$

*Condition* 2: From (2) we have:

$$g(\psi) = \psi - 6G_1\sin(\psi) - 0.28 \tag{5}$$

$$\therefore \quad g'(\psi) = 1 - 6G_1\cos(\psi).$$

$\therefore$ If a solution $\phi_{ss}$ of (5) exists, then $g'(\phi_{ss}) = 1 - 6G_1\cos(\phi_{ss})$.
*Fixed point analysis* says that (Tables):

$$\phi_{ss}\text{exists if}|g'(\phi_{ss})|<1 \Rightarrow |1 - 6G_1\cos(\phi_{ss})|<1 \tag{6}$$

From Tables: $\cos(\phi_{ss}) = \pm\sqrt{1 - \sin^2(\phi_{ss})}$.
From (3) we have

$$|\phi_{ss}|<\pi/2 \Rightarrow \cos(\phi_{ss}) > 0 \tag{7}$$

Hence, using (3) and (7) we have:

$$\cos(\phi_{ss}) = +\sqrt{1 - \sin^2(\phi_{ss})} = \sqrt{1 - (0.047/G_1)^2}. \tag{8}$$

Now substituting (8) in (6) we get:

$$\begin{aligned}
|1 - 6G_1\sqrt{1 - 0.0022/G_1^2}|<1 &\Rightarrow |1 - \sqrt{36G_1^2 - 0.08}|<1 \\
&\Rightarrow -1<1 - \sqrt{36G_1^2 - 0.08}<1 \\
&\Rightarrow -2< - \sqrt{36G_1^2 - 0.08}<0 \\
&\Rightarrow 0<\sqrt{36G_1^2 - 0.08}<2 \\
&\Rightarrow 0<36G_1^2 - 0.08<4 \\
&\Rightarrow 0.047<G_1<0.33
\end{aligned} \tag{9}$$

4. We have $\phi(0) = \pi/4$, $x(0) = \sin(\pi/4) = 0.7$.
   Using (9), let $G_1 = 0.1$. Hence:

$$\begin{aligned}
y(0) &= G_1x(0) = 0.1(0.7) = 0.07, \\
T(1) &= T_o - y(0) = 1 - 0.07 = 0.93\,\text{s}, \\
t(1) &= T(1) = 0.93\,\text{s}, \\
\phi(1) &= \omega.t(1) + \pi/4 \equiv 0.08\,\text{rad}, \\
x(1) &= \sin[\phi(1)] = 0.08, \\
y(1) &= G_1x(1) = 0.1(0.08) = 0.008, \\
T(2) &= T_o - y(1) = 1 - 0.008 = 0.99, \\
t(2) &= T(1) + T(2) = 1.92, \\
\phi(2) &= \omega t(2) + \pi/4 \equiv -0.26\,\text{rad}.
\end{aligned}$$

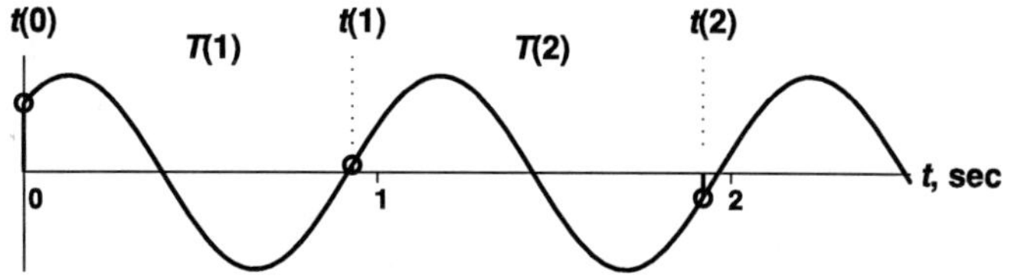


## *Tutorial 50*

**Q**: The phase equation of a first-order sinusoidal DPLL is shown below. Find approximately the steady-state phase error $\phi_{ss}$.

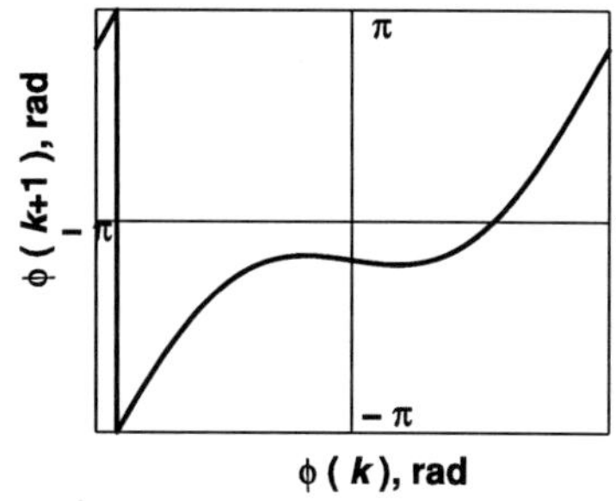


**Solution**: First, we plot the curve $y = x$. If there is a steady-state, this curve will have two intersections with phase equation curve; one of them is the steady-state point, where its projection on the y-axis gives $\phi_{ss}$. To determine which one is the steady-state point, we choose randomly some value for the initial phase $\phi$ (0), and then plot the phase plane diagram by successive projections on the equation and $y = x$. For the above figure we choose $\phi(0) = -1.5$ and we find $\phi_{ss} \approx -0.5$.

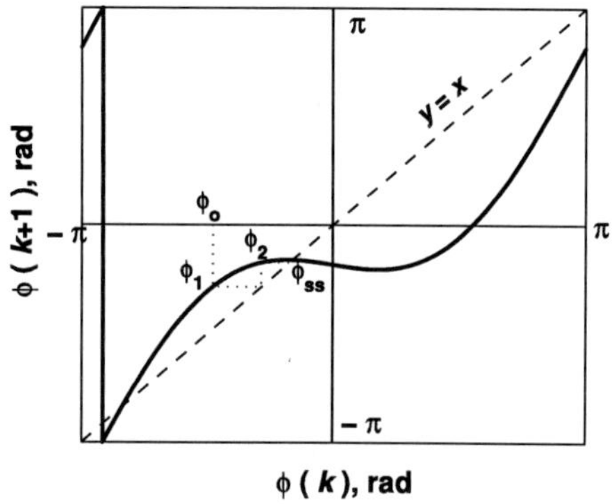

## *Tutorial 51*

**Q**: Design a digital controlled oscillator (DCO) with a clock rate of 1 MHz and a center frequency of 10 kHz. Draw the block diagram and the output signal for an input sequence of 0, 30, −30, 70,.... Indicate sampling times.

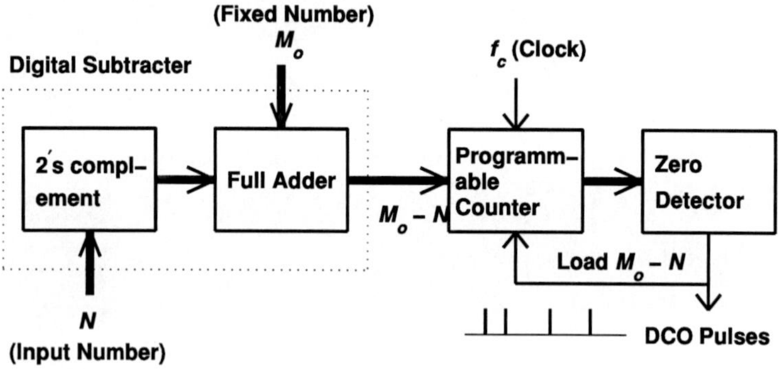


**Solution**: We have $f_c = 1$ MHz, hence, $T_c = 1/f_c = 1\mu$ s.
The center frequency is $f_o = f_c/M_o = 10$ kHz.
Hence, the free-running input number is $M_o = f_c/f_o = 100$.
For the given sequence, the values of the counter initial number are $R_o = 100 - 0 = 100$; $100 - 30 = 70$; $100 - (-30) = 130$; and $100 - 70 = 30$; respectively.
Hence, the sampling times are
0 (initial), $100\ T_c = 100\mu$ s, 170 $\mu$ s, 300 $\mu$ s, and 330 $\mu$ s.
An illustrative (not to scale) diagram is shown below.

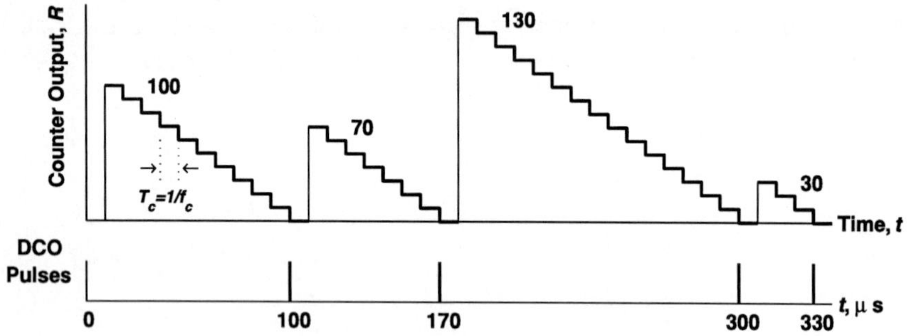

## *Tutorial 52*

**Q**: A block diagram of a delta modulation system is shown below, along with the input signal. The step of the integrator is arranged to be 0.2 V, while the sampling frequency is $f_s = 1$ kHz. Plot on the same figure the estimated signal and the output waveform. Assume that the integrator output was initially zero and the quantizer output is 1. Suggest a demodulation circuit.

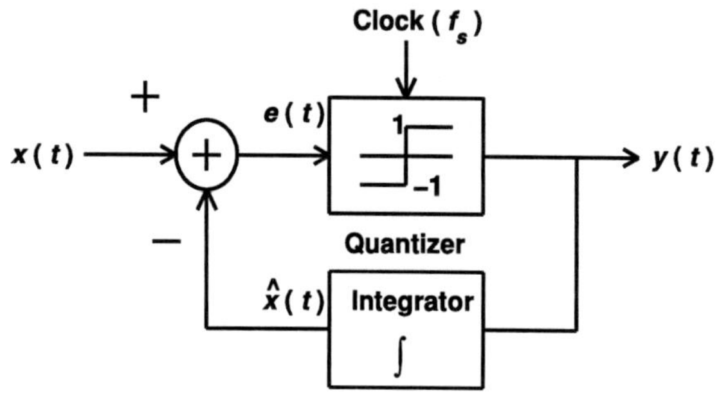


**Solution**: The sampling period is $T_s = 1/f_s = 1/1000 = 1$ ms. The plots are as shown below.

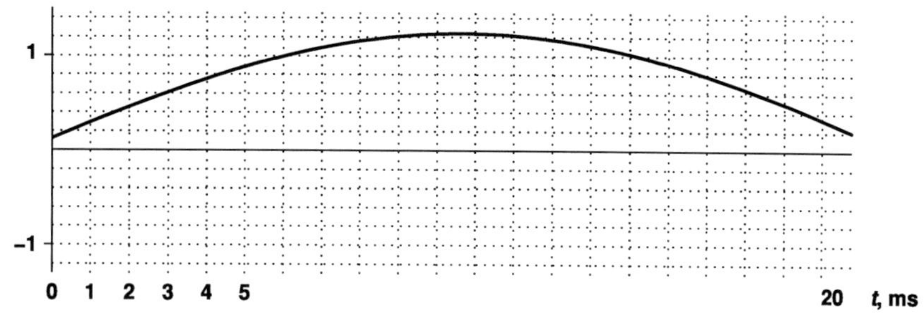


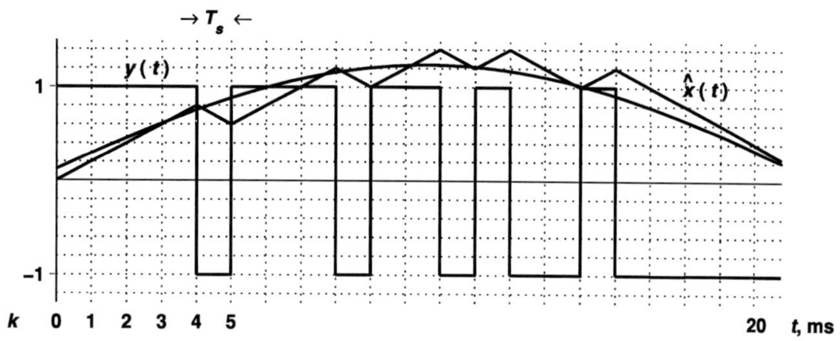

## *Tutorial 53*

**Q**: Use MATLAB to design a fourth -order elliptic lowpass filter with the following specifications:

Pass-band peak-to-peak ripple, $R_p = 0.6$ dB
Minimum stop-band attenuation $R_s = 18$ dB
The normalized pass-band-edge frequency $w_p = 0.63$ Then, implement this filter by changing the filter properties to fixed-point representation.

**Solution**: We may use the `dfilt` object for this purpose.
The first step is to design the filter using the Matlab default double-precision format:

```
M = 4; Rp = .6; Rs = 18; wp = .63; [Den, Num] = ellip(M, Rp, Rs, wp);
```

Digital filter implementation using direct form—I:

```
H = dfilt.df1(Den, Num)
```

Second step is to convert the filter coefficients to fixed point format:

```
H.Arithmetic = 'fixed'
get(H)
H.FilterInternals = 'SpecifyPrecision'
```

*Note*: "SpecifyPrecision" enables you to set your own wordlength and fraction length for the output and accumulator-related properties.
Now set the hardware specifications:

```
set(H,'InputWordlength', 16,...
'InputFracLength', 13,...
'ProductWordLength', 24,...
'AccumulatorWordLength', 28,...
'OutputWordLength', 16);
H
```

Verify that the object `dfilt` has made an appropriate auto-scaling (Hint: check the range of the filter coefficients).

## *Tutorial 54*

**Q**: Verify the equivalence of the identities shown in the figure below. What is the potential of these identities?

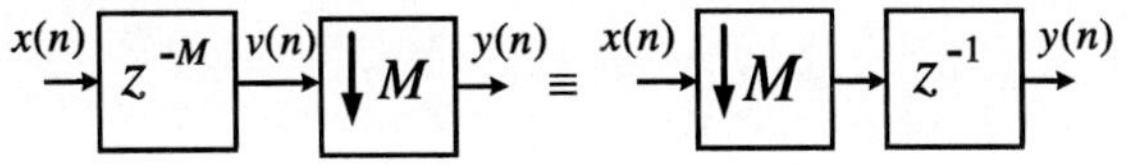


(a)

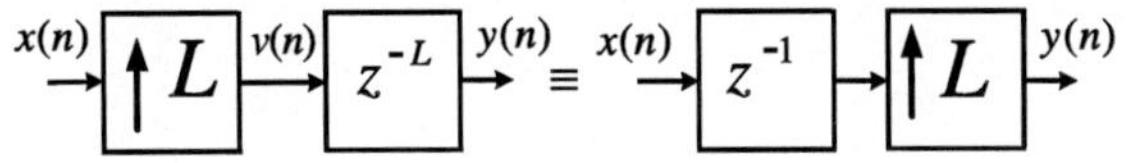


(b)

**Solution**: In both cases $H(z) = z^{-1}$, that is a time delay. In Fig. a, the right-hand side can be expressed as

$$y(n+1) = x(Mn)$$
$$y(n) = x(M(n-1)) = x(Mn - M)$$

while the left-hand side is

$$v(n) = x(n - M)$$
$$y(n) = v(Mn)) = x(Mn - M)$$

that is, both sides are of Fig. a are equivalent. The up-sampler shown in Fig. b can be verified similarly. In both cases the right-hand side implementation is more computationally efficient since after the down-sampler or before the up-sampler the filter $H(z)$ will operate at lower sampling rate. On the other hand, the left-hand side implementation is wasteful as it processes a lot of zero terms.

# Appendix B: Miscellaneous Exercises

## *Miscellaneous DSP Exercises—A*

**Q1:** Explain the meaning of a "signal". Give five examples of real-life signals.

**Q2:** State five classes of signals with brief explanations.

**Q3:** Draw a block diagram for an analog/digital signal processing system.

**Q4:** Show whether the Hilbert transform [that gives constant 90° phase shift for all sinusoidal signals of the form $x(t) = \sin(\omega t + c)$] is:
(a) Memoryless, (b) causal, (c) linear, (d) time-invariant, (e) BIBO stable.

**Q5:** Define the Dirac delta function. How can we approximate it in applications?

**Q6:** How can we represent an analog system and its I/O relationship?

**Q7:** Both Fourier and Laplace Transforms are used to represent analog systems. Which one is the more general?

**Q8:** State the conditions in the time-domain that an analog system is BIBO stable. What are the equivalent conditions in the frequency domain?

**Q9:** Explain what is the physical meaning of the cross-correlation integral. State an application for this integral.

**Q10:** What kind of signals can Fourier series represent? Is the Fourier series a frequency transform? Can it reveal the frequency content of the signal?

**Q11:** How does the trigonometric Fourier series of an odd periodic signal look like?

**Q12:** Can we use Fourier series to represent energy signals?

**Q13:** Can we use Fourier transform to represent periodic signals? How?

**Q14:** From the basic definition of Fourier transform, find and plot the amplitude and phase spectra of the signal $x(t) = \exp(-5t)u(t)$. Is this an energy or power signal? Why?

**Q15:** What is meant by the duality of the Fourier transform?

**Q16:** Using Tables, find the Fourier transform of the following signals and plot their magnitude spectra:
1. $\sin(2t + 1)$, 2. $\cos(5t)$, 3. 1, (4) $\delta(t)$, 5. $u(t)$, 6. $\mathrm{sgn}(5t)$, 7. $\Pi_1(t)$, 8. $\mathrm{sinc}(t - 5)$, 9. $\mathrm{sinc}(t)\cos(20t)$.

**Q17:** State Parseval's theorem for periodic and non-periodic analog signals and define the power and energy spectra and spectral densities.

**Q18:** Explain why the single-sided Laplace transform is sufficient for engineering applications.

**Q19:** From the basic definition, find the Laplace transform and its ROC for the signal $x(t) = \exp(t)u(t)$.

**Q20:** Plot the pole-zero diagram of the system $H(s) = s(s + 1)/[(s2 + 5s + 6)(s + 5)]$. Is the system stable? Justify your answer.

**Q21:** What is the sample space of the die tossing experiment? How can we represent outcomes of this experiment using a random variable? Is this random variable continuous or discrete? Plot the pdf of this random variable.

**Q22:** Two dies are tossed. What is the probability that the first is 1 and the second is 6?

**Q23:** Explain the physical meaning of the statistical mean and variance of a random variable.

**Q24:** Plot and find the mean and the variance of the following Gaussian pdf $p(x) = \frac{1}{\sqrt{\pi}}e^{-x^2}$.

**Q25:** Two Gaussian noise processes $n_1(t)$ and $n_2(t)$, where the first has more power than the second. Approximately plot the pdf's of the two noise random variables. Are $n_1$ and $n_2$ energy or power signals?

**Q26:** Find the mean and variance of the random signal $x(t) = \sin(wt) + n(t)$, where $n(t)$ is Gaussian noise with variance = 0.1.

**Q27:** State the Wiener–Kinchin theorem for WSS random signals.

**Q28:** Define the autocorrelation function of a random variable $X(t)$.

**Q29:** Explain the meaning of white noise. What is the effect of an ideal LPF on the correlation between white noise samples?

**Q30:** Outline the principle of a matched filter. State an application for matched filters in binary communications (Plot the block diagram).

**Q31:** Show that the matched filter is essentially a correlator.

**Q32:** Show that the ideal analog LPF is a non-causal system. How can we modify this theoretical system to be practically realizable?

**Q33:** What are the advantages of active filters over passive filters?

**Q34:** What is the importance of impedance matching between successive stages of an electrical system? How can we attain this aim when designing analog passive filters?

**Q35:** What does the 3-dB point of a filter mean?

**Q36:** If a sixth-order analog BPF is required, what is the order of the prototype normalized LPF? Why?

**Q37:** Find the transfer function and draw the circuit diagram of a second-order Butterworth LPF with cutoff frequency $f_c = 50$ Hz, maximum gain $G_m = 1$, and $RL = 10\ \Omega$.

**Q38:** What is the function of the following circuit? Find the transfer function and plot approximately its magnitude frequency response (Fig. B.1).

**Q39:** What does uniform sampling mean?

**Fig. B.1** Problem Q38

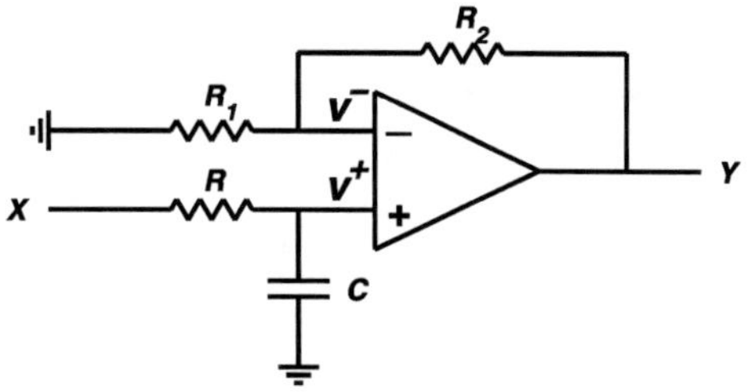


**Q40:** The signal $x(t) = \text{sinc}(t - 5)$ is sampled with $f_s = 2$ Hz. Plot the magnitude spectrum of the digitized signal $x(n)$. Is there aliasing? How can we reconstruct $x(t)$ from $x(n)$?

**Q41:** Explain what frequency aliasing means.

**Q42:** Find the convolution of the two signals: $x(n) = \{\hat{2}, 3, 4\}$ and $y(n) = \{\hat{1}, -5\}$.

**Q43:** Explain the meaning of the DTFT and the DFT. How can we represent digital systems and their I/O relationships in the time and the frequency domains using these transforms? Which transform is more useful in practice?

**Q44:** Which is more general: the DFT or the ZT? How can we represent digital systems and their I/O relationships using the ZT? How is the ZT related to DTFT?

**Q45:** State the conditions in the time-domain that a system is BIBO stable. What are the equivalent conditions in the z-domain?

**Q46:** What is the relationship between the s-plane and the z-plane?

**Q47:** Explain why the single-sided z-transform is sufficient for engineering applications.

**Q48:** Explain why the digital subsystem $H(z) = z^{-1}$ represents a unit delay for causal systems. How can we implement this system using digital hardware?

**Q49:** What is meant by the FFT?

**Q50:** Find the difference equation, the transfer function, the impulse response, and the frequency response of the digital system shown in figure below, noting that fs = 1 Hz. Is it FIR or IIR filter? LP or HP? What do you expect the function of this system? Does it have an exact linear phase? Plot the pole-zero diagram. Is the system stable?

**Q51:** a) What is the energy of the digital signal $x(n) = 0.5\, n\, T_s\, u(n)$, knowing that $f_s = 100$ Hz? (Ans. 0.726 J) (Hint: use Tables for summation.)

**Q52:** What is the energy of the analog signal $x(t) = 0.5\, t\, u(t)$? (Ans. 0.721 J)

**Q53:** How can we make the circular convolution of two finite digital sequences $x(n)$ (length $N_1$) and $h(n)$ (length $N_2$) equal to their linear convolution?

**Q54:** State the advantages of DSP over ASP.

**Q55:** Explain the function of the moving average digital filter. Is it FIR or IIR? State a real-life application for this filter.

**Q56:** Design a digital integrator. Is it FIR or IIR? LP or HP?

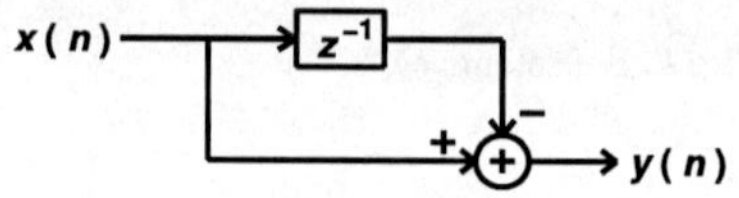


**Fig. B.2** Problem Q50

**Q57:** Implement the following digital system in cascade:

$$H(z) = \frac{(2z-5)(z+1)}{(z^3+5z^2+6z)}.$$

**Q58:** Why do we normally choose complex poles as conjugates in digital systems?

**Q59:** How can we equalize the distorting effect of a communication channel?

**Q60:** What are the advantages of FIR over IIR filters?

**Q61:** Explain Gibbs phenomenon and how it can be dealt with in digital systems.

**Q62:** Design an ideal FIR HPF with cutoff frequency = 20 Hz, knowing that the sampling frequency is $f_s = 100$ Hz. (Hint: use Tables *to find h(n) for LPF, then convert to HPF*).

**Q63:** Plot the pole diagram of the sinusoidal digital oscillator. How can we deduce the frequency of oscillation from these poles?

**Q64:** Simulate the following DSP systems using MATLAB (find the output time signal for arbitrary causal input $x(n)$ and the frequency response of the system):
(a) The system shown in Fig. B.3 (b) The system shown in Fig. B.2.

**Q65:** Design a digital DC blocker with $f_s = 20$ kHz that attenuates the frequency $f = 20$ Hz by $-0.1$ dB.

**Q66:** A signal $x(t) = \sin(t)$ is corrupted by AWGN $n(t)$ with SNR = 0 dB. Simulate the noisy signal $s(t) = x(t) + n(t)$ using MATLAB and find the spectra of all signals. Design a Butterworth analog LPF to reduce noise and find the output of this filter.

**Q67:** A random signal $n(t)$ is received and saved as a data file with $f_s = 1$ kHz and length $N = 1000$ samples. Write a MATLAB code to find the pdf of this signal.

**Q68:** An analog signal has a time duration of 3 ms. How long would be the frequency range of this signal?

*Answer to Q4 (Hilbert Transformer)*:

1. We have $y(t) = x(\omega t - \pi/2) = x\left[\omega\left(t - \frac{\pi/2}{\omega}\right)\right]$. Hence, there is a time-delay of $\frac{\pi/2}{\omega}$, where the new time axis is $t_n = t - \frac{\pi/2}{\omega}$. Therefore, the system has memory.
2. It is causal since there is no time-advance.
3. Consider only positive frequencies, hence the phase shift is $-\pi/2$ [see Tutorial 33]. Consider sinusoids in complex exponential form.
   Let $z(t) = ae^{j(\omega_1 t + c)} + be^{j(\omega_2 t + d)}$, where $a$, $b$, $c$ and $d$ are constants. We have:

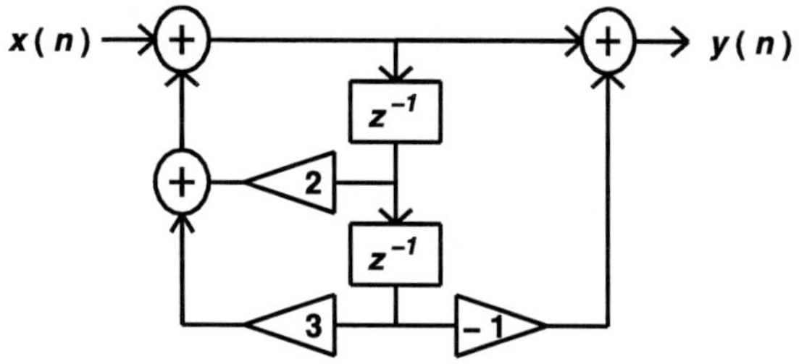


**Fig. B.3** Problem Q64

$$\mathbf{T}[z(t)] = z(t)e^{-j\pi/2}.$$

$$\mathbf{T}\left[ae^{j(\omega_1 t+c)}\right] + \mathbf{T}\left[be^{j(\omega_2 t+d)}\right] = ae^{j(\omega_1 t+c)}e^{-j\pi/2} + be^{j(\omega_2 t+d)}e^{-j\pi/2}$$

$$= \left[ae^{j(\omega_1 t+c)} + be^{j(\omega_2 t+d)}\right]e^{-j\pi/2} = \mathbf{T}[z(t)].$$ Hence, it is linear.

4. Let $y(t) = \mathbf{T}[x(t)] = e^{j(\omega t + c - \pi/2)}$ (output of HT). Now

$$\mathbf{T}[x(t - t_o)] = \mathbf{T}\left[e^{j[\omega(t-t_o)+c]}\right] = e^{j[\omega(t-t_o)+c-\pi/2]} = y(t - t_o).$$

Hence, it is time-invariant.
5. Clearly it is BIBO stable, as it does not affect the magnitude of the signal.

## *Miscellaneous DSP Exercises—B*

Q1: Explain why we cannot use the impulse invariance method to design high pass digital filters (Hint: due to aliasing).
Q2: Show that the bilinear transformation preserves stability between the analog prototype filter and the digital filter (Hint: show that the LHS of the s-plane is transformed inside the unit circle in the z-plane, as shown in the Lecture Notes).
Q3: Can we use the bilinear transformation to design a digital HPF from an analog prototype HPF? (Hint: yes, since there is no aliasing).
Q4: Explain how the analog frequency is transformed using the bilinear transformation (see Tables).
Q5: Can we use the global average to find the trend of prices in a stock market? Why? What kind of filter do you suggest for this purpose? Why? (Hint: no, since the price trend is non-stationary, i.e., time-varying. We use either an FIR moving average filter, or an alpha filter. We prefer the alpha filter for its simple structure).
Q6: Explain how we estimate a communication channel transfer function using an FIR filter.

Q7: Where do you expect the poles of a digital resonator to be located in the z-plane?

Q8: (a) Draw a block diagram for a generic FIR filter of order N.
(b) Plot the impulse response of this filter.
(c) What is the condition on this filter to have a linear phase transfer function?
(d) Plot the impulse response of a linear FIR filter and draw an efficient block diagram assuming N is odd.

Q9: Two digital filters have impulse responses given by $h_1(n) = \{1, 2, -1, -1, 2, 1\}$ (starting at $n = 9$) and $h_2(n) = \{1, 2, -1, 5, -1, 2, 1\}$ (starting at $n = 0$).

(a) Which one is an FIR filter? (Ans. both)
(b) Which one have a linear phase transfer function? (Ans. both, as they have symmetric impulse response.)
(c) Plot the above impulse responses.
(d) Draw the efficient implementation diagrams of these filters.

Q10: (a) Plot the transfer function of an ideal digital LPF with cutoff frequency 10 Hz and sampling frequency 50 Hz. Use the frequency range [–100, 100 Hz].
(b) Roughly plot the impulse response of this filter.
(c) repeat part (a) for a high-pass filter with cutoff 10 Hz.

Q11: Explain Gibbs phenomenon. How can we alleviate the effect of this phenomenon? (Hint: when we truncate the infinite impulse response of an ideal digital filter using a time window, we get magnitude ripples in the filter transfer function, with maxima at the ends of the transfer function. This ripple is significant when we use a rectangular widow.)

Q12: Explain why we need just one matched filter at the receiver of a binary communication system with antipodal signals (symbols).

Q13: Explain the basic operation of the optimal receiver in a binary communication system with orthogonal signals. [Hint: it consists of two matched filters, one matched to the symbol that represents logic "0", the other to "1" (explain how we choose the impulse responses). If "0" was transmitted, the output of the first matched filter (which is matched to "0") will be higher than that of the second (which is matched to "1"). This is because the matched filter is a correlator. Hence, the comparator will decide that "0" was transmitted. Similar reasoning if "1" was transmitted, where the output of the second filter will be higher this time. Draw a block diagram.]

Q14: A signal $x(t) = \sin(\omega\, t)$ was corrupted by a white Gaussian noise $n(t)$ with variance 0.1. Find the statistical mean, time mean, and variance of the signal $s(t) = x(t) + n(t)$. [*Ans. Statisticalmean* = *mean* $\{x(t) + n(t)\}$ = *mean* $\{x(t)\}$ + *mean* $\{n(t)\} = x(t) + 0 = x(t)$, *thetimemean* = *mean*$\{x(t)\}$ + *mean* $\{n(t)\} = 0 + 0 = 0$. Note that since $n(t)$ is ergodic, its time mean = its statistical mean.]

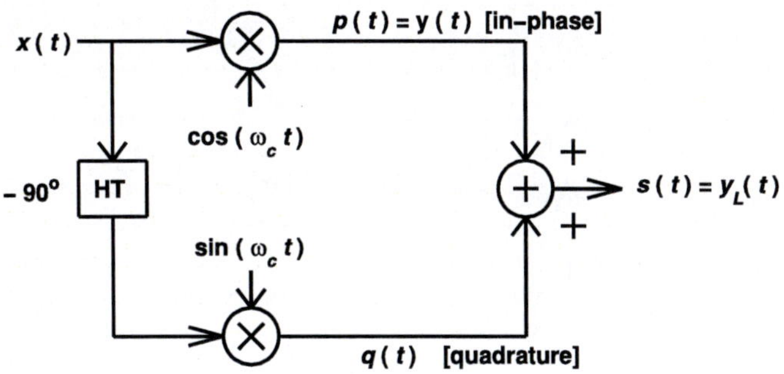


**Fig. B.4** Problem Q18

Q15: The analog Hilbert transform is a linear system that shifts the phase of the input signal by –90°. Explain the operation of this system in the frequency domain and in the time domain.

Q16: Define the analytic signal and explain how it can be used for spectral economy.

Q17: Explain with an example why we modulate signals before radio transmission.

Q18: The figure below shows SSBSC AM generator. Show how this structure is used for spectral economy. Suggest a demodulation circuit (Fig. B.4).

Q19: Explain the operation of the first-order sinusoidal DPLL with a sinusoidal input. State and comment on the system equations. Draw the block diagram and the sampling process.

Q20: What are the main advantages of digital PLLs over analog PLLs?

Q21: Draw the block diagrams of a generic analog PLL and a generic DPLL.

Q22: Explain the meaning of the locking range of a DPLL.

Q23: Explain how can the DPLL track the frequency of a sinusoidal signal corrupted by noise. [Hint: the peak of the PLL frequency pdf will estimate the input frequency, if the original frequency is inside the loop locking range.]

Q24: How do you expect the relation between the variance of the PLL frequency and the input SNR? [Ans. the variance decreases when SNR increases; it approaches zero (hence the frequency pdf will approach a spike over the true input frequency) if the SNR is very high.]

Q25: For a sinusoidal input signal $x(t) = A \sin[\omega_o t + \theta(t)]$, the phase equation in a first-order SDPLL is given by $\phi(k) = \theta(k) - \omega_o \sum_{i=0}^{k-1} y(i)$, where $y(i)$ is the output of the digital filter at the $i$th sampling instant, and $\theta(k)$ is the information-bearing phase. Explain (with a block diagram) how to demodulate PM signals using SDPLL.

Q26: Find the system equation for the second-order SDPLL.

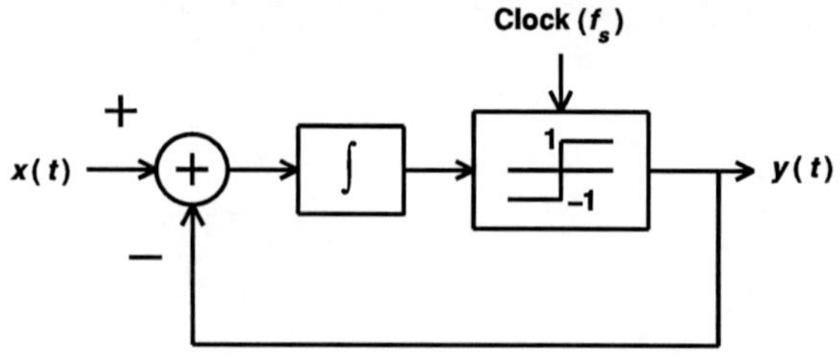


**Fig. B.5** Problem Q37

Q27: Explain the basic operation of an adaptive filter. State how can we choose the reference signal in communications and noise reduction. Draw a block diagram for an adaptive noise canceler.

Q28: The algorithm for adaptive Wiener filter is as follows:

Define:
$\mathbf{h}(k) = [h_o(k)h_1(k)\ h_2(k) \cdots h_M(k)]$, the filter coefficients at the $k$th instant.
$\mathbf{y}(n) = [y(n)y(n-1) \cdots y(n-M)]$, the observed signal vector at the time instant $n$. The algorithm can be described in vector form (MATLAB-like code) as follows:

**h**(0) = 0; % Initialize the filter coefficients.
For n = 1: N % N = length(y);
$\hat{x}$(n) = **h**(n − 1) * **y**T(n); % Filter output (this is matrix multiplication).
e(n) = d(n) − $\hat{x}$(n);
**h**(n) = **h**(n − 1) + $\mu$ * e(n) **y**(n); % $\mu$ is the convergence-accuracy coefficient.
end Show how to implement this algorithm using DSP hardware.

Q29: What is the basis of finding the coefficients of the Wiener filter? [Ans. minimizing the mean-squared error between the estimated signal and the reference signal.]

Q30: What is meant by the intersymbol interference (ISI)? Explain how an adaptive filter with a training sequence can be used for ISI reduction.

Q31: What are the mean and variance of uniform quantization noise?

Q32: Explain why we need non-uniform quantization for audio applications, and how to achieve this kind of quantization.

Q33: How can we improve the signal-to-quantization noise ratio? Explain [Hint: by oversampling.]

Q34: Explain how the oversampling process can assist in reducing the order of the anti-aliasing filter of the compact disc system.

Q35: With a block diagram, explain the operation of the analog delta modulation system. Suggest a demodulation circuit. What is the slope overload? What is the granular noise? How can we improve the accuracy of this system? Implement the system in the digital domain.

Q36: Why do we need an adaptive delta modulation system? An adaptive algorithm for the step size is given by: $\Delta_n = \Delta_{n-1} K^{y(n)y(n-1)}$, where $K > 1$ is a constant. Implement this algorithm using DSP hardware. Draw the block diagram of the adaptive modulator and the demodulator.

Q37: A block diagram of a sigma-delta modulator is shown below. Suggest a decoder. Represent this system in Laplace domain and show how it can be used for noise shaping (Fig. B.5).

Q38: Why is the sigma-delta modulator more appropriate for audio applications than the delta modulator?

Q39: Design a circuit for DC blocking.

Q40: Design a digital SD modulator and demodulator.

## *Miscellaneous DSP Exercises: C*

Q1: A fifth-order elliptic low-pass filter with the following specifications:

Pass-band peak-to-peak ripple = 0.5 dB
Minimum stop-band attenuation = 40 dB
The pass-band-edge frequency = $0.3\pi$ Use Matlab to implemented this filter with second-order sections.

Q2: A 12-bit A/D converter is used to sample an analog signal. The sampled signal $x(n)$ is stored in the lower bits of DSP processor such that the corresponding maximum dynamic range will be 1/16. Then $x(n)$ is to be passed to a 16-bit IIR filter whose transfer function is

$$H(z) = \frac{1}{1 - 0.98z^{-1}}.$$

Scale the output signal such that its upper limit does not exceed unity.

Q3: Consider the first-order IIR filter described by the infinite-precision difference equation

$$y(n) = -0.625y(n-1) + x(n).$$

Show whether this system would trap into limit cycle when implemented using 4-bit (a) rounding arithmetic, (b) truncation arithmetic. Let $x(n) = 0$ and $y(0) = 3/8$. Find the magnitude and frequency of the oscillation for each case, if any.

Q4: A second-order IIR filter described by the following infinite-precision difference equation

$$y(n) = -a_1 y(n-1) - a_2 y(n-2) + x(n).$$

Derive the dead band bound that govern the occurrence of limit cycles in this structure when implemented using rounding arithmetic.

## Miscellaneous DSP Exercises: D

Q1: Prove that $c_M(n) = \frac{1}{M}\sum_{k=0}^{M-1} e^{-j2\pi kn/M}$.

Q2: Figure B.6 shows the spectrum of a band-pass signal. Show that down-sampling this signal by $M = 3$ produces an alias-free spectrum.

Q3: Write a MATLAB code to change the sampling rate from 48 to 44.1 kHz as explained in Example (2), Chap. 5.

Q4: Consider the sinusoidal signal $x(n) = \cos(0.1\,\pi n)$. Determine the frequency spectrum of the 3-fold down-sampled version.

Q5: Prove the decimation and interpolation identities shown in Fig. 5.13.

Q6: Given the following specifications of the sampling rate conversion DSP system shown in Fig. 5.11: $L = 2$, $M = 3$ and the input audio signal is sampled at 6 kHz, whereas the output signal sampling rate should be 9 kHz. Determine the filter length and cutoff frequencies for the required filter (window design method could be used).

Q7: The signal $x(n)$ was sampled at a frequency $f_s = 10$ kHz. Consider the following two cases: (a) Resample $x(n)$ at a new sampling frequency $f_{s1} = 22$ kHz. (b) The signal signal $x(n)$ is to be resampled to a new sampling frequency $f_{s2} = 8$ kHz.

Q8: Consider the transfer function of a FIR filter: $H(z) = 0.25 + 0.5\,z^{-1} + 0.75\,z^{-2} + z^{-3} + 1.25\,z^{-4} + 1.5\,z^{-5}$. Use polyphase decomposition technique to implement a factor of $M = 3$ decimator.

Q9: A single stage decimator structure is used to reduce the sampling rate of a signal from 12000 to 500 Hz. The specifications of the single-stage decimator low-pass FIR filter $H(z)$ are: pass-band edge = 225 Hz, stop-band edge = 250 Hz, pas-band ripple = 0.004, and stop-band attenuation = 0.001. Assume $H(z)$ as an equiripple linear-phase FIR filter.

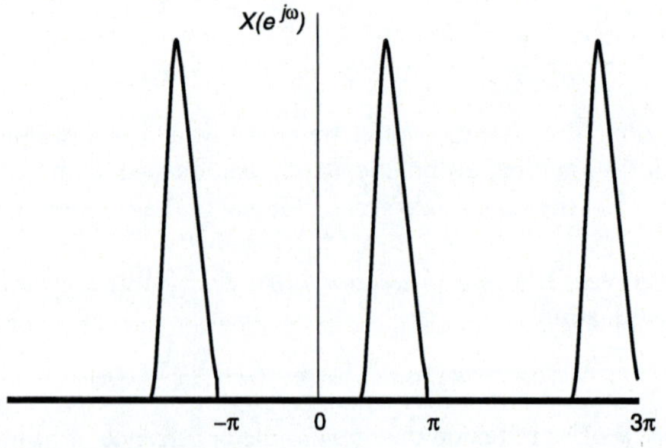


**Fig. B.6** Problem 2

Q10: To convert the CD audio at the sampling rate of 44.1 kHz to the MP3 sampling rate of 48 kHz, a conversion factor of $L/M = 160/147$ is needed. Single-stage scheme would require an unfeasible FIR filter sizes for interpolation and decimation. How would you tackle this problem?

# Appendix C: Tables and Formulas

## *Basic Definitions*

The rectangular pulse: $\Pi(t/T) = \Pi_T(t) = \begin{Bmatrix} 1, & |t| \leq T/2 \\ 0, & \text{elsewhere} \end{Bmatrix}$

The triangular pulse: $\Lambda(t/T) = \Lambda_T(t) = \begin{Bmatrix} 1 - 2|t|/T, & |t| \leq T/2 \\ 0, & \text{elsewhere} \end{Bmatrix}$

The unit step function: $u(t) = \begin{Bmatrix} 1, & t \geq 0 \\ 0, & t < 0 \end{Bmatrix}$,

The sinc function: $\text{sinc}(t) = \frac{\sin(\pi t)}{(\pi t)}$

The signum function: $\text{sgn}(t) = \begin{Bmatrix} 1, & t \geq 0 \\ -1, & t < 0 \end{Bmatrix}$

## *Useful Formulas*

1. $e^{\pm j\phi} = \cos(\phi) \pm j\sin(\phi)$ (Euler);
   $|e^{\pm j\phi}| = 1$;
   $\cos(x) = \cosh(jx) = (e^{jx} + e^{-jx})/2$;
   $\sin(x) = (1/j)\sinh(jx) = (e^{jx} - e^{-jx})/(2j)$;

2. $\cos^2(\phi) + \sin^2(\phi) = 1$;
   $\tan(x) = \sin(x)/\cos(x)$;
   $\cos^2(\phi) = (1/2) + (1/2)\cos(2\phi)$;
   $\sin^2(\phi) = (1/2) - (1/2)\cos(2\phi)$

3. $\cos(x \pm y) = \cos(x)\cos(y) \mp \sin(x)\sin(y);$
$\sin(x \pm y) = \sin(x)\cos(y) \pm \cos(x)\sin(y);$
$\tan(x \pm y) = [\tan(x) \pm \tan(y)]/[1 \mp \tan(x)\tan(y)];$
$\sin(2x) = 2\sin(x)\cos(x); \cos(2x) = 2\cos^2(x) - 1.$

4. $\cos(x)\cos(y) = (1/2)[\cos(x-y) + \cos(x+y)];$
$\sin(x)\sin(y) = (1/2)[\cos(x-y) - \cos(x+y)];$
$\sin(x)\cos(y) = (1/2)[\sin(x-y) + \sin(x+y)];$

5. $\tan(x/2) = \sin(x)/[1+\cos(x)] = [1-\cos(x)]/\sin(x);$
$\sin(x/2) = \pm\sqrt{[1-\cos(x)]/2};$
$\cos(x/2) = \pm\sqrt{[1+\cos(x)]/2}.$

6. $\sinh(x) = (e^x - e^{-x})/2;$
$\cosh(x) = (e^x + e^{-x})/2;$
$\tanh(x) = \sinh(x)/\cosh(x);$
$\sinh^{-1}(x) = \ln(x + \sqrt{x^2 - 1});$
$\cosh^{-1}(x) = \ln(x + \sqrt{x^2 - 1});$
$\tanh^{-1}(x) = \frac{1}{2}\ln\frac{1+x}{1-x};$
$\cosh^2(x) - \sinh^2(x) = 1;$

7. $df(y)/dx = [df(y)/dy][dy/dx];$
$$d\left[\int_a^t f(u)du\right]/dt = f(t), \quad \text{where } a \text{ is constant}$$

8. $dx^a/dx = ax^{a-1}$;

$$\int x^a dx = x^{a+1}/(a+1), (a \neq -1);$$

$d\ln(x)/dx = 1/x$;

$$\int x^{-1} dx = \ln(x);$$

9. $df(y)/dx = [df(y)/dy][dy/dx]$;
$d\sin(x)/dx = \cos(x)$;
$d\cos(x)/dx = -\sin(x)$;
$d\tan(x)/dx = -1/\cos^2(x)$;
$de^x/dx = e^x$;
$dc^x/dx = \ln(c)c^x$;
$d\tan^{-1}(x)/dx = 1/(1+x^2)$;
$d\tanh^{-1}(x)/dx = 1/(1-x^2)$;
$d\sinh^{-1}(x)/dx = 1/\sqrt{x^2+1}$;
$d\cosh^{-1}(x)/dx = 1/\sqrt{x^2-1}$;
$d\sin^{-1}(x) = -d\cos^{-1}(x) = 1/\sqrt{1-x^2}$.

10. $$\sum_{n=0}^{N-1} r^n = \left\{ \begin{array}{ll} (1-r^N)/(1-r), & r \neq 1 \\ N, & r = 1 \end{array} \right\};$$

$$\sum_{n=0}^{N-1} (a+nr) = \frac{N}{2}[2a+(N-1)r] = \frac{N}{2}\ [\text{first} + \text{last}].$$

11. If $\log_b(y) = x \Leftrightarrow y = b^x$;
$\log_b(xy) = \log_b(x) + \log_b(y)$;
$\log_b(x/y) = \log_b(x) - \log_b(y)$;
$\log_b(x^n) = n\log_b(x)$; $x^0 = 1$;
$\log_b b = 1$;
$\log_b(y) = \log_c(y)/\log_c(b)$;
$\log_a(a^x) = x$.

Natural logarithm: $\ln(x) = \log_e(x), e \approx 2.718281828459\cdots$;
Common logarithm: $\log_{10}(x)$.

12. Power series expansion: $1/(1-x) = 1 + x + x^2 + x^3 + \cdots$ ( for $|x| < 1$);

$\sin(x) = x - x^3/3! + x^5/5! - \cdots$;
$\sin^{-1}(x) = x + x^3/6 + 3x^5/40 + 5x^7/112 + 35x^9/1152 + \cdots$;
$\cos(x) = 1 - x^2/2! + x^4/4! - \cdots$;
$\cos^{-1}(x) = \pi/2 - x - x^3/6 - 3x^5/40 - 5x^7/112 - \cdots$ (for $|x| < 1$);
$\tan(x) = x + x^3/3 + 2x^5/15 + 17x^7/315 + 62x^9/2835 + \cdots$ (for $|x| < \pi/2$)
$\tan^{-1}(x) = x - x^3/3 + x^5/5 - \cdots$ (for $|x| < 1$);
$\tan^{-1}(1+x) = \pi/4 + x/2 - x^2/4 + x^3/12 + x^5/40 + \cdots$
$e^x = x + x^2/2! + x^3/3! + \cdots$ (for $|x| < \infty$);
$\ln(1+x) = x - x^2/2 + x^3/3 - x^4/4 + \cdot$ (for $|x| < 1$)

13. The convolution integral: $x(t) * h(t) = \int_{-\infty}^{\infty} x(\lambda)h(t-\lambda)d\lambda$;
14. The correlation integral: $R_{xy}(\tau) = \int_{-\infty}^{\infty} x(t)y(\tau + t)dt$;
Autocorrelation: $R_x(\tau) = \int_{-\infty}^{\infty} x(t)x(\tau + t)dt$.
15. Discrete convolution: $x(n) * h(n) = \sum_{k=-\infty}^{\infty} x(k)h(n-k)$;
16. Dirac delta function, $\delta(t)$: It is defined by the integral

$$\int_{-\infty}^{\infty} x(t)\delta(t - t_o)dt = x(t_o),$$

where $x(t)$ is any continuous function.
Note that: $\delta(at) = (1/|a|)\delta(t); \delta(-t) = \delta(t); \int_{-\infty}^{\infty} \delta(t)dt = 1$.

17. $x(t) * \delta(t - t_o) = x(t - t_o); x(t)\delta(t - t_o) = x(t_o)\delta(t - t_o)$

18. Gaussian distribution: $p(\lambda) = \frac{1}{\sigma\sqrt{2\pi}} e^{-(\lambda - m)^2/2\sigma^2}$, $m$ = mean, $\sigma^2$ = variance
19. Schwarz's inequality: $|\int g_1(x)g_2(x)dx|^2 \leq (\int |g_1(x)|^2dx)(\int |g_2(x)|^2dx)$
*(equality holds only when $g_1(x) = kg_2^*(x)$, $k$ being a constant.)*
20. Roots of the quadratic equation: $ax^2 + bx + c = 0$ are given by

$$r_{1,2} = \frac{-b \pm \sqrt{b^2 - 4ac}}{2a}$$

21. Complex cubic roots of 1: Call them $\{u_k | k = 0, 1, 2\}$. Let $z = 1$. Then:

$$z = 1 = 1e^{j2\pi} \Rightarrow \sqrt[3]{Z} = (e^{j2k\pi})^{1/3}\{k = 0, 1, 2, \ldots\}$$
$$= e^{jk2\pi/3} = \left\{1, e^{j2\pi/3}, e^{j4\pi/3}\right\} = \left\{1, \frac{-1 \pm j\sqrt{3}}{2}\right\} = \{u_k | k = 0, 1, 2\}.$$

Complex cubic roots of a real number $a = h^3$ are $\{r_k = hu_k \mid k = 0, 1, 2\}$.

22. Roots of the cubic equation: $x^3 + ax^2 + bx + c = 0$.

Let $p = b - \frac{a^2}{3}$;

$q = c + \frac{2a^3 - 9ab}{27}$;

$h = \sqrt[3]{\frac{q}{2} \pm \sqrt{\frac{q^2}{4} + \frac{p^3}{27}}}$ ($\pm$ : take either $+$ or $-$; but choose carefully when $p = 0$ so that $h \neq 0$;

$r_k = hu_k\{k = 0, 1, 2\}$.

Then: $x_k = \frac{p}{3r_k} - \frac{a}{3} - r_k$

23. $(x + y)^n = \sum_{k=0}^{n} C(n, k)x^{n-k}y^k$ where $C(n, k) = n!/[k!(n - k)!]$.

**Laplace transform pairs and theorems**

| $x(t)$ | $X(s)$ | Property | Transform pair |
|---|---|---|---|
| $\delta(t)$ | 1 | Linearity | $ax(t) + by(t) \overset{\mathcal{L}}{\longleftrightarrow} aX(s) + bY(s)$ |
| $u(t)$ | $\frac{1}{s}$ | Scaling | $x(at) \overset{\mathcal{L}}{\longleftrightarrow} X(s/a)/\lvert a\rvert$ |
| $\frac{t^n}{n!}u(t)$ | $\frac{1}{s^{n+1}}$ | Time shift (delay) | $x(t - t_o) \overset{\mathcal{L}}{\longleftrightarrow} X(s)e^{-st_o}, t_o > 0$ |
| $e^{-at}\, u(t)$ | $\frac{1}{s+a}$ | s-Shift | $x(t)e^{-at} \overset{\mathcal{L}}{\longleftrightarrow} X(s + a)$ |
| $\frac{t^n}{n!}e^{-at}u(t)$ | $\frac{1}{(s+a)^{n+1}}$ | Convolution | $x(t) * y(t) \overset{\mathcal{L}}{\longleftrightarrow} X(s)Y(s)$ |
| $\sin(\omega_o\, t)u(t)$ | $\frac{\omega_o}{s^2+\omega_o^2}$ | Conjugation | $x^*(t) \overset{\mathcal{L}}{\longleftrightarrow} X^*(s^*)$ |
| $\cos(\omega_o\, t)u(t)$ | $\frac{s}{s^2+\omega_o^2}$ | Initial value theorem | $\lim_{t\to 0} x(t) = \lim_{s\to\infty} sX(s)$ |
| $t\,\sin(\omega_o\, t)u(t)$ | $\frac{2\omega_o s}{(s^2+\omega_o^2)^2}$ | Final value theorem | $\lim_{t\to\infty} x(t) = \lim_{s\to 0} sX(s)$ |
| $t\,\cos(\omega_o\, t)u(t)$ | $\frac{\omega_o}{(s^2+\omega_o^2)^2}$ | Time differentiation | $dx(t)/dt \overset{\mathcal{L}}{\longleftrightarrow} sX(s) - x(0^-)$ |
| $e^{-at}\,\sin(\omega_o\, t)u(t)$ | $\frac{\omega_o}{(s+a)^2+\omega_o^2}$ | Time integration | $\int_{0^-}^{t} x(u)du \overset{\mathcal{L}}{\longleftrightarrow} \frac{X(s)}{s}$ |
| $e^{-at}\,\cos(\omega_o\, t)u(t)$ | $\frac{s+a}{(s+a)^2+\omega_o^2}$ | Multiplication by $t^n$ | $t^n x(t) \overset{\mathcal{L}}{\longleftrightarrow} (-1)^n \frac{d^n X(s)}{ds^n}$, $n=$ integer |

**Fourier transform pairs and theorems**

| $x(t)$ | $X(f)$ | Properties |
|---|---|---|
| $\Pi(t/T) = \Pi_T(t)$ | $T\,\mathrm{sinc}(Tf)$ | $ax(t) + by(t) \overset{\mathcal{F}}{\longleftrightarrow} aX(f) + bY(f)$ |
| $\Lambda(t/T) = \Lambda_T(t)$ | $\frac{T}{2}\,\mathrm{sinc}^2(\frac{T}{2}f)$ | $x(at) \overset{\mathcal{F}}{\longleftrightarrow} X(f/a)/\lvert a\rvert$ |
| $\mathrm{sinc}(Lt)$ | $\frac{1}{L}\Pi_L(f)$ | $x(-t) \overset{\mathcal{F}}{\longleftrightarrow} X(-f)$ |
| $e^{-at}\,u(t),\ a > 0$ | $\frac{1}{a+j2\pi f}$ | $x^*(t) \overset{\mathcal{F}}{\longleftrightarrow} X^*(-f),\ x^*(-t) = X^*(f)$ |
| $te^{-at}\,u(t),\ a > 0$ | $\frac{1}{(a+j2\pi f)^2}$ | $X(t) \overset{\mathcal{F}}{\longleftrightarrow} x(-f)\, X(t) \overset{\mathcal{F}}{\longleftrightarrow} x(f)$, $x$ even |
| $e^{-a\lvert t\rvert},\ a > 0$ | $\frac{2a}{a^2+(2\pi f)^2}$ | $x(t - t_o) \overset{\mathcal{F}}{\longleftrightarrow} X(f)e^{-j2\pi f t_o}$ |
| $e^{-a^2t^2}$ | $\frac{\sqrt{\pi}}{a}e^{-(\pi f/a)^2}$ | $x(t)e^{j2\pi f_o t} \overset{\mathcal{F}}{\longleftrightarrow} X(f - f_o)$ |
| $1$ | $\delta(f)$ | $x(t)\cos(2\pi f_o t) \overset{\mathcal{F}}{\longleftrightarrow} \frac{1}{2}X(f - f_o) + \frac{1}{2}X(f + f_o)$ |
| $\delta(t)$ | $1$ | $d^n x(t)/dt^n \overset{\mathcal{F}}{\longleftrightarrow} (j2\pi f)^n X(f)$ |
| $\mathrm{sgn}(t)$ | $\frac{1}{j\pi f}$ | $\int_{-\infty}^{t} x(u)du \overset{\mathcal{F}}{\longleftrightarrow} \frac{X(f)}{j2\pi f} + \frac{1}{2}X(0)\delta(f)$ |
| $u(t)$ | $\frac{1}{2}\delta(f) + \frac{1}{j2\pi f}$ | $x(t) * y(t) \overset{\mathcal{F}}{\longleftrightarrow} X(f)Y(f)$ |
| $\cos(2\pi f_o\,t + \theta_o)$ | $\frac{1}{2}e^{j\theta_o}\delta(f - f_o) + \frac{1}{2}e^{-j\theta_o}\delta(f + f_o)$ | $x(t)y(t) \overset{\mathcal{F}}{\longleftrightarrow} X(f) * Y(f)$ |
| $\sin(2\pi f_o\,t + \theta_o)$ | $\frac{1}{2j}e^{j\theta_o}\delta(f - f_o) - \frac{1}{2j}e^{-j\theta_o}\delta(f + f_o)$ | If $x(t)$ is real, then $X(-f) = X^*(f)$ |
| $\sum_{n=-\infty}^{\infty}\delta(t - nT_s)$ | $f_s\sum_{k=-\infty}^{\infty}\delta(f - kf_s)$, where $f_s = \frac{1}{T_s}$ | |
| $e^{\pm j2\pi f_o t}$ | $\delta(f \mp f_o)$ | |

**z-Transform pairs and theorems**

| $x(n)$ | $X(z)$ | Property | Transform pair |
|---|---|---|---|
| $\delta(n)$ | $1$ | Linearity | $ax(n) + by(n) \overset{\mathcal{Z}}{\longleftrightarrow} aX(z) + bY(z)$ |
| $u(n)$ | $\frac{z}{z-1}$ | z-Scaling | $a^n x(n) \overset{\mathcal{Z}}{\longleftrightarrow} X(z/a)$ |
| $a^n\,u(n)$ | $\frac{z}{z-a}$ | Convolution | $x(n) * y(n) \overset{\mathcal{Z}}{\longleftrightarrow} X(z)Y(z)\, x(n)\cdot$ |
| | | | $y(n) \overset{\mathcal{Z}}{\longleftrightarrow} X(z) * Y(z)$ |
| $na^n\,u(n)$ | $\frac{az}{(z-a)^2}$ | Conjugation | $x^*(n) \overset{\mathcal{Z}}{\longleftrightarrow} X^*(z^*)$ |
| $(n + 1)a^n\,u(n)$ | $(\frac{z}{z-a})^2$ | z-Differentiation | $nx(n) \overset{\mathcal{Z}}{\longleftrightarrow} -zdX(z)/dz$ |
| $\sin(bn)u(n)$ | $\frac{\sin(b)z}{z^2-2\cos(b)z+1}$ | Time integration | $\sum_{k=-\infty}^{n} x(k) \overset{\mathcal{Z}}{\longleftrightarrow} \frac{1}{1-z^{-1}}X(z)$ |
| $\cos(bn)u(n)$ | $\frac{z^2-\cos(b)z}{z^2-2\cos(b)z+1}$ | Time differentiation | $x(n) - x(n - 1) \overset{\mathcal{Z}}{\longleftrightarrow} (1 - z^{-1})X(z)$ |
| | | Time shift (delay) | $x(n - m) \overset{\mathcal{Z}}{\longleftrightarrow} z^{-m}X(z)$ |
| | | Initial value theorem | If $x(n) = 0$ for $n < 0$, then $x(0) = \lim_{z\to\infty} X(z)$ |

**Denominator polynomial coefficients for normalized LPF's**

| $n$ | $b_o$ | $b_1$ | $b_2$ | $b_3$ | $b_4$ | $b_5$ | $b_6$ |
|---|---|---|---|---|---|---|---|
| *Butterworth* | | | | | | | |
| 1 | 1.0 | | | | | | |
| 2 | 1.0 | 1.4142 | | | | | |
| 3 | 1.0 | 2.0 | 2.0 | | | | |
| 4 | 1.0 | 2.6131 | 3.4142 | 2.6131 | | | |
| 5 | 1.0 | 3.2361 | 5.2361 | 5.2361 | 3.2361 | | |
| 6 | 1.0 | 3.8637 | 7.4641 | 9.1416 | 7.4641 | 3.8637 | |
| 7 | 1.0 | 4.4940 | 10.0978 | 14.5918 | 14.5918 | 10.0978 | 4.4940 |
| *0.5-dB passband ripple Chebychev* ($\varepsilon^2 = 0.1220$) | | | | | | | |
| 1 | 2.8628 | | | | | | |
| 2 | 1.5162 | 1.425 | | | | | |
| 3 | 0.7157 | 1.5349 | 1.2529 | | | | |
| 4 | 0.3791 | 1.0255 | 1.7169 | 1.1974 | | | |
| 5 | 0.1789 | 0.7525 | 1.3096 | 1.9374 | 1.1725 | | |
| 6 | 0.0948 | 0.4324 | 1.1719 | 1.5898 | 2.1718 | 1.1592 | |
| 7 | 0.0447 | 0.2821 | 0.7557 | 1.6479 | 1.8694 | 2.4127 | 1.1512 |
| *1.0-dB passband ripple Chebychev* ($\varepsilon^2 = 0.2589$) | | | | | | | |
| 1 | 1.9652 | | | | | | |
| 2 | 1.1025 | 1.0977 | | | | | |
| 3 | 0.4913 | 1.2384 | 0.9883 | | | | |
| 4 | 0.2756 | 0.7426 | 1.4539 | 0.9528 | | | |
| 5 | 0.1228 | 0.5805 | 0.9744 | 1.6888 | 0.9368 | | |
| 6 | 0.0689 | 0.3071 | 0.9393 | 1.2021 | 1.9308 | 0.9283 | |
| 7 | 0.0307 | 0.2137 | 0.5486 | 1.3575 | 1.4288 | 2.1761 | 0.9231 |
| *3-dB passband ripple Chebychev* ($\varepsilon^2 = 0.9953$) | | | | | | | |
| 1 | 1.0024 | | | | | | |
| 2 | 0.7079 | 0.6449 | | | | | |
| 3 | 0.2506 | 0.9283 | 0.5972 | | | | |
| 4 | 0.1770 | 0.4048 | 1.1691 | 0.5816 | | | |
| 5 | 0.0626 | 0.4080 | 0.5489 | 1.4150 | 0.5745 | | |
| 6 | 0.0442 | 0.1634 | 0.6991 | 0.6906 | 1.6628 | 0.5707 | |
| 7 | 0.0157 | 0.1462 | 0.3000 | 1.0518 | 0.8314 | 1.9116 | 0.5684 |

**Normalized low-pass filter LC element values**

| $n$ | $R_s$ | $C_1$ | $L_2$ | $C_3$ | $L_4$ | $C_5$ | $L_6$ | $C_7$ |
|---|---|---|---|---|---|---|---|---|
| *LPF Butterworth LC element values* $(R_L = 1\,\Omega)$ | | | | | | | | |
| 2 | 1.0 | 1.4142 | 1.412 | | | | | |
| 3 | 1.0 | 1.0 | 2.0 | 1.0 | | | | |
| 4 | 1.0 | 0.7654 | 1.8478 | 1.8478 | 0.7654 | | | |
| 5 | 1.0 | 0.6180 | 1.6180 | 2.0 | 1.6180 | 0.6180 | | |
| 6 | 1.0 | 0.5176 | 1.4142 | 1.9319 | 1.9319 | 1.4142 | 0.5176 | |
| 7 | 1.0 | 0.4450 | 1.2470 | 1.8019 | 2.0 | 1.8019 | 1.2470 | 0.4450 |
| *LPF 0.5-dB ripple Chebychev LC element values* $(R_L = 1\,\Omega)$ | | | | | | | | |
| 2 | 1.9841 | 0.9827 | 1.9497 | | | | | |
| 3 | 1.0 | 1.8636 | 1.2804 | 1.8636 | | | | |
| 4 | 1.9841 | 0.9202 | 2.5864 | 1.3036 | 1.8258 | | | |
| 5 | 1.0 | 1.8068 | 1.3025 | 2.6914 | 1.3025 | 1.8068 | | |
| 6 | 1.9841 | 0.9053 | 2.5774 | 1.3675 | 2.7133 | 1.2991 | 1.7961 | |
| 7 | 1.0 | 1.7896 | 1.2961 | 2.7177 | 1.3848 | 2.7177 | 1.2961 | 1.7896 |
| *LPF 1-dB ripple Chebychev LC element values* $(R_L = 1\,\Omega)$ | | | | | | | | |
| 2 | 3.0 | 0.5723 | 3.1317 | | | | | |
| 3 | 1.0 | 2.2160 | 1.0883 | 2.2160 | | | | |
| 4 | 3.0 | 0.6529 | 4.4110 | 0.8140 | 2.5346 | | | |
| 5 | 1.0 | 2.2072 | 1.1279 | 3.1025 | 1.1279 | 2.2072 | | |
| 6 | 3.0 | 0.6729 | 3.8725 | 0.7706 | 4.7107 | 0.9692 | 2.4060 | |
| 7 | 1.0 | 2.2043 | 1.1311 | 3.1472 | 1.1942 | 3.1472 | 1.1311 | 2.2043 |

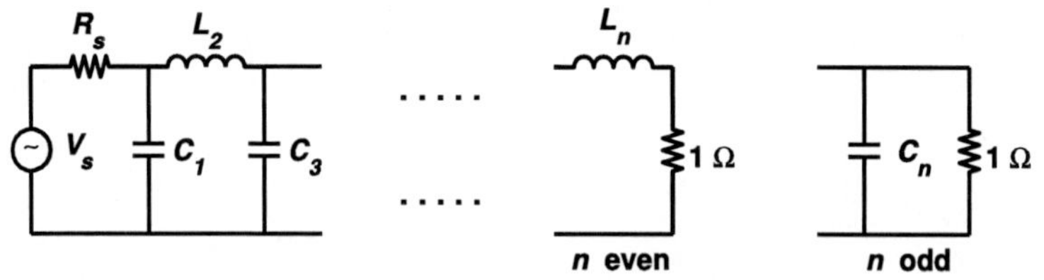


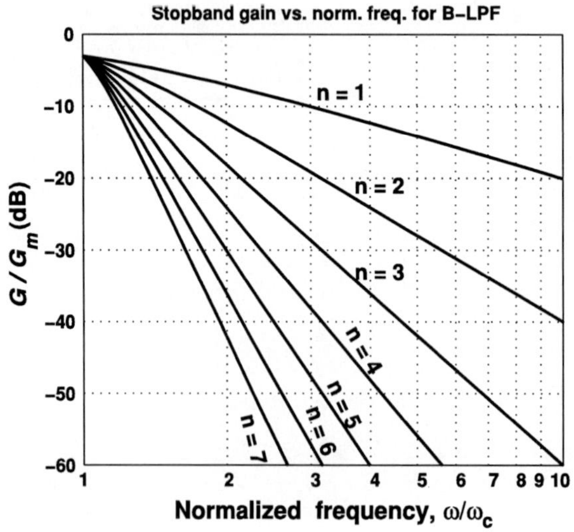

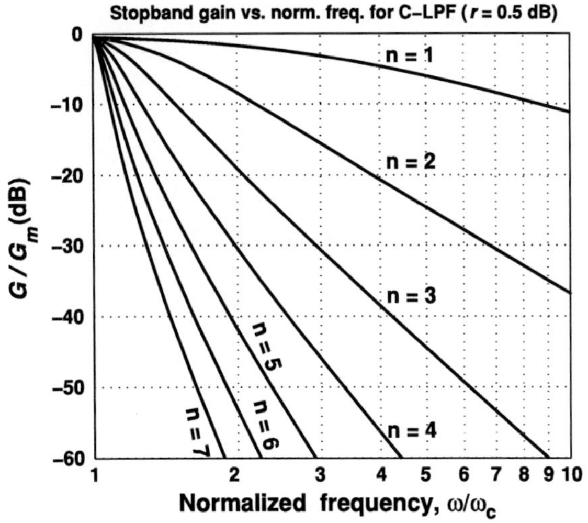
Stopband gain vs. norm. freq. for C–LPF (r = 0.5 dB)
G / G_m (dB)
0
–10
–20
–30
–40
–50
–60
n = 1
n = 2
n = 3
n = 4
n = 5
n = 6
n = 7
1
2
3
4
5
6
7
8
9
10
Normalized frequency, ω/ω_c

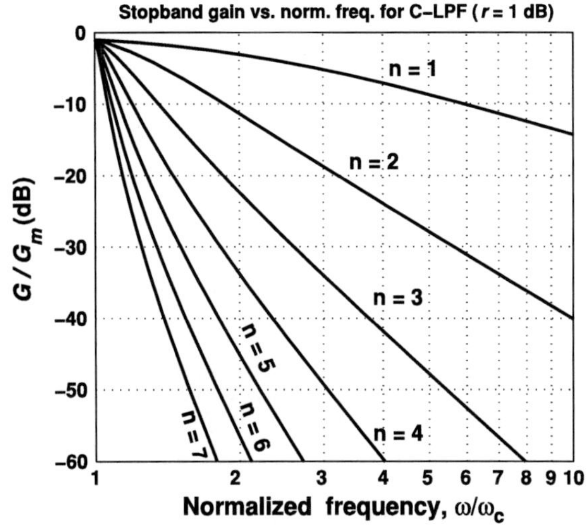
Stopband gain vs. norm. freq. for C–LPF (r = 1 dB)
G / G_m (dB)
0
–10
–20
–30
–40
–50
–60
n = 1
n = 2
n = 3
n = 4
n = 5
n = 6
n = 7
1
2
3
4
5
6
7
8
9
10
Normalized frequency, ω/ω_c

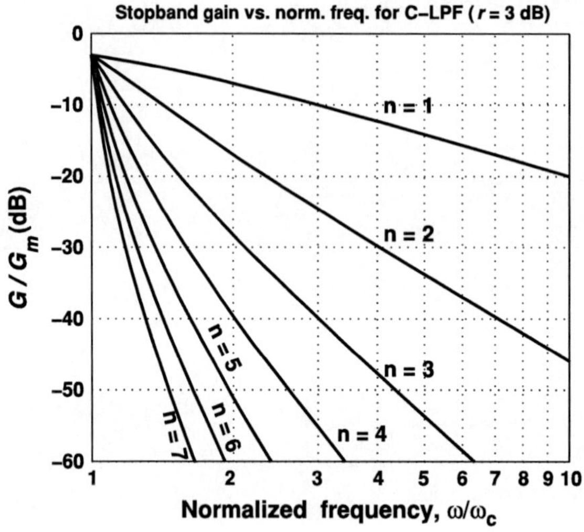


## *Useful Definitions and Relations*

Complex Fourier Series (FS):

$$x(t) = \sum_{k=-\infty}^{\infty} X_k e^{+j2\pi k f_o t}, \text{ where:}$$

$$X_o = \frac{1}{T_o} \int_0^{T_o} x(t)dt,$$

$$X_k = \frac{1}{T_o} \int_0^{T_o} x(t)e^{-j2\pi k f_o t} dt$$

Trigonometric FS:

$$x(t) = a_o + \sum_{n=1}^{\infty} a_n \cos(2\pi n f_o t) + b_n \sin(2\pi n f_o t), \text{ where:}$$

$$a_o = X_o, a_n = X_n + X_{-n}, b_n = j(X_n - X_{-n}).$$

[Note that $X_n$'s are the coefficients of the Complex FS.]

Fourier Transform (FT) pair:

$$X(f) = \int_{-\infty}^{\infty} x(t)e^{-j2\pi ft}dt,$$

$$x(t) = \int_{-\infty}^{\infty} X(f)e^{+j2\pi ft}df$$

(Single-sided) Laplace Transform (LT):

$$X(s) = \int_{0^-}^{\infty} x(t)e^{-st}dt, \text{ where:}$$

$$s = \sigma + j\omega, \quad \omega = 2\pi f.$$

Discrete-Time Fourier Transform (DTFT) pair:

$$X_s(f) = \sum\nolimits_{n=-\infty}^{\infty} x(n)e^{-j2\pi nT_s f},$$

$$x(n) = \frac{1}{f_s}\int_{0}^{f_s} X_s(f)e^{+j2\pi nT_s f}df$$

Some important DTFT pairs:

1. $\frac{2f_c}{f_s}$ sinc $\left(\frac{2f_c}{f_s}n\right) \overset{DTFT}{\longleftrightarrow} \Pi_{2f_c}(f), -\frac{f_s}{2} < f < \frac{f_s}{2}$
   *(Periodic rectangular pulse, period* $= f_s$*)*
2. $\Pi_N(n) \overset{DTFT}{\longleftrightarrow} N\text{sinc}(Nf/f_s)/\ \text{sinc}(f/f_s)$ [*a sinc-like function*]

The DTFT modulation property:
If $x(n) \overset{DTFT}{\longleftrightarrow} X_s(f)$, then: $x(n)\cos(2\pi f_o/f_s) \overset{DTFT}{\longleftrightarrow} X_s(f - f_o) + X_s(f + f_o)$
Discrete Fourier Transform (DFT) pair:

$$X(k) = \sum\nolimits_{n=0}^{N-1} x(n)e^{-j2\pi kn/N},$$

$$x(n) = \frac{1}{N}\sum\nolimits_{k=0}^{N-1} X(k)e^{+j2\pi kn/N}$$

(Single-sided) z-Transform (ZT):

$$X(z) = \sum\nolimits_{n=0}^{\infty} x(n)z^{-n}$$

Low-pass to band-pass transformations

$$\omega_{LN} = \left(\omega^2 - \omega_g^2\right)/(\omega\omega_b), \quad s_{LN} = \left(s^2 + \omega_g^2\right)/(s\omega_b), \quad \text{where:}$$

$$\omega_b 7 = \omega_u - \omega_l, \quad \omega_g = \sqrt{\omega_u\omega_l}.$$

Low-pass to high-pass transformations

$$\omega_{LN} = \omega_c/\omega; \quad s_{LN} = \omega_c/s.$$

Low-pass to band-stop transformations

$$\omega_{LN} = \omega\omega_b/\left(\omega^2 - \omega_g^2\right), \quad s_{LN} = s\omega_b/\left(s^2 + \omega_g^2\right), \quad \text{where:}$$

$$\omega_b = \omega_u - \omega_l, \quad \omega_g = \sqrt{\omega_u\omega_l}.$$

Circuit denormalization

$R \rightarrow RZ; \quad L \rightarrow L\frac{Z}{W}; \quad C \rightarrow \frac{C}{ZW}$, where:
$Z$ = Impedance scaling factor (ISF)
$W$ = Frequency scaling factor (FSF)

Hardware analog filter transformations

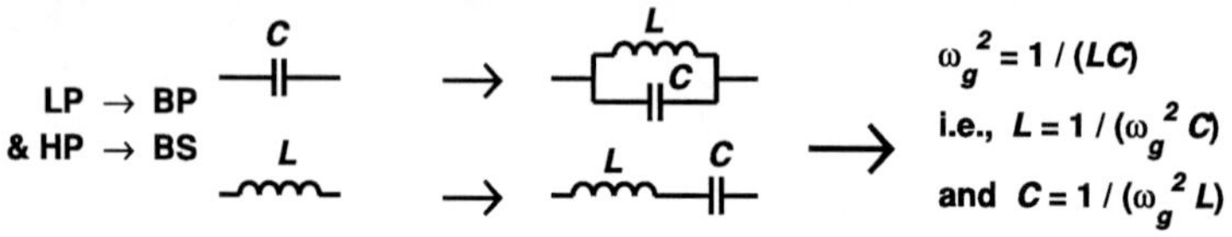


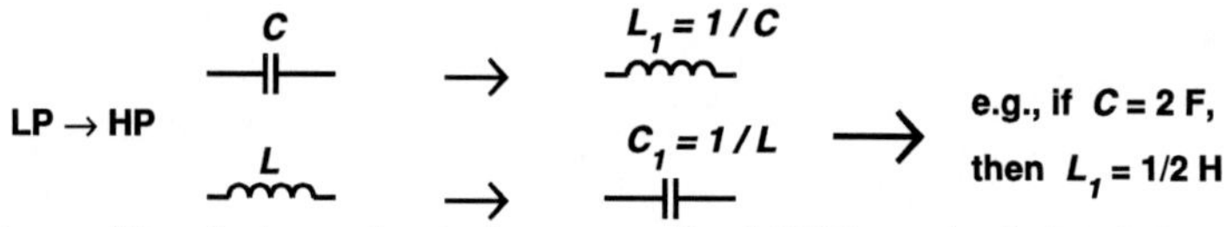


Hardware filter design rules (using normalized LPF standard circuits)
HPF design:

1. Transform $LP_N \rightarrow HP_N$ circuit components
2. Denormalize $HP_N \rightarrow HP$

BPF design:

1. Denormalize $LP_N \rightarrow LP$
2. Transform $LP \rightarrow BP$ circuit components

BSF design:

1. Transform $LP_N \rightarrow HP_N$ circuit components
2. Denormalize $HP_N \rightarrow HP$
3. Transform $HP \rightarrow BS$ circuit components

Magnitude response of Butterworth LPF:

$$|H(\omega)| = G_m/\sqrt{1 + (\omega/\omega_c)^{2n}};$$

where $G_m$ = maximum gain.

Butterworth DC gain: $G_{dc} = G_m$

Magnitude response of Chebychev-I LPF:

$$|H(\omega)| = G_m \Big/ \sqrt{1 + \varepsilon^2 C_n^2\left(\frac{\omega}{\omega_c}\right)}$$

where $C_o(x) = 1$, $C_1(x) = x$, $C_{n+1}(x) = 2xC_n(x) - C_{n\text{-}1}(x)$.

Chebychev DC gain:

$$G_{dc} = \begin{cases} G_m & n \text{ odd} \\ G_m/\sqrt{1+\varepsilon^2} & n \text{ even} \end{cases}$$

Matched filter: $h(t) = s(T - t)$.

Inverting and non-inverting amplifiers:

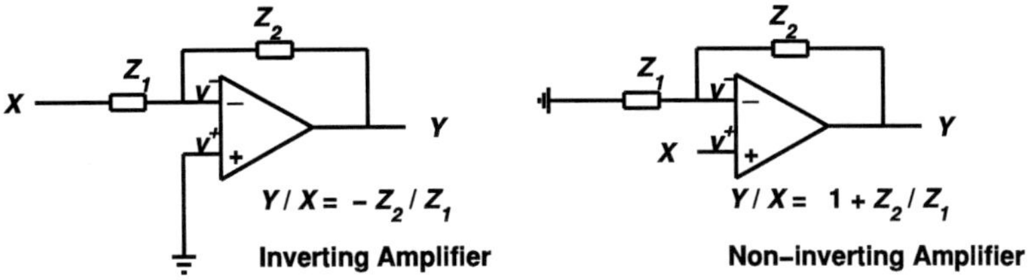


Hilbert Transformer (HT): $H(f) = -j\text{sgn}(f)$ with $h(t) = 1/(\pi \cdot t)$.

Hilbert Transform of $x(t)$ is: $\hat{x}(t) = x(t) * h(t)$ or: $\hat{X}(f) = -j\ \text{sgn}(f)X(f)$.

The analytic associate of $x(t)$ **is:** $z(t) = x(t) + j\hat{x}(t)$

Digital HT:

$$H(e^{j\Omega}) = -j\ \text{sgn}(\Omega), |\Omega| < \pi; h(n) = \left\{ \begin{matrix} \frac{2}{\pi}\frac{\sin^2(n\cdot\pi/2)}{n}, & n \neq 0 \\ 0 & \text{elsewhere} \end{matrix} \right\}$$

Bilinear Transform: $s \Rightarrow \frac{z-1}{z+1}$ with $\Omega = 2\tan^{-1}(\omega) = 2\tan^{-1}(2\pi f)$.

Impulse invariant transform:

$$H_a(s) = \sum_{m=1}^{M} \frac{c_m}{s - p_m} \Rightarrow H(z) = T_s \sum_{m=1}^{M} c_m \frac{z}{z - z_m} \quad \text{where} \quad z_m = e^{p_m T_s}.$$

Digital domain frequency: $\Omega = 2\pi \cdot v = 2\pi \cdot f/f_s = \omega/f_s = \omega T_s$

Sinusoidal PLL system equation: $\phi(k) = \theta(k) - \omega_0 \sum_{i=0}^{k-1} y(i)$

Sinusoidal PLL: 1st-order system equation:

$$\phi(k+1) = \phi(k) - K_2 \sin[\phi(k)] + \Lambda_o$$

where $\Lambda_o = 2\pi(\omega - \omega_o)/\omega_o$ and $K_2 \omega G_1 A$.

If $K_1 \omega G_1 A$ and $W = \omega_o/\omega$, then $K_2 = K_1(\omega_o/\omega) = K_1/W$.

Note that $G_2 = 0$ in this case.
Sinusoidal PLL: 2nd-order system equation:

$$\phi(k+2) - 2\phi(k+1) + \phi(k) = K_2 \sin[\phi(k)] - rK_2 \sin[\phi(k+1)],$$
$$\text{where } r = 1 + G_2/G_1.$$

Fixed point analysis: $g(x) = x$ has a solution $x^*$ only if $|g'(x^*)| < 1$.

Jacobian matrix: $\frac{\partial(u,v)}{\partial(x,y)} = \begin{bmatrix} \partial u/\partial x & \partial u/\partial y \\ \partial v/\partial x & \partial v/\partial u \end{bmatrix}$ *where u, v are functions of x, y.*

Eigenvalues $\{\lambda\}$ of a matrix X:

$$|\lambda I - X| = 0; \quad \text{where } I \text{ is the identity matrix.}$$

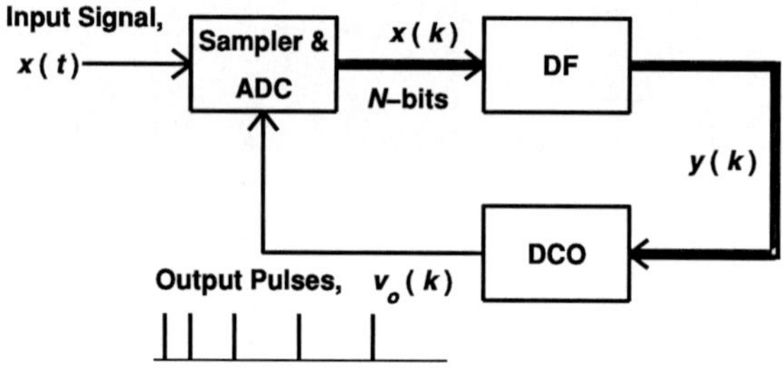


Integrator $H(z) = 1/(1 - z^{-1})$; Differentiator $H(z) = (1 - z^{-1})/T_s$
Error functions:

$$\text{erf}(x) = \frac{2}{\sqrt{\pi}} \int_0^x e^{-v^2} dv$$

$$Q(x) = \frac{1}{\sqrt{2\pi}} \int_x^\infty e^{-v^2/2} dv = \frac{1}{2} \text{ erfc} \left( \frac{x}{\sqrt{2}} \right) = \frac{1}{2} - \frac{1}{2} \text{ erf} \left( \frac{x}{\sqrt{2}} \right)$$

For $x > 0$ we have:

$$Q(x) \approx \frac{1}{\sqrt{2\pi}} e^{-x^2/2} \sum_{k=1}^{5} b_k g^k; \quad \text{where}$$

$g = 1/(1 + 0.2316419x)$,
$b_1 = 0.3198, \quad b_2 = -0.3565, \quad b_3 = 1.7814,$
$b_4 = -1.8212, \quad b_5 = 1.3302$

Direct forms of IIR implementation:

$$y(n) = b_o x(n) + b_1 x(n-1) + \cdots + b_M x(n-M) - a_1 y(n-1) - \cdots - a_N y(n-N)$$

$$H(z) = \frac{\sum_{i=0}^{M} b_i z^{-i}}{1 + \sum_{k=1}^{N} a_k z^{-k}} = \left( \sum_{i=0}^{M} b_i z^{-i} \right) \left( \frac{1}{1 + \sum_{k=1}^{N} a_k z^{-k}} \right) = H_1(z) H_2(z) = H_2(z) H_1(z).$$

**Direct Form – I**

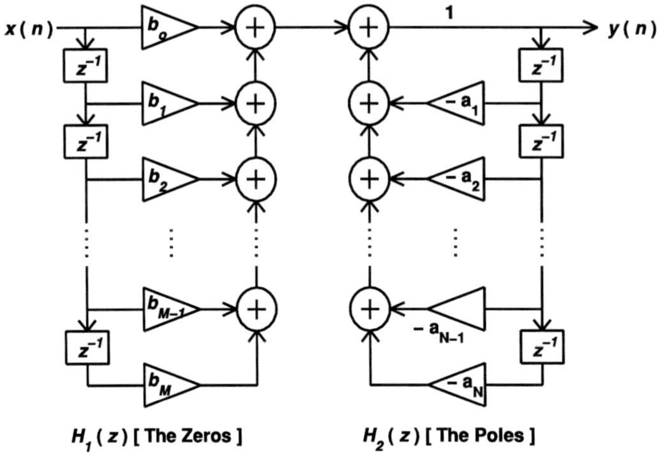


**Direct Form – II**

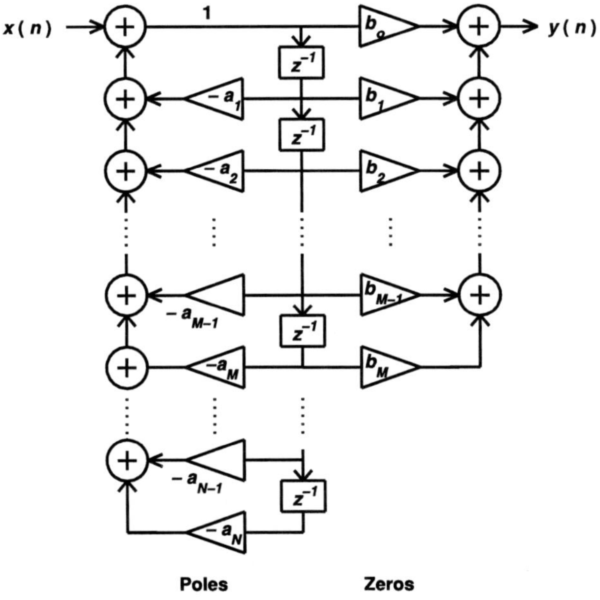

# Appendix D: DSP Lab Experiments

## *Experiment # 1: Computing the Convolution Integral*

### Introduction

The convolution integral of two functions $x(t)$ and $h(t)$ is one of the most significant topics in signal processing. This is so because the output $y(t)$ of any linear time-invariant system is given by the convolution integral of the input signal $x(t)$ and the system impulse response $h(t)$ as follows:

$$y(t) = x(t) * h(t) = \int_{-\infty}^{\infty} x(\lambda)h(t-\lambda)d\lambda = h(t) * x(t) = \int_{-\infty}^{\infty} h(\lambda)x(t-\lambda)d\lambda.$$

Steps of finding the convolution can be found in the tutorials. In this experiment we discuss the numerical implementation of this integral.

Due to the finite word-length and memory on the computer, we can only deal with finite length discrete-time signals. Analog signals and systems should be approximated by their discrete counterparts. This approximation should be mathematically well founded.

First, we should note that if $x(t)$ and $h(t)$ are finite-length functions defined over the intervals $a_1 \le t \le b_1$ and $a_2 \le t \le b_2$, respectively, then the interval of their convolution $y(t)$ will be $a_1 + a_2 \le t \le b_1 + b_2$.

From **Calculus** we have:

$$y(t) = \int_{-\infty}^{\infty} x(\lambda)h(t-\lambda)d\lambda = \lim_{\Delta\lambda \to 0} \sum_{k=-\infty}^{\infty} x(k\Delta\lambda)h(t-k\Delta\lambda)\Delta\lambda.$$

Hence, we can approximate the above continuous-time convolution as follows:

$$y(t) \approx \Delta\lambda \left[ \sum_{k=-\infty}^{\infty} x(k\Delta\lambda)h(t-k\Delta\lambda) \right].$$

The variable $t$ is considered as a constant in the above integration and summation. It has the characteristics of the dummy variable $\lambda$, hence, it can be approximated as $t \approx n\Delta\lambda$, for some integer $n$. The above equation will be better written as follows:

$$y(n) \approx \Delta\lambda \left[ \sum_{k=-\infty}^{\infty} x(k)h(n-k) \right],$$

where $\Delta\lambda$ has been removed from expressions of the form $k\Delta\lambda$ for easier notation and suitability for implementation on the digital computer.

**MATLAB®** is currently the best tool for mathematical modeling. In addition to the basic mathematical features, it has some dedicated toolboxes like the signal processing toolbox, the communications toolbox, and the control toolbox. Also, symbolic solutions are possible through the MATLAB symbolic toolbox. In this book we adopt MATLAB for DSP simulations.

Straightforward calculation of the above discrete convolution using for-next loops is time-consuming and not suitable for DSP purposes, especially for real-time processing of long signals (like speech signals). As we will see later in this course, this problem is closely related to that of calculating the Fourier transform of a signal. Scientists have discovered efficient algorithms for calculating the above summation by making use of symmetric terms in the discrete Fourier transform. This algorithm is called the Fast Fourier Transform (FFT), which is available on MATLAB as fft and ifft (for inverse Fourier Transform).

The above discrete convolution can be implemented on MATLAB as follows:

```
y = Ts * conv(x,h);
```

where `y` is the output vector, `Ts` is the time step (equivalent to $\Delta\lambda$), `x` and `h` are two vectors representing the two signals to be convolved. Note that * on MATLAB means scalar or matrix multiplication, while .* means vector multiplication (*i.e., multiplication of corresponding elements in the two vectors*).

The accuracy of computation would be dependent on `Ts`. However, the less `Ts` the more computation time. Therefore, there is a compromise between speed and accuracy.

Using `y = Ts * conv(x,h)` gives a function whose length is larger than both `x` and `h`. In many applications we need to fit all signals using the same word length. In this case, we can use the following fft-based algorithm:

```
y = Ts * ifft(fft(x) * fft(h));    % If we do not allow negative time,
```

or

```
y = Ts * fftshift(ifft(fft(x) * fft(h)));    % If we allow negative time.
```

## MATLAB Simulation

Task 1

Write a MATLAB code to find the convolution of the two continuous-time finite-length signals *x*(*t*) and *h*(*t*) as shown in Fig. D.1.

- Open a new m-file and give it a name (e.g., DSP_Exp1.m) and save it in your directory.
- Choose a unified interval for x and h (i.e., `a1` = `a2` = `-10`, `b1` = `b2` = `10`).
- Take `Ts` = `0.01` and define the time vector `t`.
- Express `x` and `h` first in terms of the unit step function, which is denoted in MATLAB as `stepfun(t,to)`. On MATLAB command line, type >`help stepfun` to see how this function works. Plot all step functions associated with the functions `x` and `h`, then plot `x`, `h`, and their convolution `y` on the same graph.
- Plot the theoretical convolution (see the *Note* below) and the numerical result on another graph.
- Define all axes using xlabel and ylabel commands.
- Limit the axes lengths using `axis([x_min x_max y_min y_max])` command.
- Define all functions using `text(a,b,'text')` command to write on figures.
- Find the convolution using the fft-based algorithm and compare with the first method.

*Note*: Using Fig. D.1 we have:

$$y(t) = h(t) * x(t) = \int_{-\infty}^{\infty} h(\lambda)x(t-\lambda)d\lambda = \int_{0}^{3} x(t-\lambda)d\lambda = \int_{0}^{0} 3e^{-2(t-\lambda)}u(t-\lambda)d\lambda$$

$$\xrightarrow{v=t-\lambda} = -\int_{t}^{t-3} 3e^{-2v}u(v)dv = \int_{t-3}^{t} 3e^{-2v}u(v)dv$$

$$= \begin{cases} 0, & t<0 [\text{Since } u(v)=0 \text{ for } v<0] \\ 3\int_{0}^{t} e^{-2v}dv = (3/2)(1-e^{-2t}) & t\geq 0 \& t-3<0 \rightarrow 0\leq t<3 \\ 3\int_{t-3}^{t} e^{-2v}dv \approx (3/2)e^{-2t+6} & t\geq 0 \& t-3\geq 0 \rightarrow t\geq 3. \end{cases}$$

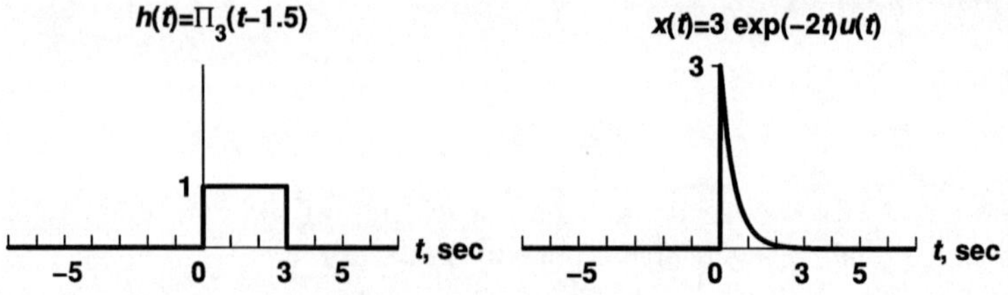


**Fig. D.1** A plot of signals $h(t) = \Pi_3\ (t - 1.5)$ and $x(t) = 3e^{-2t}\ u(t)$

Task 2

Write a code to invert the function $x(t)$ around the y-axis to get $x(-t)$. Then a code to find the shifted versions $x(t + t_o)$ and $x(-t + t_o)$. Take $t_o = 3$ and $-3$. Plot all functions on the same graph. Comment on these operations.

Task 3

The Dirac delta function $\delta(t)$ is an important tool in signal processing. Roughly speaking, it is equivalent to a pulse of infinite height and very narrow width (approaching zero). It is defined by the following integral:

$$\int_{-\infty}^{\infty} g(t)\delta(t - t_o)dt = g(t_o)$$

where $g(t)$ is a continuous function, $t_o$ is a constant. The delta function has the following properties:

P1: $\int_{-\infty}^{\infty} \delta(t)dt = 1$ (unit area),
P2: $\delta(t) = \delta(-t)$ (even),
P3: $x(t) * \delta(t) = x(t)$, or, generally, $x(t)*\delta(t - t_o) = x(t - t_o)$, where $t_o$ is a constant.

The Dirac delta function can also be defined as the limit of several even functions that can satisfy the above properties in the limit. These definitions include:

1. Limit of the weighted rectangular pulse (box), $\Pi_{2a}(t)$ (see Fig. D.2, left):

$$\delta(t) = \lim_{a \to 0} \frac{1}{2a} \Pi_{2a}(t) = \lim_{(a \to 0)} \frac{1}{2a} \begin{Bmatrix} 1, & |t| \le a \\ 0, & |t| > 0 \end{Bmatrix}$$

2. Limit of the weighted absolutely-decaying exponential:

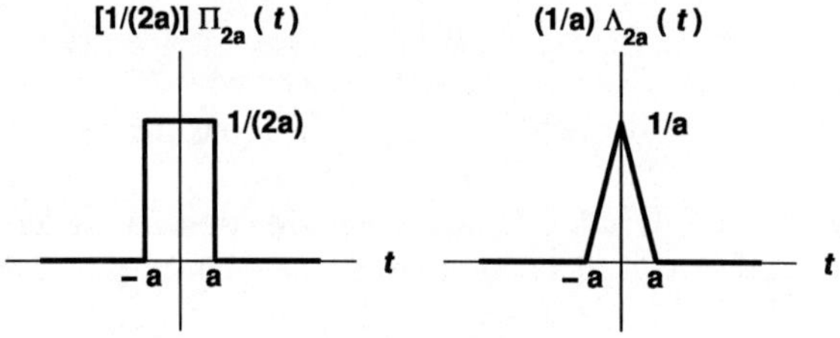


**Fig. D.2** The weighted rectangular and triangular pulses

$$\delta(x) = \lim_{(a \to 0)} \frac{1}{2a} e^{-|x|/a}$$

3. Limit of the weighted triangular pulse $\Lambda_{2a}(t)$ (see Fig. D.2, right):

$$\delta(t) = \lim_{(a \to 0)} \frac{1}{a} \Lambda_{2a}(t) = \lim_{(a \to 0)} \frac{1}{a} \left\{ \begin{array}{ll} 1 - \frac{|t|}{a}, & |t| \le a \\ 0, & |t| > 0 \end{array} \right\}$$

The above definitions are of practical importance since they can be used for approximating the delta function.

On MATLAB, use `stepfun` to define a weighted rectangular pulse that approximates $\delta(t)$ and call it `d`. Choose `a = 11 * Ts`. Replace `h` in the above code with `d`. To see whether this approximation is successful, convolve `d` with `x`. According to P3 above, you should obtain y ≈ x. Plot your results and comment.

Task 4

In Task 3 we chose the pulse width as `a = 11 * Ts`. This choice will decide the accuracy of approximating the delta function. To see the effect of this choice, consider different widths for the rectangular pulse (hence, different approximations for the delta function). Let the width parameter `a = r * Ts`, where `r` is an integer ranging from 1 to `R`. Take `R = 100`. For each value of `r` (hence, for each approximation of the delta function), find the mean squared error (MSE), which is defined as follows:

$$e(r) = \frac{1}{N} \sum_{n=1}^{N} [y_r(n) - x(n)]^2$$

where $N$ is the total number of samples in each signal. Use the function `mean` in MATLAB (type `>help mean`) to find the above summation. If you use `y = conv(d,x)`, the interval of `y` is larger than that of `x`, you should extend `x` range (by zero padding) to be able to find the above summation. Plot MSE versus `a = r * Ts` and comment on the shape of the resulting curve.

Task 5

In the above code, use the other two approximations for the delta function and repeat the procedure in Task 4. Plot the three MSE curves on one figure and compare your results. Decide which approximation is better for practical implementation of the delta function. Take `Ts = 0.005`, repeat the same steps, and see whether the MSE is reduced.

## *Experiment # 2: Generation and Demodulation of AM Signals*

### Introduction

Information-bearing signals are normally of low-frequency nature. For example, the frequency content of the human speech signal ranges from 200 to 4 kHz, while audio signals (in general) can reach up to 20 kHz. For long-range transmission of signals, we should convert them to radio waves and transmit them through antennas. The antenna dimensions [length or diameter (for dish antenna)] should be proportional to $\lambda/4$ or $\lambda/2$, $\lambda$ being the wavelength, where $c = \lambda f = 3 \times 10^8$ m/s. Hence, for baseband radio transmission of a speech signal through, we need an antenna of length $c/(4 \times 4000) \approx 18.7\,\text{km}$. Therefore, we should increase the frequency of the signal (without changing the information) before radio transmission. This is called modulation. In fact, high frequency is also more resistive to noise in space. Mobile phones operate near 900 MHz, UHF TV operates in the range 300–800 MHz, and satellite links operate in the frequency range from few hundred MHz to 40 GHz or more.

Amplitude modulation (AM) is widely used in communication systems. The standard AM modulator multiplies the message $m(t)$ by a scaling factor $\beta$ (called the *modulation index*) and a high-frequency sinusoid called the *carrier*, $c(t) = A_c \cos(\omega_c t)$, and then add the result to the carrier (for synchronization purposes) to get the standard AM signal

$$x(t) = A_c[1 + \beta m(t)]\cos(\omega_c t) \tag{1}$$

The magnitude spectrum of this signal is given by:

$$X(f) = A_c\left[\frac{1}{2}\left\{\delta(f - f_c) + \delta(f + f_c)\right\} + \frac{\beta}{2}\left\{M(f - f_c) + M(f + f_c)\right\}\right] \tag{2}$$

For demodulation of the AM signal at the receiver, an envelope detector is usually used. However, if the AM signal is weak and the modulation index is small, it is better to use a square-law device followed by a low-pass filter as shown in Fig. D.3 below, where $r(t)$ is the received AM signal and $s(t)$ is the demodulated signal.

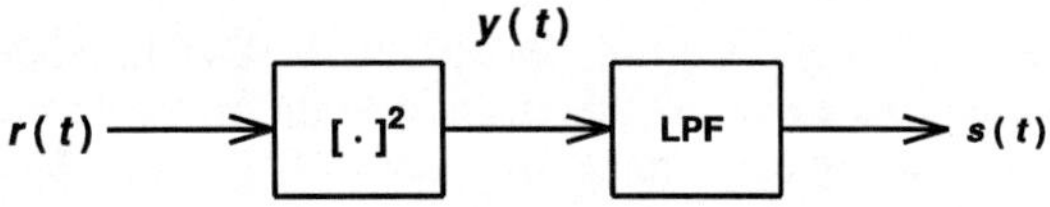


**Fig. D.3** Demodulation of standard AM signals using a square-law device

Under noise-free conditions, we have $r(t) = x(t)$. Hence, the squarer output will be as follows:

$$y(t) = x^2(t) = A_c^2[1 + \beta m(t)]^2[\cos(\omega_c t)]^2$$
$$= A_c^2\left[1 + 2\beta m(t) + \beta^2 m^2(t)\right]\left[\frac{1}{2} + \frac{1}{2}\cos(2\omega_c t)\right].$$

The LPF will remove the high frequency components. Hence, the output signal will be given by

$$s(t) = 0.5A_c^2 + A_c^2\beta.m(t) + 0.5A_c^2\beta^2 m^2(t),$$

which contains a d.c. term, the original message (scaled by $A_c^2\ \beta$), and an error term $0.5{A_c}^2\ \beta^2\ m^2\ (t)$, which we cannot get rid of by filtering. The relative error is given by:

$$e = \frac{0.5A_c^2\beta^2 m^2(t)}{A_c^2\beta m(t)} = \beta m(t)/2 \qquad (3)$$

To reduce error, we should have $|\beta m(t)/2| << 1 \quad \forall t$. Hence, this method is efficient only if the message is weak and the modulation index is small.

**MATLAB Simulation**

In this experiment we will simulate the generation and demodulation of standard AM signals as explained above.

Task 1

Write a MATLAB code to generate a standard AM signal. First, generate a sinusoid with frequency $f_o = 2$ Hz and a carrier with frequency 10 Hz. Take all amplitudes to be 1. Select a value for the modulation index and generate the modulated signal that would be transmitted. Plot all time signals and their spectra.

Task 2

Write a code to simulate the function of the receiver. First comes the square-law device, followed by the LPF. Use an ideal LPF with a carefully chosen cutoff frequency $f_1$ (Take $f_1 = 3f_o$). Filter the squared received signal, then plot the demodulated signal in the time and frequency domains. Compare with the original signal. Change the modulation index and comment on the results.

Task 3

Repeat Tasks 1 and 2 above for the rectangular pulse $\Pi_2(t-2)$, using the same carrier.

## *Experiment # 3: Random Signal Analysis*

### Introduction

A random process is a function of two variables: an event and time. A realization of the random process is called a "sample function". All sample functions constitute an "ensemble". At a specific time instant $t = t_o$, the values of sample functions are represented by a "random variable". For example, noise $n(t)$is a random process.

If we have a signal $x(t) = a\cos(\omega_o t)$ and this signal is corrupted by noise $n(t)$, then the result would be the random signal $y(t) = a\cos(\omega_o t) + n(t)$.
To study the statistical properties of noise and noisy signals, we may repeat the realization of the random signal $y(t)$ many times. If we have three realizations (repetitions) of the noise process $n(t)$ with a given noise power, we get the sample functions or "Realizations" as shown in Fig. D.4.

The "ensemble average" at $t = t_o$ is given by:

$$n_{av} = [n_1(t_o) + n_2(t_o) + n_3(t_o)]/3$$

If we have a large number of realizations and the process is stationary, then we may have:

$$\text{Ensemble average } (m) = \text{Time average } (m_t)$$

In this case we call the process "ergodic". For ergodic processes we can calculate the ensemble mean using the time mean, which is much easier and does not require more than one realization. On MATLAB, this is obtained by using the instruction "mean". Note that the ensemble mean for the above signal at the time instant $t$ is given by:

$$m = \mathcal{E}\{n(t)\} = \int_{-\infty}^{\infty} np(n)dn$$

where $p(n)$ is the probability density function (pdf) of noise, and $\mathcal{E}\{.\}$ is the statistical expectation.

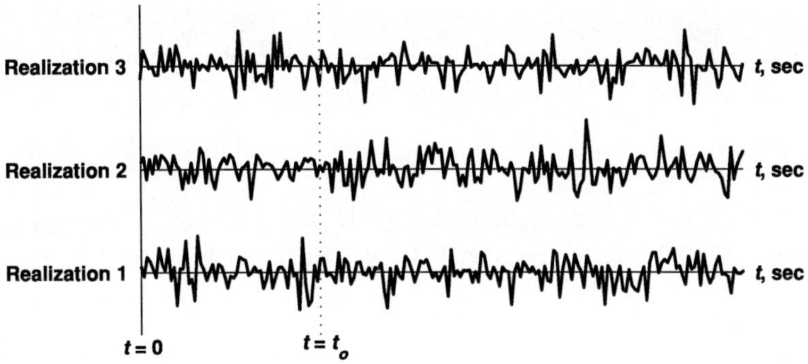


**Fig. D.4** Sample functions of a noise process

## MATLAB Simulation

In this experiment we will study the statistical properties of a sinusoidal signal imbedded in AWGN (additive white Gaussian noise) at different SNRs (signal-to-noise-ratios).

Task 1

Write a MATLAB code to generate $M = 10$ realizations of a sinusoidal signal $x(t) = a\cos(\omega_o t)$ corrupted by AWGN process $n(t)$ to give the noisy signal $y(t) = a\cos(\omega_o t) + n(t)$. Take $f_o = 0.2$ Hz, $a = 1$, and SNR $= 1$ dB. Show that the ensemble mean of noise is approximately zero, and the ensemble mean of the noisy signal is the deterministic signal $s(t)$. This approximation is improved if we take a larger number of realizations $M$ (e.g., 50, 100). Plot the time signals and their spectra for the first realization; also plot the ensemble means.
Repeat the above steps for different SNR values (e.g., −5, −1, 0, 1, 3, 10).

Task 2

For each realization in **Task 1**, find the pdf of noise $p_n(n)$ and take the average of all realizations. Compare this pdf with the theoretical pdf given by:

$$p_n(n) = \frac{1}{\sqrt{2\pi\sigma}} e^{-n^2/2\sigma^2}$$

Show practically that

$$\int_{-\infty}^{\infty} p_n(n)dn \approx 1$$

Compare this result with the theoretical result, which is 1 exactly.
Find the mean and variance of one realization and compare with the theoretical values. As in **Task 1**, consider different SNRs.

Task 3

Find the autocorrelation function of the signals $x(t)$, $n(t)$, and $y(t)$. Find the cross-correlation $R_{xn}(\tau)$ between the signal $x(t)$ and the noise process $n(t)$.

## *Experiment # 4: Filter Design with Application to Noise Reduction*

### Introduction

MATLAB provides built-in codes for designing analog and digital filters. In this experiment we will consider Butterworth and Chebychev-I filters. You can type on the command line `>help butter` and `> help cheby1` to know the design parameters of these filters. As an application of filtering, we will design a filter for the purpose of reducing noise that corrupts a narrowband signal of known frequency (if the signal frequency is unknown, then this approach fails and we need an adaptive filter for noise reduction).
*White Gaussian noise* $n(t)$ is a broadband signal since it is uncorrelated (i.e. its *autocorrelation* function is a weighted delta function, hence its *power spectral density* is constant for all frequencies as shown in Fig. D.5.
Normally, we are interested in a specific frequency band ($-B < f < B$) for practical applications. For example, in speech signal processing, the important frequency band is about 4 kHz, and the whole audible spectrum is less than 20 kHz. Since noise power is given by

$$p_n = \int_{-\infty}^{\infty} G_n(f)df,$$

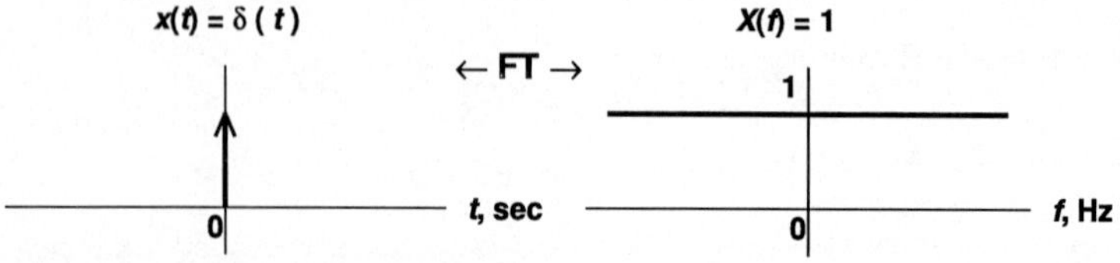


**Fig. D.5** Spectrum of a time delta function

then we can reduce this power by applying a filter (LPF or BPF) to extract only the area of interest in the spectrum.

## MATLAB Simulation

In this experiment we will design filters for the purpose of noise reduction in narrowband signals.

Task 1

Consider a signal $x(t)$ corrupted by additive Gaussian noise $n(t)$, where $x(t) = a\sin(\omega_o t)$, with $a = 1, f_o = 1\,\text{Hz}, 0 \leq t \leq 10\,\text{s}$, and SNR = 2 dB. Plot the time signals $x(t)$ and $s(t) = x(t) + n(t)$ with their spectra. Design a LPF (Butterworth and Chebychev 3 dB ripple) of minimum possible order K such that the attenuation is less than −80 dB for $f > 10$ Hz. You should carefully specify the appropriate cut-off frequency $f_c$. Plot the filter transfer function given by

$$H(f)_{dB} = 20\log_{10}|H(f)|,$$

and vary $K$ until you reach the proper order. Then filter $s(t)$ and plot the time output signal $y(t)$ with its spectrum and compare with $s(t)$.

Task 2

In this task we study the effect of the sampling frequency on a recorded signal, and we conclude that if we want to process a recorded signal, we should know its sampling frequency. Listen to any test audio signal, e.g., Ohno.wav (can be downloaded from http://free-loops.com/download-free-loop-3022.html), using a media player (just double-click on the file icon). On MATLAB use `x = wavread('Ohno')` to convert the audio file to a vector of numerical values (voltages) suitable for mathematical processing. The sampling frequency of Ohno can be read from the media player; it is `fs = 22 kHz`. Try changing fs to 35 kHz then re-write the signal on another file using the statement

```
Ohno1 = wavwrite(x, fs, 'Ohno1.wav');
```

Now listen to the new file. Change the sampling frequency to 15 kHz and repeat the process.

Task 3

In this task we consider audio effects as explained in the lecture notes. Implement an FIR filter with 4 coefficients to simulate echoes, and then listen to the resulting audio signal. Change the magnitude of the coefficients and listen again.

Task 4

Now listen to a single-tone test audio signal saved as "stone" (e.g., 440 Hz sound from http://www.mediacollege.com/audio/tone/download/). Find its sampling frequency `fs` (it is 44.1k Hz). Read the signal as x. Add Gaussian noise n (of power = −20 dB) to the signal and write the result as `s` = `x` + `n` in your directory as an audio signal using `wavwrite(s,fs,'stonen.wav')`. Listen to the corrupted signal and compare with the original one. Plot the time signals `x` and `s` with their spectra versus the normalized frequency ranges `fn` = `f/fs` and `fN` = `f/(fs/2)`. Now design a digital filter (LPF Butterworth) of order 10 and normalized cut-off frequency `wcr=wc/(fs/2)`, where wcr is the cut-off frequency which should be chosen carefully to reduce noise. Plot the frequency response of the digital filter using freqz. Then filter the discrete signal `s` to get the signal `y`. Plot `y` and its spectrum `Y`, then compare with the original and noisy signals.

## *Experiment # 5: A Sinusoidal Digital Oscillator*

### Introduction

A digital oscillator is a system that generates an output waveform (like a sinusoid) without a need for an input signal, except for a D.C. supply and perhaps a trigger (like a delta function) at the starting time. The theory of operation is based on the fact that a digital system with poles on the circumference of the unit circle in the z-plane is neither stable nor divergent, but oscillatory (or, *marginally stable*).
To design a digital sinusoidal oscillator, we need a transfer function in the z-domain whose impulse response is a sinusoid. Using Tables we can reach at the following z-transform pair:

$$h(n) = \sin[(n+1)b] \overset{z}{\longleftrightarrow} H(z) = \frac{\sin(b)z^2}{z^2 - 2\cos(b)z + 1} \qquad (1)$$

Since $n$ represents the time count, $b$ would represent the normalized radian frequency $\Omega_o = \omega_o T_s$, hence the frequency of oscillation is $f_o = \omega_o/2\pi = (b/T_s)/2\pi = (b/2\pi)f_s$ Hz, and we should have $|b| < \pi$.
Note that the two poles of this system are the roots of $z^2 - 2\cos(b)z + 1 = 0$, which are given by:

$$p_{1,2} = \cos(b) \pm \sqrt{\cos^2(b) - 1} = \cos(b) \pm j\sin(b) = e^{\pm jb} \qquad (2)$$

Hence, the poles are *exactly on the circumference* of the unit circle, and the system is neither stable nor unstable (oscillatory). Figure D.6 shows the implementation

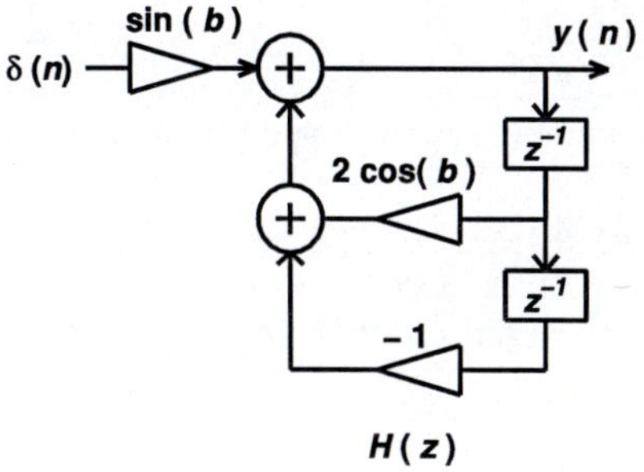


**Fig. D.6** Implementation of the sinusoidal digital oscillator

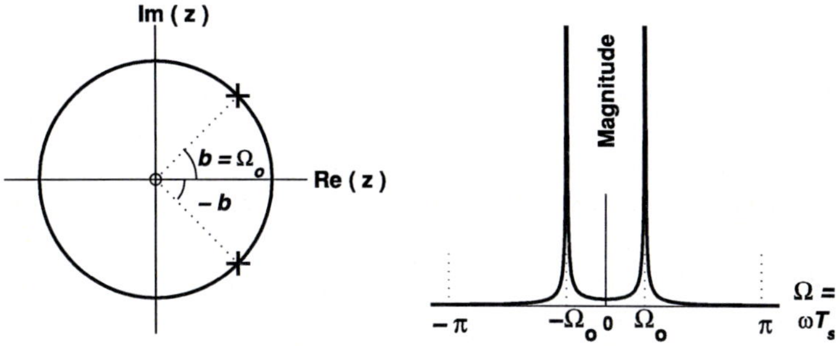


**Fig. D.7** PZ-diagram and frequency response of the sinusoidal digital oscillator

diagram of this system. The pole-zero diagram and frequency response are shown in Fig. D.7.

## MATLAB Simulation

In this experiment we will design a second-order IIR filter to perform as a sinusoidal oscillator.

Task 1

The generated sinusoid has an amplitude $A = 1$ and frequency $f_o = 1$ Hz. Choose the sampling frequency of the system as $f_s = 100$ Hz ($f_s \gg f_o$). Write the transfer function of the system as in Eq. 1 above, from which find the frequency response of the system and plot it.

Task 2Now analyze the system using `zplane` function on MATLAB. Find the poles of the system using `roots`. Find the angles of these poles, from which find the frequency of the oscillator.

Task 3

Define the system true time and plot the theoretical impulse response as in Eq. 1.

Task 4

Analyze the system using `tf`, `impulse`, and `freqz`. Plot all results.

Task 5

Now simulate the oscillator circuit as shown in Fig. D.6. Enter a delta signal and find the output. Plot and compare with previous methods of finding the impulse response.

Task 6

Find the effect of changing the sampling frequency on the system. Reduce the sampling frequency to 10 Hz and compare.

Task 7

With $f_s = 100$ Hz and $A = 1$, try to generate 2, 10, 45, 60, and 70 Hz sinusoids. See what happen if $f_o > f_s/2$. Change the amplitude and plot the resulting output signals.

### *Experiment # 6: Sampling and Reconstruction*

## Introduction

To process analog (i.e., continuous-time) signals using digital technology, we should convert these signals into *digital signals*through *sampling* and A/D conversion. In most applications, sampling is uniform, i.e., the sampling interval $T_s$ and the sampling frequency, $f_s = 1/T_s$, are constant. Ideal sampling can be formulated as multiplication of the analog signal $x(t)$ by the a train of time impulses $p(t) = \sum_{n=-\infty}^{\infty} \delta(t - nT_s)$ as shown in Fig. D.8 The spectrum of this time impulse train is the frequency impulse train $P(f) = f_s \sum_{k=-\infty}^{\infty} \delta(f - kf_s)$. The above multiplication in the time domain would be a convolution in the frequency domain between the frequency impulse train and the spectrum of the analog signal, $X(f)$, which results in the scaling (by $f_s$) and repetition (every $f_s$) of this spectrum as shown in Fig. D.9, where we used the normalized frequency $v = f/f_s$. The new spectrum is called the discrete-time Fourier transform (DTFT). Reconstruction of the sampled signal (back to the analog signal) is possible

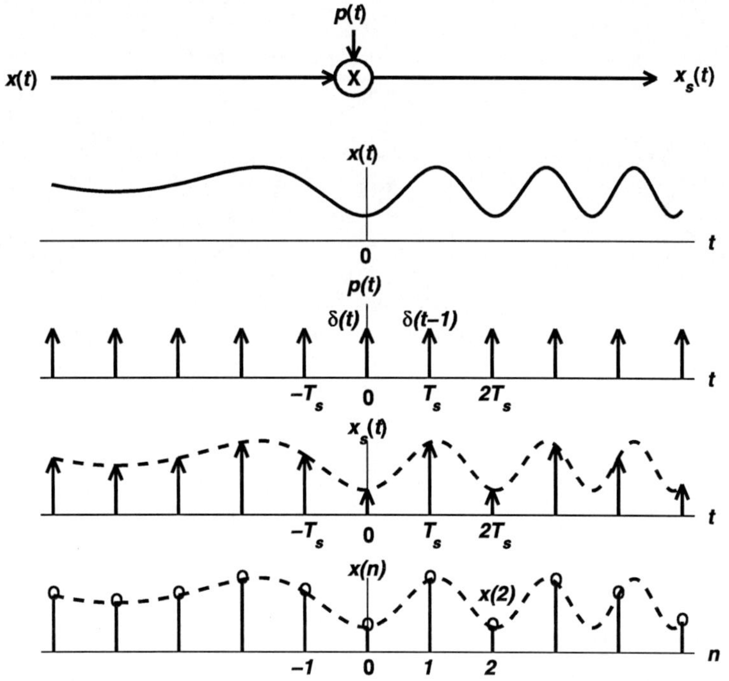


**Fig. D.8** Sampling process in the time-domain

through an analog low-pass filter on the condition that the sampling frequency is more than twice the signal bandwidth $B$, i.e.,

$$B \leq \frac{f_s}{2}.$$

The sampling rate $f_s = 2B$ is called the ***Nyquist rate***, which is the minimum rate for safe sampling. If $B > f_s/2$, replicas of the signal spectrum $X(f)$ that constitute the DTFT will overlap, and part of the information is damaged. This overlap is called *frequency aliasing*. Hence, in practice, we normally bandlimit the signal before sampling to remove the unimportant frequency content that may cause aliasing. This can be achieved using a LPF (which is called *anti-aliasing filter*). If the sampling frequency is near the Nyquist rate, the anti-aliasing filter should have a sharp cutoff at $f = B$, otherwise aliasing may occur. If $f_s \gg 2B$, we can relax this condition on the anti-aliasing filter. This is important in hardware implementation to reduce the LPF complexity.

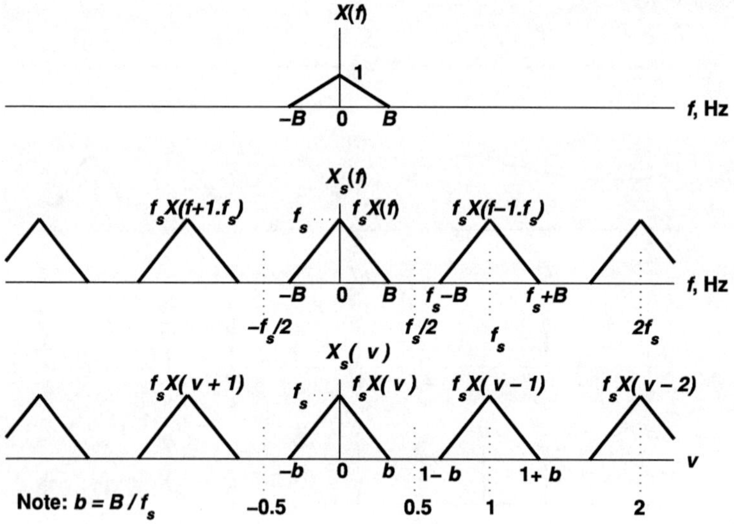


**Fig. D.9** Spectra of an analog signal and its digitized version

## MATLAB Simulations

Task 1

Write a MATLAB code to simulate the sampling process of an analog signal. Since analog signals are approximated on the computer, you should distinguish between the simulation sampling period (frequency) and the actual sampling period (frequency). Take the system sampling period to be T = 1e−4 s. With a time limit of 10 s, generate the global time and frequency vectors. Sinusoidal and linear FM signals are important in applications. Simulate a sinusoid of frequency 20 Hz (hence, to avoid aliasing, the sampling frequency should be more than 40 Hz). Plot the time signal and its spectrum.

Task 2

Use the MATLAB function `square` to simulate the time impulse train. Its duty cycle (the "ON"/"OFF" duration ratio) should be very small to simulate impulses. Its frequency would be the actual sampling frequency. Plot it in the time and the frequency domains and verify the theoretical formulas stated above.

Task 3

Now multiply the analog signal by the sampling signal (the impulse train). Plot the resulting spectrum for a sampling frequency of 25, 100, and 500 Hz and check for aliasing.

Task 4

Design an analog 9th-order Butterworth LPF with cutoff frequency of 50 Hz and filter the sampled signal. Plot the output signal and its spectrum and compare with the original analog signal for sampling frequency of 50, 100, and 500 Hz and check for aliasing.

Task 5

Now generate a LFM signal of initial frequency 10 Hz and modulation index of 0.7. Note that the sinusoid is a periodic function (hence, its spectrum is impulses) while the LFM is non-periodic (hence, its spectrum is a continuous function of frequency). From the spectrum we see that the maximum frequency in the LFM is about 50 Hz, hence, we expect the sampling frequency to be at least 100 Hz to avoid aliasing. Try `fs = 50, 100`, and `500` Hz then compare the results.

Task 6

In this task we study the audio effect of **aliasing**. Download and listen to an audio signal such as Ohno.wav which was used before in Experiment 4. On MATLAB use `[x,fs,bits] = wavread('Ohno')` to convert the audio file Ohno.wav to a vector of numerical values (and know its original sampling frequency and number of quantization bits). The sampling frequency of Ohno can be read `fs = 22` kHz. Try downsampling Ohno to `fk = fs/k` (k integer) then re-write the signal on another file using the statement `wavwrite(x,fs1,'Ohno1.wav')`. Now listen to `Ohno1.wav`. At what value of `k` does aliasing start to occur? Why?

## *Experiment # 7: Binary Signal Transmission*

### Introduction

In binary communication systems, binary data (which is a sequence of 0's and 1's) are often transmitted through a channel using two *orthogonal signals*, $s_0$ $(t)$ and $s_1$ $(t)$. One possible signal configuration is shown in Fig. D.10.
If data is large, then 0's and 1's are equally probable ($p(1) = p(0) = 1/2$) and *statistically independent*. The AWGN channel adds Gaussian noise (wideband, approximately white) with power spectral density (PSD) $= \eta/2$ (W/Hz). Hence, the received signal will be in the following form:

$$r(t) = s_i(t) + n(t), i \in \{0, 1\}, 0 \leq t \leq T.$$

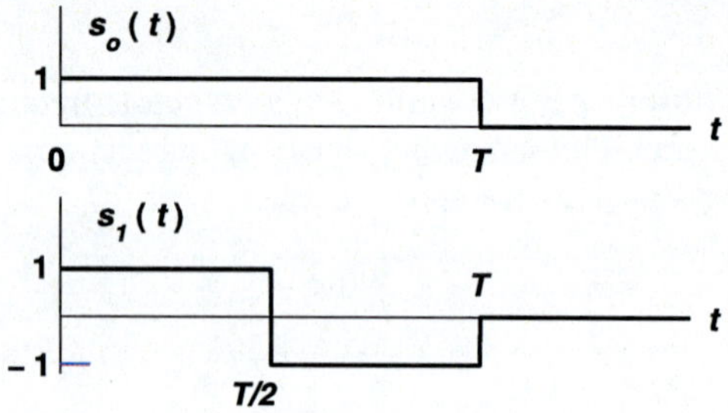


**Fig. D.10** Orthogonal signals for binary transmission

Assuming that $s_0$ and $s_1$ are as shown in Fig. D.10, and that $s_0$ was transmitted, the outputs of the two matched filters at the time instant $t = T$ are as follows:

$$r_0 = \int_0^T r(t)s_0(t)dt = \int_0^T [s_0(t) + n(t)]s_0(t)dt$$

$$= \int_0^T s_0^2(t)dt + \int_0^T n(t)s_0(t)dt = E + n_0$$

$$r_1 = \int_0^T r(t)s_1(t)dt = \int_0^T [s_0(t) + n(t)]s_1(t)dt$$

$$= \int_0^T s_0(t)s_1(t)dt + \int_0^T n(t)s_1(t) = 0 + \int_0^T n(t)s_1(t) = n_1$$

Both $n_0$ and $n_1$ are Gaussian and have zero means. The variances of $n_0$ and $n_1$, $\sigma_i^2(i = \in \{1, 2\})$, are given by:

$$\sigma_i^2 = \frac{\eta}{2}E,$$

where $E$ is the energy of the signals $s_0$ and $s_1$ [Note that ${s_1}^2\ (t) = {s_0}^2\ (t)$]. Now if $s_1\ (t)$ was transmitted, then $r_0 = n_0$, $r_1 = E + n_1$ with same statistics as above.

## Probability of Error

The matched filter compares $r_1$ and $r_0$. It will decide that a "0" was transmitted if $r_0 > r_1$, and that "1" was transmitted if $r_0 < r_1$. If "0" was transmitted, then error will occur only if $r_1 > r_0$.
It can be shown that the above probability of error can be expressed as follows:

$$P_e = \frac{1}{2} - \frac{1}{2}\text{erf}\left(\sqrt{\frac{SNR}{2}}\right)$$

where the error function $\text{erf}(y) = \frac{2}{\sqrt{\pi}}\int_0^y e^{-u^2}du$ is a reference function in MATLAB, and SNR is the signal-to-noise ratio defined by:

$$\text{SNR} = E/\eta,$$

and is normally given in dB as $\text{SNR}_{dB}$, where $\text{SNR}_{dB} = 10\log_{10}(\text{SNR})$.
The probability of error $P_e$ is *the basis for performance evaluation of communication systems.*
The same probability of error $P_e$ is obtained if "1" was transmitted. Hence, the *average* probability of error is given by:

$$P_{e,av} = P_e.$$

Figure D.11 shows the general shape of $P_e$ against SNR.

### Binary Transmission Using Antipodal Signals

Two signals $s_0$ $(t)$ and $s_1$ $(t)$ are said to be *antipodal* if $s_0$ $(t) = -s_1$ $(t)$. One possible configuration is to use two voltage levels, $\pm V$. Figure D.12 shows another configuration.
Using orthogonal signals, we need a bank of two matched filters for optimum reception. However, if we use two antipodal signals, we need *only one* matched filter. The received signal is $r(t) = \pm s(t) + n(t)$. Following the same analysis as for orthogonal signals, we find the following results:

- Output of the matched filter $z = \pm E + n_a$ (where $n_a = \int_0^T n(t)s(t)dt$),
- Variance of $n_a = \frac{\eta}{2}E$ (same as that of $n_1$ and $n_o$ for orthogonal transmission).
- Probability of error $= P_e = \frac{1}{2} - \frac{1}{2}\text{erf}(\sqrt{\text{SNR}})$. The decision will be as follows: if $z > 0$, then $s(t)$ was transmitted (which may represent "1"), otherwise $-s(t)$ was transmitted.

Hence, $P_e$ using antipodal signals is less than $P_e$ using orthogonal signals for the same SNR. Therefore, if no modulation is used later in the system, antipodal signal transmission is more efficient for baseband binary communications.

### MATLAB Simulation

In this experiment we will consider simulation of baseband binary signal generation and transmission, as well as performance analysis using the error

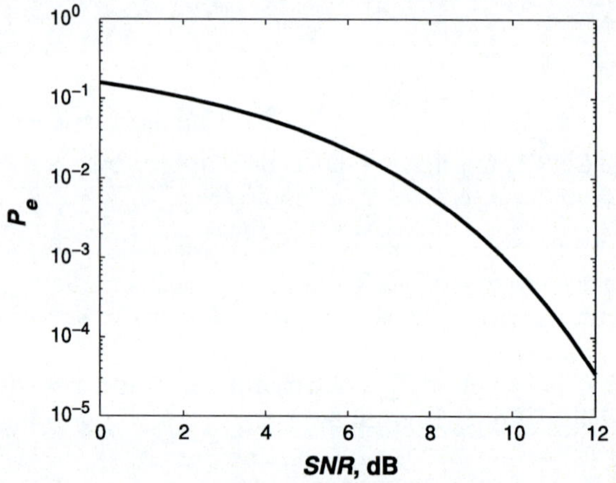


**Fig. D.11** Probability of error versus SNR in baseband binary communications

function. This is important in many applications like communications between computers operating in the same neighborhood.

Task 1

Simulate a binary communication system with two orthogonal signals. Use the flowchart shown in Fig. D.13. Take $E = 1$ and generate $N = 1000$ bits data sequence {d(n)}. To generate a sequence of equiprobable 1's and 0's, use "rand" function on MATLAB, which gives you a random number uniformly distributed over the interval (0,1). Use the simple rule: if rand >0.5, then d(n) is "1", otherwise "0". However, you should use a smart algorithm for data generation that avoids the use of loops. Mix data with noise to simulate transmission and matched filter reception, then find the probability of error as a function of the SNR. Better to use a separate function for the probability of error calculation. Compare with the theoretical curve.

**Fig. D.12** Two antipodal signals

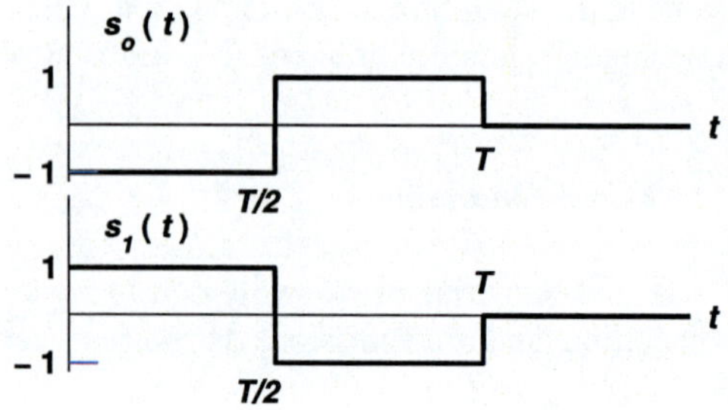

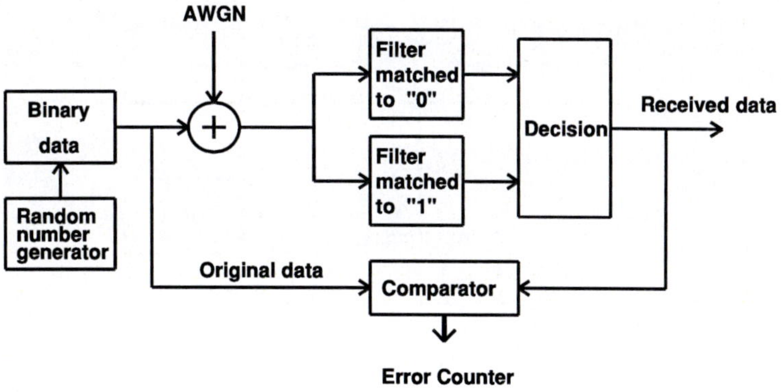


**Fig. D.13** Flow chart of a binary communication system simulator

Task 2

Using the same approach (but one matched filter), simulate a binary communication system with two antipodal signals. Find the probability of error versus SNR. Compare with the theoretical result and with orthogonal signal transmission.

Task 3

Simulate the above system without matched filters and compare with results in Tasks 1 and 2 above.

## *Experiment # 8: Simulation of the Sinusoidal Digital Phase-Locked Loop*

### Introduction

The sinusoidal DPLL (SDPLL) is an important system in signal processing and communications. Like other PLLs, the SDPLL is a feedback system that arranges its local frequency to be equal to the input frequency. It can be used for signal detection, frequency tracking, and synchronization. Unless the incoming frequency $\omega$ is equal to the center frequency $\omega_o$, the first-order SDPLL (which utilizes a multiplication constant $G_1$ only in its filter) has always a non-zero steady-state phase error, $\phi_{ss}$. The 2nd-order SDPLL [which utilizes a first-order digital filter $H(z) = G_1 + G_2/(1 - z^{-1})$] always locks on zero $\phi_{ss}$. A block diagram of the SDPLL is shown in Figure D.14 below. The sampler here operates as phase error detector (PED).

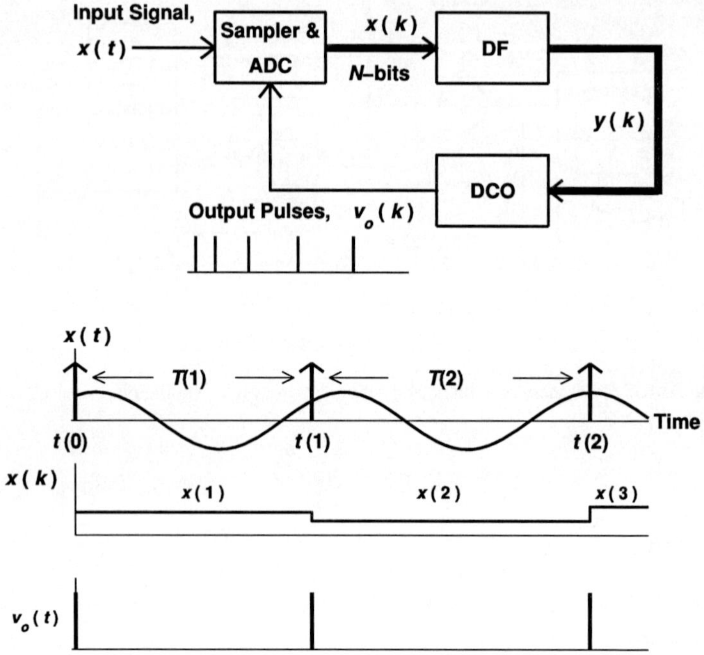


**Fig. D.14** Block diagram of the SDPLL with associated waveforms

The input $x(t)$ is assumed to be a sinusoidal signal as follows:

$$x(t) = A\sin(\omega t + \theta_o) + n(t).$$

The phase difference equation for the 1st-order SDPLL is given by:

$$\phi(k+1) = \phi(k) - K_2 \sin[\phi(k)] + \Lambda_o$$

where $\Lambda_o = 2\pi(\omega - \omega_o)/\omega_o$ and $K_2 = \omega\, G_1 A$. If we define $K_1 = \omega_0 G_1\, A$, and the frequency ratio $W = \omega_o/\omega$, then we have $K_2 = K_1\, (\omega/\omega_o) = K_1/W$.

The locking range is determined by the conditions:

$$K_1 > 2\pi|1 - W|$$

and

$$K_1 < \sqrt{(4 + 4\pi^2)W^2 - 8\pi^2 W + 4\pi^2}.$$

It should be noted that extreme points in this range does not ensure locking for all values of the initial phase error. The steady-state phase error is given by:

$$\phi_{ss} = \sin^{-1}(\Lambda_o/K_2).$$

## MATLAB Simulations

Task 1

Write a MATLAB code to simulate the locking process of the 1st-order SDPLL. First, plot the locking range. Then, let $A = 1$, $\omega_o = 2\pi$ (rad/s), $W = 0.9$, $K_1 = 0.8$ and take $\theta_o = \phi\,(0) = 0$ rad. Hence, the incoming frequency is $f = f_o/W = 1/0.9 = 1.1$ Hz. Make sure that the loop is inside the lock range. As we expect locking normally in less than 50 cycles, consider only 50 samples. Plot the input signal and the sampled signal. Also plot the phase $\phi(k)$ and the instantaneous frequency $[\omega(k) = 2\pi/T(k)]$ as functions of time. Check with the theoretical value of $\phi_{ss}$. Vary the initial phase $\phi\,(0) = \theta_o$ to take on the values −3, −2, −1, 0, 1, 2, 3 and see the difference in the locking process. Does the initial phase affect $\phi_{ss}$?

Task 2

Repeat Tasks 1 and 2 for various combinations of $(W,K_1)$ as follows: (0.9,1.5), (1.2,1.7), and (1.4,3). Let $(W,K_1)$ be outside the locking range and plot the phase and frequency transients.

Task 3

Plot the phase plane diagram of the 1st-order SDPLL for Tasks 1 and 2.

## *Experiment # 9: Adaptive Wiener Filter for Noise Reduction and Channel Estimation*

### Introduction

Wiener filter is an optimum filter for estimation or prediction signals corrupted by noise or distorted by the transmission channel. Adaptive Wiener filter is a programmable filter whose coefficients [i.e., its impulse response non-zero values, $\{h(k)\}$] are changed (adapted) according to the current available samples of the observed signal $\{y(n)\}$ and a desired (reference) signal $\{d(n)\}$, to give an optimal estimate $\hat{x}(n)$ of the original signal $\{x(n)\}$ at the time instant $n$ [see Fig. D.15]. An adaptive filter utilize s a feedback algorithm to update the filter coefficients at each time instant $n$; hence, it can compensate for time-varying channel conditions.

The filter coefficients are adapted according to an algorithm, which can be implemented by hardware or simulated on a microprocessor or a computer. The adaptive FIR Wiener filter algorithm is a least mean-squared (LMS) error

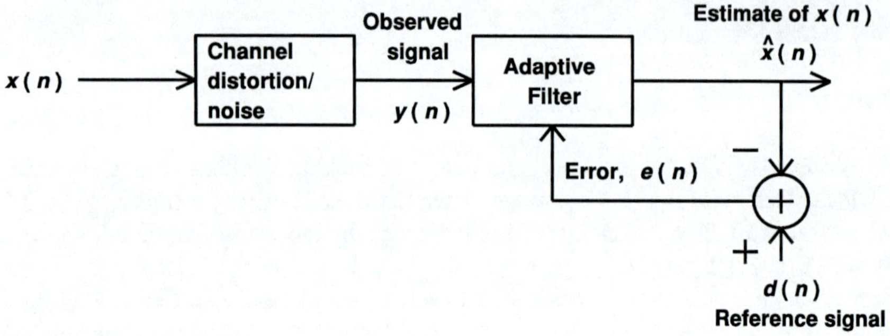


**Fig. D.15** Adaptive filter configuration

algorithm, based on minimizing the MSE error $e_{mse} = \mathcal{E}\{[e(n)]^2\} = \mathcal{E}\{[d(n) - \hat{x}(n)]^2\}$ at every time instant $n$ as we have shown earlier.

Define:

$\mathbf{h}(k) = [h_o(k)\, h_1(k)\, h_2(k) \cdots h_M(k)]$, the filter coefficients at the $k$th instant.

$\mathbf{y}(k) = [y(k)\, y(k-1)\, y(k-2) \cdots y(k-M)]$, observed signal vector at $k$th instant.The algorithm can be described in vector form **(MATLAB-like code)** as follows:

$\mathbf{h}(0) = \mathbf{0}$; % Initialize the filter coefficients.
for $n = 1 : \mathrm{N}$ % N = length(y);
$\hat{x}(n) = \mathbf{h}(n-1)\mathbf{y}^T(n)$; % Filter output (this is matrix multiplication).
$e(n) = d(n) - \hat{x}(n)$;
$\mathbf{h}(n) = \mathbf{h}(n-1) + \mu * e(n)y(n)$; % $\mu$ is the step-size.
end

The choice of $\mu$ will affect the estimation accuracy and the convergence speed of the algorithm. Small values of $\mu$ will give better accuracy but slower convergence. Large values will do the contrary. Very small or very large values for $\mu$ will cause significant errors. Hence, a compromise would be optimal.

Larger filter length $M + 1$ gives better estimation, but more delay.

*Application 1: Noise reduction in narrowband signals*: For estimation of narrowband signals (like single-tone sinusoids) with known frequency band, we can use a normal LPF for noise reduction, but for unknown frequency band, we use adaptive Wiener filter with $d(n) = y(n)$ and input sequence $\{y(n-1)\}$, as shown in Fig. D.16.

*Application 2: Channel estimation*: In mobile communications, a "training sequence" is sent before transmission of data. The receiver knows this signal and utilizes a copy of it as the desired signal $d(n)$. The adaptive Wiener filter can arrange its optimal coefficients during the short period of transmitting the training sequence before actual data are transmitted. Hardware or software implementation of the above algorithm is possible.

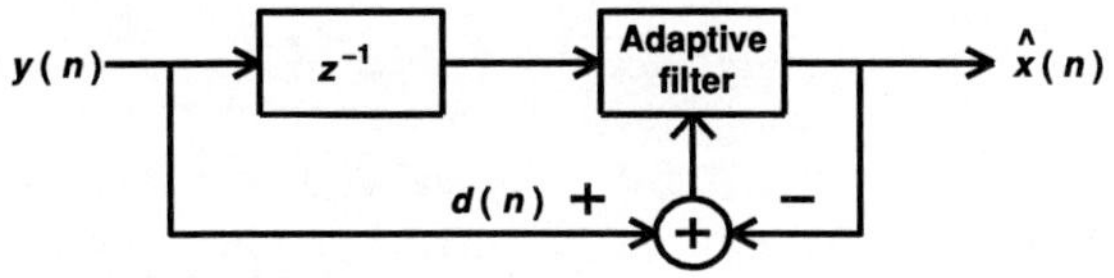


**Fig. D.16** Noise reduction using an adaptive filter

## MATLAB Simulation

Task 1

On MATLAB simulate the signal $y(k) = x(k) + n(k)$, where $x(k) = \sin(\omega_o kT_s)$, $T_s = 0.01$, the time vector is $0 < t < 10$ (s), $f_o = 2$ Hz, $n(t)$ is Gaussian noise, $SNR = 2$ dB. Assume that you know $f_o$ and try to reduce noise using a LPF (choose the cutoff frequency carefully).

Task 2

Now assume you don't know the signal frequency. Implement the adaptive Wiener filter for the purpose of noise reduction as shown in Fig. 2. Choose the number of taps $M + 1 = 101$, $\mu = 0.001$ and plot all signals. Vary $M + 1$ to 5, 10 and compare.

Task 3

With $M + 1 = 101$, vary $\mu$ to take the values 1e−4, 2e−4, 5e−4, 1e−3, 2e−3, 5e−3, 1e−2, 2e−2 and find the mean-squared error (MSE) as a function of $\mu$. Find $\mu_{\min}$, the value of $\mu$ that gives minimum MSE, and plot the corresponding signals and spectra.

Task 4

Assume that the communication channel causes ISI that spans 11 symbols, where the channel transfer function is

$$\mathrm{Hc} = [.05 - .063.088 - .126 - .25.9047.250.0.126.038.088];$$

Assume channel noise of variance −40 dB. Generate a random sequence of length N = 20 binary symbols of 0's and 1's (take number of realizations R = 20 and R = 200) and use them as the transmitted signal which will pass through the channel Hc, also use them as a reference signal (the receiver knows these sequences). The channel estimated transfer function will be given by the inverse of the filter updated transfer function, $H_e(z)$. Plot $|1/H_c(f)|$ and $|H_e(f)|$ and compare. Plot the mean-squared error signal (MSE) used by the filter versus the number of realizations. Compare this with the MSE in $H_c$ estimation.

Task 5

Re-run **Task 4** several times (for R = 20 and 200) and check how this affects the MSE of the adaptive filter error signal. Change channel noise to −20 dB and check the MSE for a given R and N.

Task 6

Take R = 200 and N = 50, 500, 1000 and compare the MSEs.

Task 7

Change the channel noise to −20 dB with R = 200, N = 500 and check the MSE. Re-run the code and check this result.

## *Experiment # 10: Delta Modulation System*

### Introduction

Delta modulation (DM) system is a single-bit differential PCM system that utilizes higher sampling rates to reduce the number of bits necessary for quantization, hence it reduces the cost of the system since increasing the sampling frequency is less expensive than increasing the word length. This makes DM suitable as ADC for audio applications, where the signal band is 20 kHz, hence the DM can use moderately high sampling frequencies.

DM system can be built in the analog domain or in the digital domain. However, it is not possible to implement an adaptive analog DM, hence we prefer the digital DM. Figure D.17 shows a digital DM system that consists of a 2-level quantizer and an integrator in the feedback loop. If the step Δ is small and the input signal has a steep slope, the DM can loose tracking and a slope overload occur. To avoid this problem, adaptive DM should be used. Figure D.18 shows an adaptive DM system.

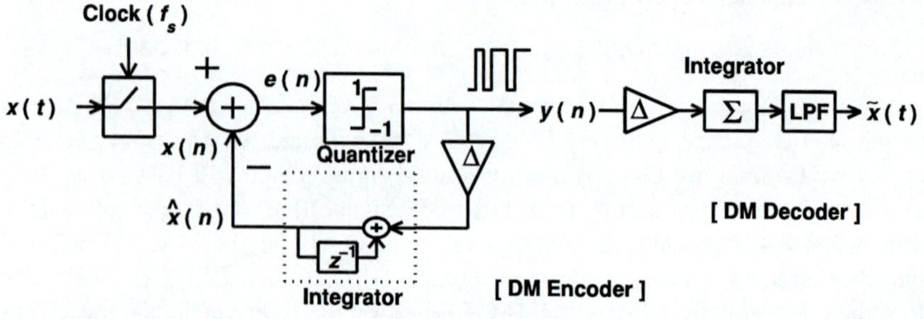


Fig. D.17 DM configuration

## MATLAB Simulation

In this experiment we will simulate DM and adaptive DM systems and compare their performance using sinusoidal and LFM input signals.

Task 1

On MATLAB generate the signal $x(k) = A\sin(\omega_o kT_s)$, with sampling period $T_s = 0.02$ s, amplitude $A = 0.5$ V, frequency $f_o = 1$ Hz, and time vector $0 < t < 10$ (s). With a step of $\Delta = 0.07$, simulate the delta modulation system in Fig. D.17. Plot the input signal, the estimated signal, and the output of the DM system along with their spectra. Can you find the effect of quantization on the output signal spectrum? The output signal is a square wave, however, its spectrum reveals a sinusoidal content.

Task 2

Implement the DM demodulator shown in Fig. D.17. Use a digital 4th-order Butterworth low-pass filter. Plot the demodulated signal and its spectrum and compare with the original signal.

Task 3

Vary the quantization step to 0.03, 0.1, and 0.2. Check the quantization noise and the slope overload.

Task 4

Repeat Tasks 1, 2, and 3 for a linear FM signal with amplitude 0.2 V, initial frequency 0.5 Hz, and modulation index of 0.5.

Task 5

Now implement the adaptive delta modulation system as shown in Fig. D.18, with an initial step of 0.03 and a step modification constant $K = 1.3$. For sinusoidal and LFM inputs, plot the input signal and its modulated version along with their spectra. Compare with the non-adaptive DM system.

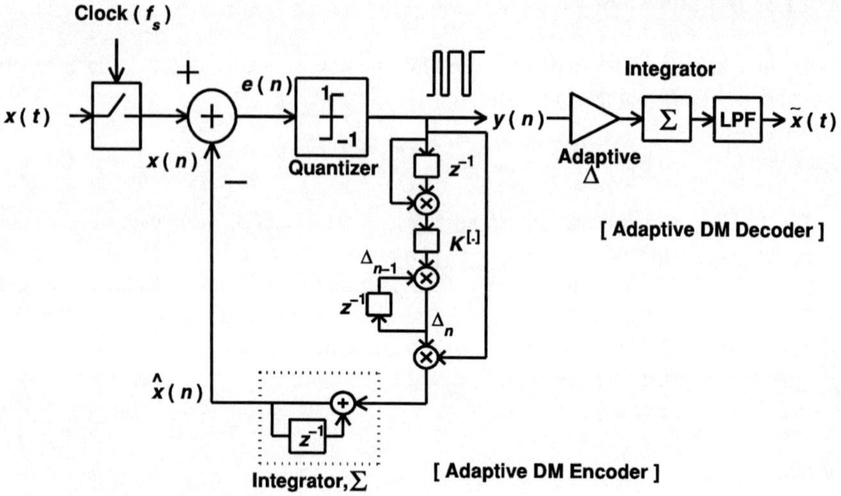


**Fig. D.18** Adaptive DM system

# DSP Project

## *Part 1: Problem Solving Using MATLAB*

(A) *Prime numbers*

(A-1) *A prime number* (p) is an integer that has no factors or divisors except 1 and itself. The first 12 prime numbers are 2, 3, 5, 7, 11, 13, 17, 19, 23, 29, 31, and 37. Note that 2 is the only even prime number. Prime numbers play a significant role in coding and computation algorithms and circuits.

(A-2) Write a MATLAB code to find whether a given integer $x$ is a prime.

(A-3) Write a MATLAB code to find the number of primes less than or equal to a given real number $x$ (not necessarily an integer). This is called "prime counting function, $\pi(x)$". For example, $\pi(10) = \pi(10.9) = 4$.

(A-4) *Prime number theorem* As $x$ becomes large, $\pi$ $(x)$ approaches $x/\ln(x)$. Prove using MATLAB simulation.

(B) *Non-linear equations*

Using MATLAB, find whether the equation $\cos(x) = x^2$ has a solution over the interval $[-3, +3]$.

(C) *3D plotting*
Plot the function $z(x, y) = \cos(x)\exp(-|y|)$ and its first derivatives. Plot a few 2D cross-sections of $z$.

(D) *Computation algorithms* are of fundamental importance in designing new generations of faster and more efficient computers and digital signal processors (DSP). Mersenne primes (MP's) are important in computation algorithms and hardware as they enable binary arithmetic while using "Number Theoretic Transforms". An integer m is **a Mersenne Prime**if it is a prime number of the form $\mathbf{m = 2p - 1}$, where **p** is a prime. There are **only 44** MP's discovered so far. The first 9 MP's are **3, 7, 31, 127, 8191, 131071, 524287, 2147483647, 2305843009213693951** (corresponding to **p = 2, 3, 5, 7, 13, 17, 19, 31, 61**), while the last known MP is **m44 = 232582657 − 1** (discovered in 2006 and is composed of 9808358 digits). Write a MATLAB code to find the first 5 MP's.

## *Part 2: System Implementation*

Choose Project-2a or Project-2b as follows.

### Project-2a: Hardware Design of Digital Systems

I. Design a digital integrating or differentiating circuit. Using ADC/DAC, apply analog sinusoids (from a signal generator) with different frequencies and find the magnitude and phase spectra. Verify the function of the circuit on the oscilloscope.

II. Design a first-order digital phase-locked loop or a digital delta modulation system and verify the circuit operation using a sinusoidal signal generator and an oscilloscope.

### Project-2b: Software Analysis of the Sinusoidal Digital Phase-Locked Loop

#### A: Frequency Tracking of a Sinusoid

Task 1: Noise-Free Analysis

Using MATLAB, simulate the 1st-order sinusoidal DPLL shown in Fig. D.19 Consider a sinusoidal input signal, plot the locking range of the loop, and study the loop behavior for different circuit parameters. Consider the phase plane, the sampling process and the transient phase and frequency. For a fixed frequency ratio $W$ and initial phase (but different combinations), find the effect of the loop gain $K_1$ on the locking speed. Locking is practically reached when the difference

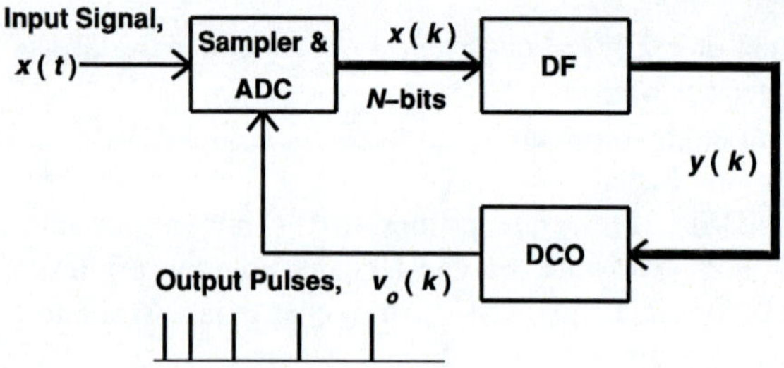


**Fig. D.19** Block diagram of the SDPLL

between successive phase errors is smaller than some positive number (e.g., 0.01). Also find the effect of the initial phase error on the locking speed assuming the frequency ratio and the loop gain are fixed.

Task 2: Noise Analysis

Consider a sinusoidal input signal corrupted by an AWGN noise. Find the pdf of the SDPLL output frequency for different SNRs. Discuss whether the SDPLL can estimate the input frequency in noisy environments. Take several combinations of the loop gain and the frequency ratio. Plot the variance of the loop frequency estimate as a function of SNR.

**B: Demodulation of PM Signals Using SDPLL**

The first-order SDPLL can demodulate PM signals as shown in Fig. D.20 below. Using MATLAB, simulate the PM demodulation circuit and study its behavior under noise-free conditio ns for different values of the modulation index.

**C: Second-Order SDPLL**

The second-order SDPLL utilizes a proportional-plus-accumulation filter and locks on zero phase. Simulate this loop and study its performance (as a frequency estimator) in Gaussian noise. Note that two initial phases should be considered.

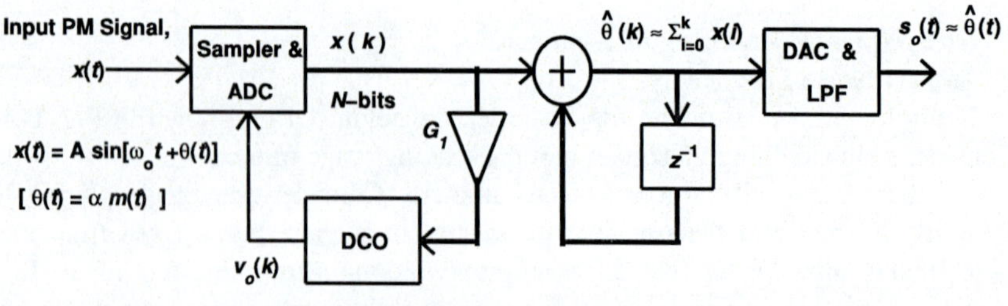


**Fig. D.20** Circuit diagram for PM demodulation using 1st-order SDPLL

# Authors' Biographies

**Zahir M. Hussain** took the first rank in Iraq in the General Baccalaureate Examinations 1979 with an average of 99%. He received the B.Sc. and M.Sc. degrees in electrical engineering from the University of Baghdad in 1983 and 1989, respectively, and the PhD degree in electrical engineering from Queensland University of Technology, Brisbane, Australia, in 2002. From 1989 to 1998 he researched and lectured on elec-trical engineering and mathematics. In 2001 he joined the School of Electrical and Computer Engineering (SECE), RMIT University, Melbourne, Australia, as a researcher then lecturer of signal process-ing and communications. He was the academic leader of a 3G commercial communications project at RMIT 2001–2002. In 2002 he was promoted to Senior Lecturer. Since 2001 he has been the senior supervisor for 15 PhD candidates at RMIT (with 9 completions between 2001 and 2009); also the second supervisor for another 7 PhD candidates. Dr. Hussain has over 200 internationally refereed technical publications on signal processing, communications, and electronics. His joint work on single-bit processing with Professor Peter O'Shea has led to an Australian Research Council (ARC) Discovery Grant (2005–2007). In 2005 he was promoted to Associate Professor of Signal Processing. He got the RMIT 2005 and 2006 Publication Awards (also shared the 2004 Publication Award with Professor Richard Harris, now Chair of Telecommunications at Massey University, NZ). In 2006 he was the Head of the Communication Engineering Discipline at SECE, RMIT. In 2007 he won the RMIT Teaching Award. Dr. Hussain is a member of Engineers Australia (formerly IEAust), IET (formerly IEE), and a senior member of IEEE and the Australian Computer Society (ACS). He worked on the technical program committees of many conferences and served as a reviewer for leading IEEE and Elsevier journals. Prof. Hussain has recently joined Kufa University, staying as Adjunct Professor at RMIT. .

**Amin Z. Sadik** received the B.Sc. (1983) and M.Sc. degrees (1988) in electrical engineering from the University of Baghdad, Iraq, and Baghdad University of Technology, Iraq, respectively. From 1989 to 1995, he was a lecturer in the School of Electrical Engineering, University of Technology, Baghdad, Iraq. From 1995 to 2001 he was a lecturer at the University of Salahuddin, Erbil, Iraq, and from 2001 to 2004 he was a lecturer at the University of Al-Balqa, Jordan. In 2006 he received his PhD degree in Electrical Engineering from the School of Electrical and Computer Engineering, RMIT University, Melbourne, Australia. Since then, he worked as a Research Fellow in RMIT University and as a Senior Research Fellow (Digital Signal Processing) in QUT. Dr. Sadik received the *Best Paper Award* in IEEE TENCON 2005 and DSPCDC 2006. He contributed to the technical committees of several international conferences and to the review process in Elsevier and other journals. In Feb. 2009 he joined the School of Engineering and Physical Sciences, Heriot-Watt University, Dubai Campus. His research interests include digital signal processing and digital communications.

**Peter O'Shea** received the B.E., Dip.Ed., and Ph.D. degrees, all from the University of Queensland, Australia. He has worked as an engineer at the Overseas Telecommunications Commission for 3 years, at University of Queensland for 4 years, at Royal Melbourne Institute of Technology (RMIT) for 7 years, and at Queensland University of Technology (QUT) for 9 years, where he is currently a Professor. He has received awards in Student Centered Teaching from the Faculty of Engineering and the University President at both RMIT and QUT. His interests are in signal processing for communications, power systems and biomedicine, and reconfigurable computing.